Engineering Analysis with SOLIDWORKS Simulation 2022

Paul M. Kurowski, Ph.D., P.Eng.

CERTIFIED
Solution
Partner

PUBLICATIONS

Design Generator, Inc.

SDC Publications
P.O. Box 1334
Mission, KS 66222
913-262-2664
www.SDCpublications.com
Publisher: Stephen Schroff

About the Cover
The image on the cover shows von Mises stresses in PIPES assembly model. The PIPES model comes with study and result plots fully defined; it can be found in Chapter 20 in the downloadable exercises that accompany this book.

ISBN-13: 978-1-63057-469-7
ISBN-10: 1-63057-469-4

Printed and bound in the United States of America.

About the cover

The image on the cover shows a displacement plot produced by nonlinear analysis. The model 2022 COVER PAGE.sldprt can be found in folder *24* in the set of exercises that accompany this book.

SOLIDWORKS Models

All models used in this book may be downloaded from www.SDCPublications.com

About the Author

Dr. Paul Kurowski obtained his MSc and PhD in Applied Mechanics from Warsaw Technical University. He completed postdoctoral work at Kyoto University. Dr. Kurowski is an Assistant Professor in the Department of Mechanical and Materials Engineering at the University of Western Ontario. His teaching includes Finite Element Analysis, Computer Aided Engineering, Product Design, Kinematics and Dynamics of Machines, Mechanical Vibrations and Reverse Engineering. His interests focus on Computer Aided Engineering methods used in mechanical design process.

Dr. Kurowski is also the President of Design Generator Inc., a consulting firm with expertise in Product Development, Design Analysis, and training in Computer Aided Engineering.

Dr. Kurowski has published many technical papers and taught professional development seminars for the Society of Automotive Engineers (SAE), the American Society of Mechanical Engineers (ASME), the Association of Professional Engineers of Ontario (PEO), the Parametric Technology Corporation (PTC), Rand Worldwide, SOLIDWORKS Corporation and others.

Dr. Kurowski can be contacted at www.designgenerator.com.

Acknowledgements

This book is in its 20th edition since "Engineering Analysis with COSMOSWorks 2003." The book takes a unique approach by bridging introductory theory with examples showing the theory's practical implementations.

The book evolves together with SOLIDWORKS Simulation software, and I hope that every year it offers better value to readers.

I thank my students for their valuable comments and questions. I thank my wife Ela for editing, project management and support.

Paul Kurowski

Table of contents

About the Author i

Acknowledgements i

Table of contents ii

Before You Start 1

Notes on hands-on exercises and functionality of SOLIDWORKS Simulation
Prerequisites
Selected terminology
Graphics

1: Introduction 5

What is Finite Element Analysis?
Finite Element Analysis used by Design Engineers
Objectives of FEA for Design Engineers
What is SOLIDWORKS Simulation?
Fundamental steps in an FEA project
Errors in FEA
A closer look at finite elements
What is calculated in FEA?
How to interpret FEA results
Units of measure
Using online help
Limitations of Static studies

2: Static analysis of a plate 31

Using the SOLIDWORKS Simulation interface
Linear static analysis with solid elements
Controlling discretization error with the convergence process
Finding reaction forces
Presenting FEA results in a desired format

3: Static analysis of an L-bracket 1

Stress singularities
Differences between modeling errors and discretization errors
Using mesh controls
Analysis in different SOLIDWORKS configurations

Nodal stresses, element stresses

4: Static and frequency analyses of a pipe support   99
Use of shell elements
Frequency analysis
Bearing load

5: Static analysis of a link   127
Symmetry boundary conditions
Preventing rigid body motions
Limitations of the small displacements theory

6: Frequency analysis of a tuning fork and a plastic part   137
Frequency analysis with and without supports
Rigid body modes
The role of supports in frequency analysis
Symmetric and anti-symmetric modes

7: Thermal analysis of a pipe connector and a heater   147
Analogies between structural and thermal analysis
Steady state thermal analysis
Analysis of temperature distribution and heat flux
Thermal boundary conditions
Thermal stresses
Vector plots

8: Thermal analysis of a heat sink   167
Analysis of an assembly
Global and local Contact conditions
Steady state thermal analysis
Transient thermal analysis
Thermal resistance layer
Use of section views in result plots

9: Static analysis of a hanger   183
Global and local Contact conditions
Hierarchy of Contact conditions

10: Thermal stress analysis of a bi-metal loop   193
Thermal deformation and thermal stress analysis
Converting Sheet Metal bodies to Solid bodies

"Parasolid" round trip
Saving model in deformed shape

11: Buckling analysis of an I-beam   201
Buckling analysis
Buckling load safety factor
Stress safety factor

12: Static analysis of a bracket using adaptive solution methods   209
h-adaptive solution method
p-adaptive solution method
Comparison between h-elements and p-elements

13: Drop test   227
Drop test analysis
Stress wave propagation
Direct time integration solution

14: Selected nonlinear problems   239
Large displacement analysis
Analysis with shell elements
Membrane effects
Following and non-following load
Nonlinear material analysis
Residual stress

15: Mixed meshing problem   283
Using solid and shell elements in the same mesh
Mixed mesh compatibility
Manual and automatic finding of contact sets
Shell Manager

16: Analysis of weldments using beam and truss elements   293
Different levels of idealization implemented in finite elements
Preparation of a SOLIDWORKS model for analysis with beam elements
Beam elements and truss elements
Analysis of results using beam elements
Limitations of analysis with beam elements

17: Review of 2D problems   321
Classification of finite elements

2D axisymmetric element
2D plane stress element
2D plane strain element

18: Vibration analysis - modal time history and harmonic   **349**
Modal Time History analysis (Time Response)
Harmonic analysis (Frequency Response)
Modal Superposition Method
Damping

19: Analysis of random vibration   **377**
Random vibration
Power Spectral Density
RMS results
PSD results
Modal excitation

20: Topology Optimization   **397**
Definition of Topology Optimization
Design space
Goals and constraints
Topology Optimization criteria
Examples of Topology Optimization

21: Miscellaneous topics – part 1   **417**
Mesh quality
Solvers and solvers options
Displaying mesh in result plots
Automatic reports
E drawings
Non uniform loads
Frequency analysis with pre-stress
Interference fit analysis
Rigid connector
Pin connector
Bolt connector
Remote load/mass
Weld connector
Bearing connector
Cyclic symmetry
Strongly nonlinear problem

Submodeling
Automated detection of stress singuylarities
Stress averaging at mid-side nodes
Terminology issues in the Finite Element Analysis

22: Miscellaneous topics – part 2   **489**
Symmetry
Antisymmerty
Displacement and stress singularities
Shell elements
2D problems

23: Implementation of FEA into the design process   **539**
Verification and Validation of FEA results
FEA driven design process
FEA project management
FEA project checkpoints
FEA reports

24: Glossary of terms   **559**

25: Resources available to FEA users   **567**

26: List of exercises   **571**

Before You Start

Notes on hands-on exercises and functionality of SOLIDWORKS Simulation

This book goes beyond a standard software manual. It introduces you to **SOLIDWORKS Simulation** software and the fundamentals of Finite Element Analysis (FEA) through hands-on exercises. We recommend that you study the exercises in the order presented in the book. You will notice that explanations and steps described in detail in earlier exercises are not repeated in later chapters. Each subsequent exercise builds on the skills, experience, and understanding gained from previously presented problems.

Exercises in this book require different levels of **SOLIDWORKS Simulation**. Information on **SOLIDWORKS Simulation** is available at:

https://www.solidworks.com/category/simulation-solutions

This book deals with structural and thermal analyses using **SOLIDWORKS Simulation**. Therefore, **SOLIDWORKS Motion** analysis will not be covered. **SimulationXpress** is a simplified version of **SOLIDWORKS Simulation** and will not be covered either.

There is no need to prepare CAD models. All exercises in this book use **SOLIDWORKS** models, which can be downloaded from www.sdcpublications.com. Most of these exercises do not contain **Simulation** studies; you are expected to create studies and analyze results yourself.

Explore each exercise beyond its description by investigating other options, other menu choices, and other ways to present results. You will soon discover that the same simple logic applies to all analysis problems in **SOLIDWORKS Simulation**.

Note on numerical results presented in this book.

Many exercises are illustrated with plots showing numerical results such as displacement, stress, frequency etc. If you notice small differences between book results and your results, those differences are caused by the use of a different software release or a different solver.

"Engineering Analysis with SOLIDWORKS Simulation 2022" is an introductory text. The focus is more on understanding Finite Element Analysis than presenting all software capabilities. This book is not intended to replace software manuals. Therefore, not all **Simulation** capabilities will be covered, especially those of fatigue, design studies, advanced nonlinear, and dynamic analyses.

Readers of "Engineering Analysis with SOLIDWORKS Simulation 2022" may wish to review the books:

- Vibration Analysis with SOLIDWORKS Simulation 2022 (Figure 25-1)

- Thermal Analysis with SOLIDWORKS Simulation 2022 and Flow Simulation 2022 (Figure 25-2)

These books are designed for readers who are familiar with topics presented in "Engineering Analysis with SOLIDWORKS Simulation 2022".

The knowledge acquired by the readers of "Engineering Analysis with SOLIDWORKS Simulation 2022" will not be software specific. The same concepts, tools and methods apply to any commercial FEA program.

Prerequisites

The following prerequisites are recommended:

- An understanding of Statics, Kinematics and Dynamics
- An understanding of Mechanics of Materials
- An understanding of Heat Transfer
- An understanding of Mechanical Vibrations
- Some experience with CAD using SOLIDWORKS
- Familiarity with Windows 10 operating system

Selected terminology

The mouse pointer plays a very important role in executing various commands and providing user feedback. The mouse pointer is used to execute commands, select geometry, and invoke pop-up menus. We use Windows terminology when referring to mouse-pointer actions.

Item	Description
Click	Self-explanatory
Double-click	Self-explanatory
Click-inside	Click the left mouse button. Wait a second, and then click the left mouse button again inside the pop-up menu or text box. Use this technique to modify the names of items in SOLIDWORKS Simulation studies.
Drag and drop	Use the mouse to point to an object. Press and hold the left mouse button down. Move the mouse pointer to a new location. Release the left mouse button.
Right-click	Click the right mouse button. A pop-up menu is displayed. Use the left mouse button to select a desired menu command.

All **SOLIDWORKS** file names appear in CAPITAL letters, even though the actual file names may use a combination of capital and small letters. Selected menu items and **SOLIDWORKS Simulation** commands appear in **bold**. **SOLIDWORKS** configurations**, SOLIDWORKS Simulation** folders, icon names and study names appear in *italics* except in captions and comments to illustrations. **SOLIDWORKS** and **Simulation** also appear in bold font. Bold font is also used to draw the reader's attention to a particular term.

Graphics

To improve readability, many plots use custom defined colors where dark blue is replaced by light grey color.

1: Introduction

What is Finite Element Analysis?

Finite Element Analysis, commonly called FEA, is a numerical technique of solving field problems described by partial differential equations. It is used in engineering disciplines such as machine design, acoustics, electromagnetism, soil mechanics, fluid dynamics, and many others.

In mechanical engineering, FEA is widely used for solving structural, vibration, and thermal problems. However, FEA is not the only available tool of numerical analysis. Other numerical methods include the Finite Volume Method, or the Boundary Element Method, to mention just a few. However, due to its versatility and numerical efficiency, FEA has come to dominate the engineering analysis software market, while other methods have been relegated to niche applications. When implemented into modern commercial software, both FEA theory and numerical problem formulation become transparent to users.

Finite Element Analysis used by Design Engineers

FEA is a powerful engineering analysis tool useful in solving many problems ranging from very simple to very complex. Design engineers use FEA during the product development process to analyze the design-in-progress. Time constraints and limited availability of product data call for many simplifications of computer models. On the other hand, specialized analysts implement FEA to solve complex problems, such as vehicle crash dynamics, hydro forming, and air bag deployment.

This book focuses on how design engineers use FEA, implemented in **SOLIDWORKS Simulation**, as a design tool. Therefore, we highlight the most essential characteristics of FEA as performed by design engineers as opposed to those typical for FEA performed by analysts.

FEA for Design Engineers: one of many design tools

For design engineers, FEA is one of many design tools that are used in the design process and include CAD, prototypes, spreadsheets, catalogs, hand calculations, textbooks, etc.

FEA for Design Engineers: Based on CAD models

Modern design is conducted using CAD, so a CAD model is the starting point for analysis. Since CAD models are used for describing geometric information for FEA, it is essential to understand how to prepare CAD geometry for analysis, and how a CAD model is different from an FEA model.

FEA for Design Engineers: Concurrent with the design process

Since FEA is a design tool, it should be used concurrently with the design process. It should drive the design process rather than follow it.

Limitations of FEA for Design Engineers

An obvious question arises: would it be better to have a dedicated specialist perform FEA and let design engineers do what they do best – design new products? The answer depends on the size of the business, type of products, company organization and culture, and many other tangible and intangible factors. A consensus is that design engineers should handle relatively simple types of analysis but do it quickly and of course reliably. Analyses that are very complex and time consuming cannot be executed concurrently with the design process and are usually better handled either by a dedicated analyst or contracted out to specialized consultants.

Objectives of FEA for Design Engineers

The ultimate objective of using FEA as a design tool is to change the design process from repetitive cycles of "design, prototype, test" into a streamlined process where prototypes are not used as design tools and are only needed for final design validation. With the use of FEA, design iterations are moved from the physical space of prototyping and testing into the virtual space of computer simulations (Figure 1-1).

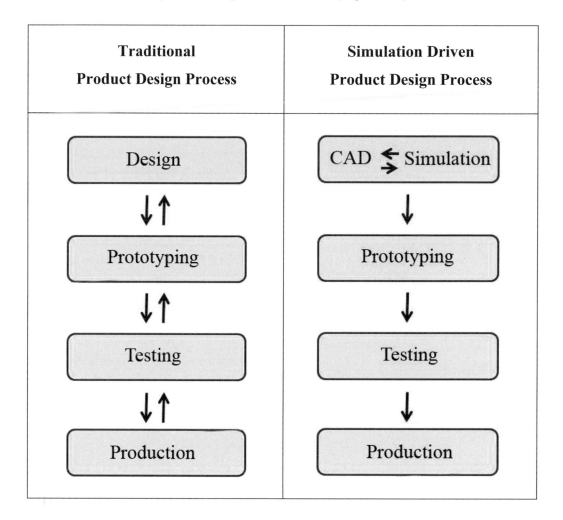

Figure 1-1: Traditional and Simulation driven product development.

Traditional product development needs prototypes to support a design in progress. The process in Simulation-driven product development uses numerical models, rather than physical prototypes, to drive development. In a Simulation driven product design process, the prototype is no longer a part of the iterative design loop.

"Simulation" means here Finite Element Analysis, but other simulation methods may be used in the product development process such as Mechanism Analysis, Computational Fluid Dynamics, and many others.

What is SOLIDWORKS Simulation?

SOLIDWORKS Simulation is a commercial implementation of FEA capable of solving problems commonly found in design engineering, such as the analysis of displacements, stresses, natural frequencies, vibration, buckling, heat flow, etc. It belongs to the family of engineering analysis software products originally developed by the Structural Research & Analysis Corporation (SRAC). SRAC was established in 1982 and since its inception has contributed to innovations that have had a significant impact on the evolution of FEA. In 1995, SRAC partnered with the **SOLIDWORKS** Corporation and created COSMOSWorks, one of the first **SOLIDWORKS** Gold Products, which became the top-selling analysis solution for the **SOLIDWORKS** Corporation. The commercial success of COSMOSWorks resulted in the acquisition of SRAC in 2001 by Dassault Systèmes, parent of **SOLIDWORKS** Corporation. In 2003, SRAC operations merged with the **SOLIDWORKS** Corporation. In 2009, COSMOSWorks was renamed **SOLIDWORKS Simulation**.

SOLIDWORKS Simulation is integrated with **SOLIDWORKS** CAD software and uses **SOLIDWORKS** for creating and editing model geometry.

Fundamental steps in an FEA project

The starting point for any **SOLIDWORKS Simulation** project is a **SOLIDWORKS** part or an assembly. First, material properties, loads, and restraints are defined. Next, as is always the case with using any FEA-based analysis tool, the model geometry is split into relatively small and simply shaped entities called finite elements. The elements are called "finite" to emphasize the fact that they are not infinitesimally small, but relatively small in comparison to the overall model size. Creating finite elements is commonly called meshing. When working with finite elements, the **SOLIDWORKS Simulation** solver approximates the sought solution (for example stress) by assembling the solutions for individual elements.

From the perspective of FEA software, each application of FEA requires three steps:

- Preprocessing of the FEA model, which involves defining the model and then splitting it into finite elements
- Solving for desired results
- Post-processing for results analysis

We will follow the above three steps in every exercise. From the perspective of FEA methodology, we can list the following FEA steps:

- Building the mathematical model
- Building the finite element model by discretizing the mathematical model
- Solving the finite element model
- Analyzing the results

The following subsections discuss these four steps.

Building the mathematical model

The starting point to analysis with **SOLIDWORKS Simulation** is a **SOLIDWORKS** model. Geometry of the model needs to be meshable into a correct finite element mesh. This requirement of mesh-ability has very important implications. We need to ensure that the CAD geometry will indeed mesh and that the produced mesh will provide the data of interest (e.g., stresses, displacements, or temperature distribution) with acceptable accuracy.

The necessity to mesh often requires modifications to the CAD geometry, which can take the form of defeaturing, idealization, and/or clean-up:

Term	Description
Defeaturing	The process of removing geometry features deemed insignificant for analysis, such as outside fillets, chamfers, logos, etc.
Idealization	A more aggressive exercise that may depart from solid CAD geometry by, for example, representing thin walls with surfaces and beams with lines.
Clean-up	Sometimes needed because geometry must satisfy high quality requirements to be meshable. To clean-up, we can use CAD quality control tools to check for problems like, for example, sliver faces, multiple entities, etc. that could be tolerated in the CAD model but would make subsequent meshing difficult or impossible.

It is important to mention that we do not always simplify the CAD model with the sole objective of making it meshable. Often, we must simplify a model even though it would mesh correctly "as is," because the resulting mesh would be large (in terms of the number of elements) and consequently, the meshing and the analysis would take too long. Geometry modifications allow for a simpler mesh and shorter solution times.

Sometimes, geometry preparation may not be required at all. Successful meshing depends as much on the quality of geometry submitted for meshing as it does on the capabilities of the meshing tools implemented in the FEA software.

Having prepared a meshable, but not yet meshed geometry, we now define material properties (these can also be imported from a CAD model), loads and restraints, and provide information on the type of analysis that we wish to perform. This procedure completes the creation of the mathematical model (Figure 1-2). Notice that the process of creating the mathematical model is not FEA specific. FEA has not yet entered the picture.

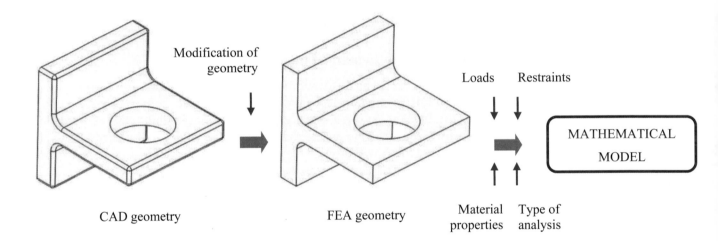

Figure 1-2: Building the mathematical model.

The process of creating a mathematical model consists of the modification of CAD geometry (here removing outside fillets), definition of loads, restraints, material properties, and definition of the type of analysis (for example linear static) that is to be performed.

You may review the differences between CAD geometry and FEA geometry using part model BRACKET_DEMO in two configurations: *01 fully featured* and *02 defeatured*.

Building the finite element model

The mathematical model now needs to be split into finite elements using the process of discretization, more commonly known as meshing. Geometry, loads, and restraints are all discretized. The discretized loads and restraints are applied to the nodes of the finite element mesh which is a discretized representation of geometry.

Solving the finite element model

Having created the finite element model, we use one of solvers provided in **SOLIDWORKS Simulation** to produce results (Figure 1-3).

Analyzing the results

Often the most difficult step of FEA is analyzing the results. Proper interpretation of results requires that we understand all simplifications
(and errors they introduce) in the first three steps: defining the mathematical model, meshing, and solving.

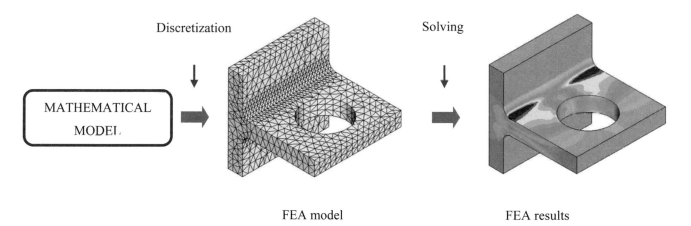

FEA model   FEA results

Figure 1-3: Building the finite element model.

The mathematical model is discretized into a finite element model. This completes the pre-processing phase. The FEA model is then solved with one of Simulation solvers. This illustration shows BRACKET DEMO model.

Errors in FEA

The process illustrated in Figure 1-2 and Figure 1-3 introduces unavoidable errors. Formulation of a mathematical model introduces modeling errors (also called idealization errors), discretization of the mathematical model introduces discretization errors, and solving introduces solution errors. Of these three types of errors, only discretization errors are specific to FEA. Modeling errors affecting the mathematical model are introduced before FEA is utilized and can only be controlled by using correct modeling techniques. Solution errors are caused by the accumulation of round-off errors.

A closer look at finite elements

Meshing splits continuous mathematical models into finite elements. The type of elements created by this process depends on the type of geometry meshed. **SOLIDWORKS Simulation** offers three types of three dimensional (3D) elements: solid elements for meshing solid geometry, shell elements for meshing surface geometry and beam elements for meshing curves (wire frame geometry). **SOLIDWORKS Simulation** also works with two dimensional (2D) elements: plane stress elements, plane strain elements, and axisymmetric elements.

Before proceeding, we need to clarify an important terminology issue. In CAD terminology, "solid" denotes the type of geometry: solid geometry to differentiate it from surface or curve (wire frame) geometry. In FEA terminology, "solid" denotes the type of element used to mesh solid CAD geometry.

Solid elements

The type of geometry that is most often used for analysis with **SOLIDWORKS Simulation** is solid CAD geometry. Meshing of this geometry is accomplished with tetrahedral solid elements. The tetrahedral solid elements in **SOLIDWORKS Simulation** can either be first order elements (called "draft quality"), or second order elements (called "high quality"). The user decides whether to use draft quality or high-quality elements for meshing. However, as we will soon prove, only high-quality elements should be used for an analysis of any importance. The difference between first and second order solid tetrahedral elements is illustrated in Figure 1-4.

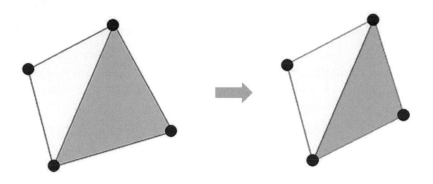

First order tetrahedral element
before deformation

First order tetrahedral element
after deformation

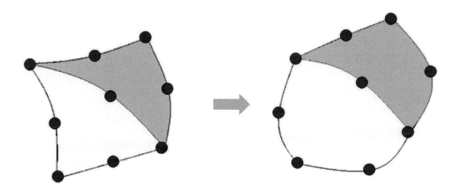

Second order tetrahedral element
before deformation

Second order tetrahedral element
after deformation

Figure 1-4: Differences between first and second order solid tetrahedral elements.

First and second order tetrahedral elements are shown before and after deformation. Notice that the first order element has corner nodes only, while the second order element has both corner and mid-side nodes; mid-side nodes are not visible. Single elements seldom experience deformations of this magnitude, which are exaggerated in this illustration.

In a first order element, edges are straight, and faces are flat. After deformation the edges and faces must retain these properties.

The edges of a second order element before deformation may either be straight or curvilinear, depending on how the element has been mapped to model the actual geometry. Consequently, the faces of a second order element before deformation may be flat or curved.

After deformation, edges of a second order element may either assume a different curvilinear shape or acquire a curvilinear shape if they were initially straight. Consequently, faces of a second order element after deformation can be either flat or curved.

First order tetrahedral elements model the linear field of displacement inside their volume, on faces, and along edges. The linear (or first order) displacement field gives these elements their name: first order elements.

If you recall from Mechanics of Materials, strain is the first derivative of displacement. Since the displacement field is linear, the strain field is constant. Consequently, the stress field is also constant in first order tetrahedral elements. This situation imposes a very severe limitation on the capability of a mesh constructed with first order elements to model the stress distribution of any complex model. To make matters worse, straight edges and flat faces cannot map properly to curvilinear geometry, as illustrated in Figure 1-5, left.

Second order tetrahedral elements have ten nodes (Figure 1-4) and model the second order (parabolic) displacement field and first order (linear) stress field in their volume, on faces and along edges. The edges and faces of second order tetrahedral elements can be curvilinear before and after deformation, therefore these elements can be mapped precisely to curved surfaces, as illustrated in Figure 1-5 right. Even though these elements are more computationally demanding than first order elements, second order tetrahedral elements are used for the majority of analyses with **SOLIDWORKS Simulation** because of their better mapping and stress modeling capabilities.

A tetrahedral solid element is the only type of solid element available in **SOLIDWORKS Simulation**.

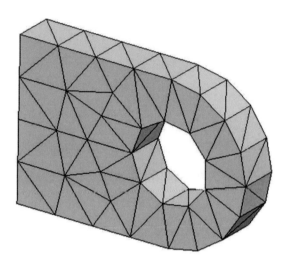
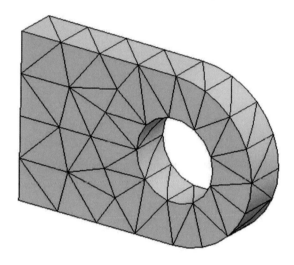

Mapping a curvilinear geometry with first order solid tetrahedral elements.

Mapping a curvilinear geometry with second order solid tetrahedral elements.

Figure 1-5: Failure of straight edges and flat faces to map to curvilinear geometry when using first order elements (left), and precise mapping to curvilinear geometry using second order elements (right).

Notice the imprecise first order element mapping of the hole; flat faces approximate the face of the curvilinear geometry. Second order elements map well to curvilinear geometry.

The mesh shows elements that are way too large for any meaningful analysis. Large elements are used only for clarity of this illustration.

This illustration shows LUG_01 model.

Shell elements

Shell elements are created by meshing surfaces or faces of solid geometry. Shell elements are primarily used for analyzing thin-walled structures. Since surface geometry does not carry information about thickness, the user must provide this information. Similar to solid elements, shell elements also come in draft and high quality with analogous consequences with respect to their ability to map to curvilinear geometry, as shown in Figure 1-6.

Mapping a cylinder with first order triangular shell elements

Mapping a cylinder with second order triangular shell elements

Figure 1-6: Mapping with first order shell elements (left) and second order shell elements (right).

The shell element mesh on the left was created with first order elements. Notice the imprecise mapping of the mesh to curvilinear geometry. The shell element mesh on the right was created with second order elements, which map correctly to curvilinear geometry. Large elements are used for clarity of this illustration.

This illustration shows model RING_01.

We need to make two important comments about Figure 1-5 and Figure 1-6. First, a mesh should never be that coarse (large size of elements compared to the model). We use a coarse mesh only to show the differences between first and second order elements clearly. Second, notice the "kinks" on the side of the second order elements; they indicate locations of mid side nodes. The second order element does map precisely to second order geometry.

As in the case of solid elements, first order shell elements model linear displacements and constant strain and stress. Second order shell elements model second order (parabolic) displacement and linear strain and stress.

The assumptions of modeling first or second order displacements in shell elements apply only to in-plane directions. The distribution of in-plane stresses across the thickness is assumed to be linear in both first and second order shell elements.

Triangular elements are the only type of shell elements available in **SOLIDWORKS Simulation**.

Certain classes of shapes can be modeled using either solid or shell elements, such as the plate shown in Figure 1-7. Often the nature of the geometry dictates what type of element should be used for meshing. For example, a part produced by casting would be meshed with solid elements, while a sheet metal structure would be best meshed with shell elements.

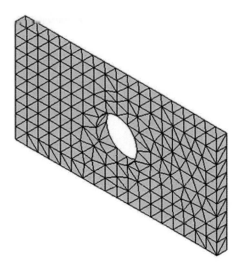
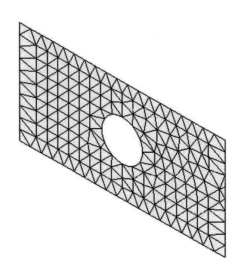

Second order tetrahedral solid elements   Second order triangular shell elements

Figure 1-7: Plate modeled with solid elements (left) and shell elements (right).

The actual choice between solids and shells depends in this case on the requirements of analysis.

Beam elements

Beam elements are created by meshing curves (wire frame geometry). They are a natural choice for meshing weldments. Assumptions about the stress distribution in two directions of the beam cross section are made.

A beam element does not have any physical dimensions in the directions normal to its length. It is possible to think of a beam element as a line with assigned beam cross section properties (Figure 1-8).

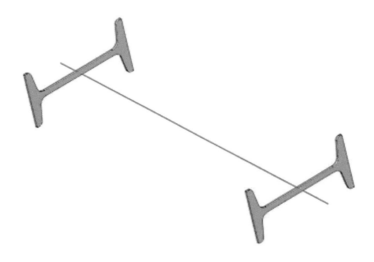

Figure 1-8: Conceptual representation of a beam element.

A beam element is a line with assigned properties of a beam cross section as required by beam theory. This illustration conceptualizes how a curve (here a straight line) defines an I-beam but does not represent actual geometry of beam cross-section.

Before we proceed with the classification of finite elements, we need to introduce the concept of nodal degrees of freedom which are of paramount importance in FEA. The degrees of freedom (DOF) of a node in a finite element mesh define the ability of the node to perform translation and rotation. The number of degrees of freedom that a node possesses depends on the element type. In **SOLIDWORKS Simulation**, nodes of solid elements have three degrees of freedom, while nodes of shell elements have six degrees of freedom.

This is because to describe the transformation of a solid element from the original to the deformed shape, we only need to know three translational components of nodal displacement. In the case of shell and beam elements, we need to know the translational components of nodal displacements and the rotational displacement components.

Using solid elements, we study how a 3D structure deforms under a load. Using shell elements, we study how a 3D structure with one dimension "collapsed" deforms under a load. This collapsed dimension is thickness which is not represented explicitly in the model geometry.

Beam elements are intended to study 3D structures with two dimensions removed from the geometry and not represented explicitly by model geometry. It is important to point out that solids, shells and beams are all 3D elements capable of deformation in 3D space.

2D elements

There are also cases where a structure's response to load can be fully described by 2D elements that have only two in-plane degrees of freedom. These are plane stress, plane strain and axisymmetric elements.

Plane stress elements are intended for the analysis of thin planar structures loaded in-plane, where out-of-plane stress is assumed to be equal to zero. Plane strain elements are intended for the analysis of thick prismatic structures loaded in-plane, where out-of-plane strain is assumed to be equal zero. Axisymmetric elements are intended for the analysis of axi-symmetric structures under axisymmetric load. In all of these special cases, the structure deformation can be fully described using elements with only two degrees of freedom per node. For plane stress and plane strain, these are two components of in-plane translation. For axisymmetric elements, these are radial and axial displacements.

2D elements are summarized in Figure 1-9. Just like solids and shells, 2D elements may be of first or second order.

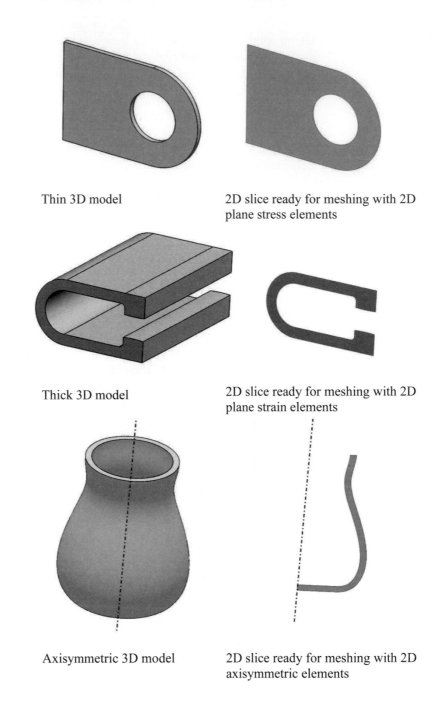

Thin 3D model

2D slice ready for meshing with 2D plane stress elements

Thick 3D model

2D slice ready for meshing with 2D plane strain elements

Axisymmetric 3D model

2D slice ready for meshing with 2D axisymmetric elements

Figure 1-9: Application of 2D elements: plane stress (top), plane strain (middle), and axisymmetric (bottom).

Figure 1-9 shows (from top to bottom) models LUG_02, CONTACT_01, VASE_01.

Figure 1-10 presents the basic library of elements in **SOLIDWORKS Simulation**. Solid elements are tetrahedral, shell elements and 2D elements are triangles and beam elements are lines. Elements such as hexahedral solids or quadrilateral shells are not available in **SOLIDWORKS Simulation**.

	3D elements			2D elements
	Solid elements	**Shell elements**	**Beam elements**	**Plate elements**
First order element Linear (first order) displacement field Constant stress field	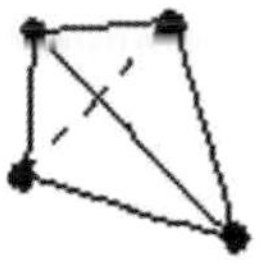		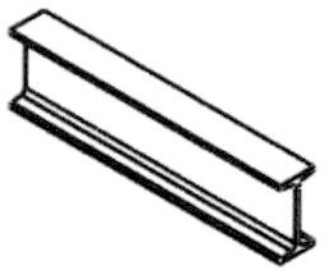	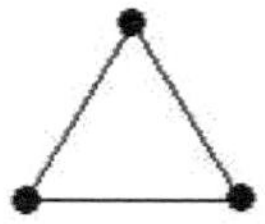
Second order element Parabolic (second order) displacement field Linear stress field	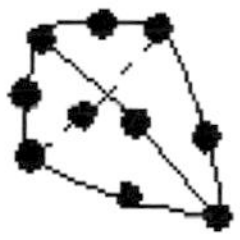	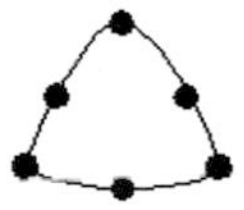		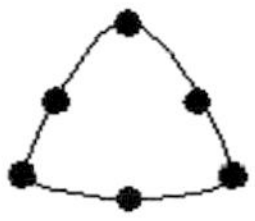

Figure 1-10: Basic element library of SOLIDWORKS Simulation.

Most analyses use the second order tetrahedral element. Element order is not applicable to beam elements. 2D elements are sometimes referred to as plate elements.

Beam element cross-section is defined by cross section and second moments of inertia of the analyzed structural member. It does not have any physical shape. The above I-beam is shown only for the ease of visualization.

What is calculated in FEA?

Each degree of freedom of a node in a finite element mesh constitutes an unknown. In structural analysis, nodal degrees of freedom represent displacement components, while in thermal analysis they represent temperatures. Nodal displacements and nodal temperatures are the primary unknowns for structural analysis and thermal analysis, respectively.

Structural analysis finds displacements, strains and stresses. If solid elements are used, then three displacement components (three translations) per node must be calculated. With shell and beam elements, six displacement components (six translations) must be calculated. 2D elements require calculations of two displacement components. Strains and stresses are calculated based on the nodal displacement results.

Thermal analysis finds temperatures, temperature gradients, and heat flow. Since temperature is a scalar value (unlike displacements, which are vectors), then regardless of what type of element is used, there is only one unknown (temperature) to be found for each node. All other thermal results such as temperature gradient and heat flux are calculated based on temperature results. The fact that there is only one unknown to be found for each node, rather than three or six, makes thermal analysis less computationally intensive than structural analysis.

How to interpret FEA results

Results of structural FEA are provided in the form of displacements and stresses. But how do we decide if a design "passes" or "fails"? What constitutes a failure?

To answer these questions, we need to establish some criteria to interpret FEA results, which may include maximum acceptable displacements, maximum stress, or the lowest acceptable natural frequency.

While displacement and frequency criteria are obvious and easy to establish, stress criteria are not. Let us assume that we need to conduct a stress analysis to ensure that stresses are within an acceptable range. To judge stress results, we need to understand the mechanism of potential failure. If a part breaks, what stress measure best describes that failure? **Simulation** can present stress results in many different forms, but it is up to us to decide which stress measures should be used to analyze results.

Discussion of various failure criteria is out of the scope of this book. Any textbook on the Mechanics of Materials provides information on this topic. Here we will limit our discussion to commonly used failure criteria: Von Mises Stress failure criterion and Maximum Tensile Stress failure criterion.

Von Mises Stress failure criterion

Von Mises stress, also known as Huber stress, is a stress measure that accounts for all six stress components of a general 3-D state of stress (Figure 1-11).

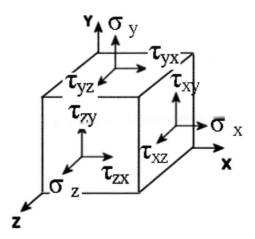

Figure 1-11: General state of stress represented by three normal stresses: σ_x, σ_y, σ_z and six shear stresses.

Two components of shear stress and one component of normal stress act on each side of this elementary cube. Due to symmetry of shear stresses, the general 3D state of stress is characterized by six stress components: σ_x, σ_y, σ_z and $\tau_{xy} = \tau_{yx}$, $\tau_{yz} = \tau_{zy}$, $\tau_{xz} = \tau_{zx}$.

Von Mises stress σ_{vm}, can be expressed by six stress components as:

$$\sigma_{vm} = \sqrt{0.5 \times \left[(\sigma_x - \sigma_y)^2 + (\sigma_y - \sigma_z)^2 + (\sigma_z - \sigma_x)^2\right] + 3 \times \left(\tau_{xy}^2 + \tau_{yz}^2 + \tau_{zx}^2\right)}$$

Von Mises stress σ_{vm}, can be also expressed by three principal stresses (Figure 1-12) as:

$$\sigma_{vm} = \sqrt{0.5 \times \left[(\sigma_1 - \sigma_2)^2 + (\sigma_2 - \sigma_3)^2 + (\sigma_3 - \sigma_1)^2\right]}$$

Notice that von Mises stress is a non-negative, scalar stress measure. Von Mises stress is commonly used to present results because the structural safety for many engineering materials showing elasto-plastic properties (for example, steel or aluminum alloy) can be evaluated using von Mises stress.

The maximum von Mises stress failure criterion is based on the von Mises-Hencky theory, also known as the shear-energy theory or the maximum distortion energy theory. The theory states that a ductile material starts to yield at a location when the von Mises stress becomes equal to the stress limit. In most cases, the yield strength is used as the stress limit. According to the von Mises failure criterion, the factor of safety (FOS) is expressed as:

$$FOS = \sigma_{limit} / \sigma_{vm}$$

where σ_{limit} is the yield strength.

Maximum Tensile Stress failure criterion

By properly adjusting the angular orientation of the stress cube in Figure 1-11, shear stresses disappear, and the state of stress is represented only by three principal stresses: σ_1, σ_2, σ_3, as shown in Figure 1-12. In **Simulation**, principal stresses are denoted as P1, P2, P3.

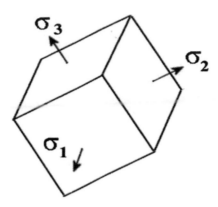

Figure 1-12: The general state of stress represented by three principal stresses: σ_1, σ_2, σ_3.

The Maximum Normal Stress Failure criterion is used for <u>brittle</u> materials. Brittle materials do not have a specific yield point. This criterion assumes that the ultimate tensile strength of the material in tension and compression is the same. This assumption is not valid in all cases. For example, cracks considerably decrease the strength of the material in tension while their effect is not significant in compression because the cracks tend to close.

This criterion predicts that failure will occur when σ_1 exceeds the ultimate tensile strength. According to the maximum principal stress failure criterion, the factor of safety FOS is expressed as:

$$FOS = \sigma_{limit}/\sigma_1$$

where σ_{limit} is the tensile strength.

Units of measure

Internally, **Simulation** uses the International System of Units (SI). However, for the user's convenience, the unit manager allows data entry in any of three systems of units: SI, Metric, and English. Results can be displayed using any of the three systems. Figure 1-13 summarizes the available systems of units.

	International System (SI)	Metric (MKS)	English (IPS)
Mass	kg	kg	lb
Length	m	cm	in
Time	s	s	s
Force	N	Kgf	lbf
Gravitational acceleration	m/s^2	G	in/s^2
Mass density	kg/m^3	kg/cm^3	lbf/in^3
Temperature	K	C	F

Figure 1-13: Unit systems available in SOLIDWORKS Simulation.

SI, Metric, and English systems of units can be interchanged when entering data and analyzing results.

As **Simulation** users, we are spared much confusion and trouble with systems of units. However, we may be asked to prepare data or interpret the results of other FEA software where we do not have the convenience of the unit manager. Therefore, we will make some general comments about the use of different systems of units in the preparation of input data for FEA models. We can use any consistent system of units for FEA models, but in practice, the choice of the system of units is dictated by what units are used in the CAD model. The system of units in CAD models is not always consistent; length can be expressed in [*mm*], while mass density can be expressed in [*kg/m³*]. Contrary to CAD models, in FEA all units must be consistent. Inconsistencies are easy to overlook, especially when defining mass and mass density, and that can lead to serious errors.

In the SI system, based on meters [*m*] for length, kilograms [*kg*] for mass, and seconds [*s*] for time, all other units are easily derived from these base units. In mechanical engineering, length is commonly expressed in millimeters [*mm*], force in Newtons [*N*], and time in seconds [*s*]. All other units must then be derived from these basic units: [*mm*], [*N*], and [*s*]. Consequently, the unit of mass is defined as a mass which, when subjected to a unit force equal to 1N, will accelerate with a unit acceleration of 1 mm/s^2. Therefore, the unit of mass in a system using [*mm*] for length and [*N*] for force is equivalent to 1000 kg or one metric ton. Therefore, mass density is expressed in metric tonnes [*tonne/mm^3*]. This is critically important to remember when defining material properties in FEA software without a unit manager. Review Figure 1-14 and notice that an erroneous definition of mass density in [*kg/m^3*] rather than in [*tonne/mm^3*] results in mass density being one trillion (10^{12}) times higher.

System SI	[m] [N] [s]
Unit of mass	kg
Unit of mass density	kg/m^3
Density of aluminum	2794 kg/m^3

System of units derived from SI	[mm] [N] [s]
Unit of mass	tonne
Unit of mass density	tonne/mm^3
Density of aluminum	2.794 x 10^{-9} tonne/mm^3

English system (IPS)	[in] [lbf] [s]
Unit of mass	slug/12
Unit of mass density	slug/12/in^3
Density of aluminum	2.614 x 10^{-4} slug/12/in^3

Figure 1-14: Mass density of aluminum in the three systems of units.

Comparison of numerical values of mass densities of 1060 aluminum alloy defined in the SI system of units with the system of units derived from SI, and with the English (IPS) system of units.

Using online help

Simulation features very extensive online Help and Tutorial functions, which can be accessed from the Help menu in the main **SOLIDWORKS** tool bar or from the **Simulation** menu. The Study advisor can be accessed from the Study drop-down menu (Figure 1-15).

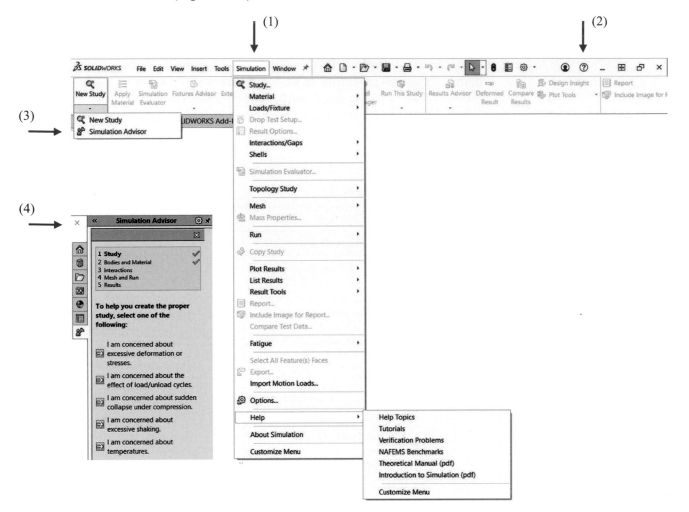

Figure 1-15: Accessing online Help, and Simulation Advisor.

Online Help and Tutorials can be accessed from the Simulation menu (1) as shown in this illustration or from the main tool bar by selecting Help (2). The Simulation Advisor can be accessed from the Study drop-down menu (3). Simulation Advisor window appears in the Task Pane (4).

Limitations of Static studies

Static study is the only type of study available in some **SOLIDWORKS** packages. Working with **Static** study we need to use important simplifications.

Linear material

Material is assumed to be linear material model. A comparison between linear and nonlinear material is shown in Figure 1-16.

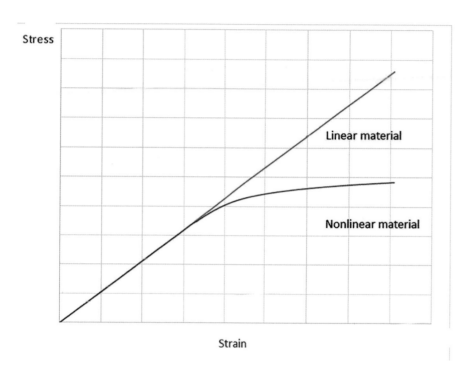

Figure 1-16: Comparison between linear and nonlinear material models.

With a linear material, stress is linearly proportional to strain. The linear range is where the linear and nonlinear material models are not significantly different.

Using a linear material model, the maximum stress magnitude is not limited at all. Material yielding is not modeled, and whether or not yield is taking place can only be established based on the stress magnitudes reported in results. Most analyzed structures experience stresses below the yield stress, and the factor of safety is most often related to the yield stress.

Static loads

All structural loads and restraints are assumed not to change with time. Dynamic loading conditions cannot be analyzed with **Static** study. This limitation implies that loads are applied slowly, and inertial effects may be ignored.

Nonlinear material analysis and loads changing with time dynamic analysis may be analyzed with **SOLIDWORKS Simulation Premium.**

Models used for illustrations

Model
BRACKET_DEMO.sldprt
LUG_01.sldprt
LUG_02.sldprt
RING_01.sldprt
CONTACT_01.sldprt
VASE_01.sldprt

2: Static analysis of a plate

Topics covered

- Using the **SOLIDWORKS Simulation** interface
- Linear static analysis with solid elements
- Controlling discretization error with the convergence process
- Finding reaction forces
- Presenting FEA results in a desired format

Project description

A steel plate is supported and loaded, as shown in Figure 2-1. We assume that the support is rigid. Rigid support is also called built-in support, fixed support, or fixed restraint. A 100000N tensile load is uniformly distributed along the end face, opposite to the supported face.

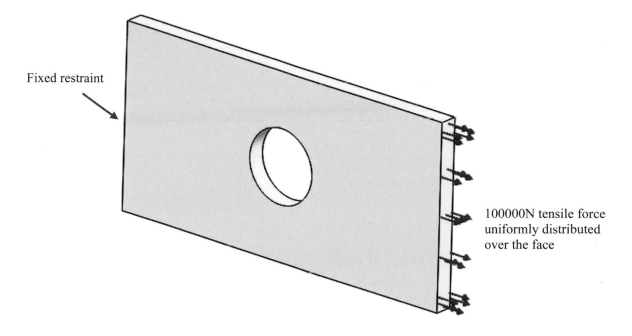

Figure 2-1: SOLIDWORKS model of a rectangular plate with a hole.

We will perform a displacement and stress analysis using meshes with different element sizes. Notice that repetitive analysis with different meshes does not represent standard practice in FEA. However, repetitive analysis with different meshes produces results which are useful in gaining more insight into how FEA works.

Procedure

In **SOLIDWORKS**, open the model file called HOLLOW PLATE. Verify that **SOLIDWORKS Simulation** is selected in the **Add-Ins** list (Figure).

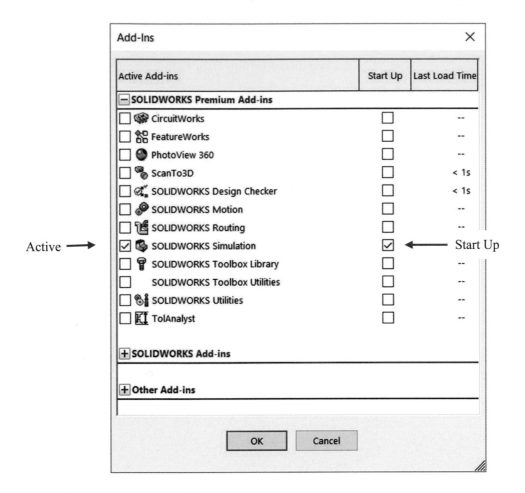

Figure 2-2: Add-Ins list in SOLIDWORKS.

Verify that SOLIDWORKS Simulation is selected as Active Add-in and Start Up Add-in.

Once **Simulation** has been added, it shows in the main **SOLIDWORKS** menu and in the Command Manager.

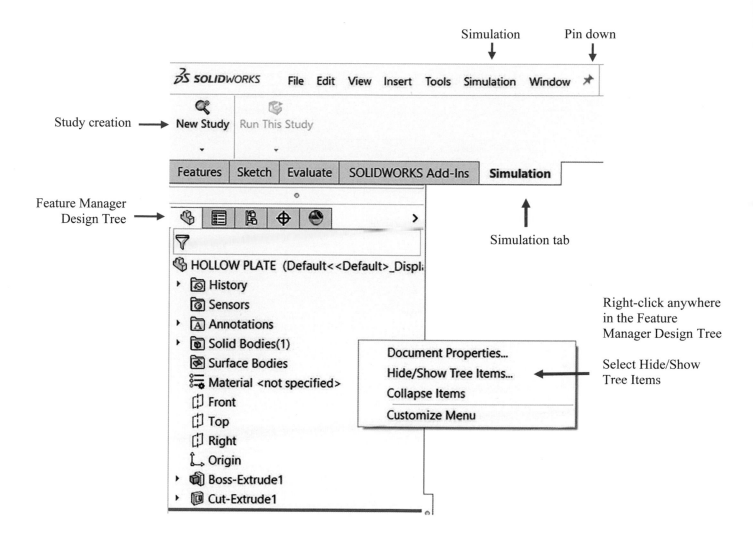

Figure 2-3: The Simulation tab is a part of the SOLIDWORKS Command Manager.

Selecting the Simulation tab in the Command Manager displays Simulation menu items (icons). Since no study has yet been created, only the Study creation icon is available; all others are grayed-out. For convenience, pin down the top tool bar as shown. Command Manager with Large Buttons is shown in this illustration.

Notice that **Feature Manager Design Tree** shown in Figure 2-3 displays **Solid Bodies** and **Surface Bodies** folders. These folders can be displayed by right-clicking anywhere in **Feature Manager Design Tree** to bring up the pop-up menu and selecting **Hide/Show Tree Items**. This will invoke **System Options-Feature Manager** (not shown here). From there, **Solid Bodies** and **Surface Bodies** folder can be selected to show. We will need to distinguish between these two different bodies in later exercises. In this exercise these two folders do not need to show.

Before we create a study, let's review the **Simulation** main menu (Figure 2-4) along with its **Options** window (Figure 2-5).

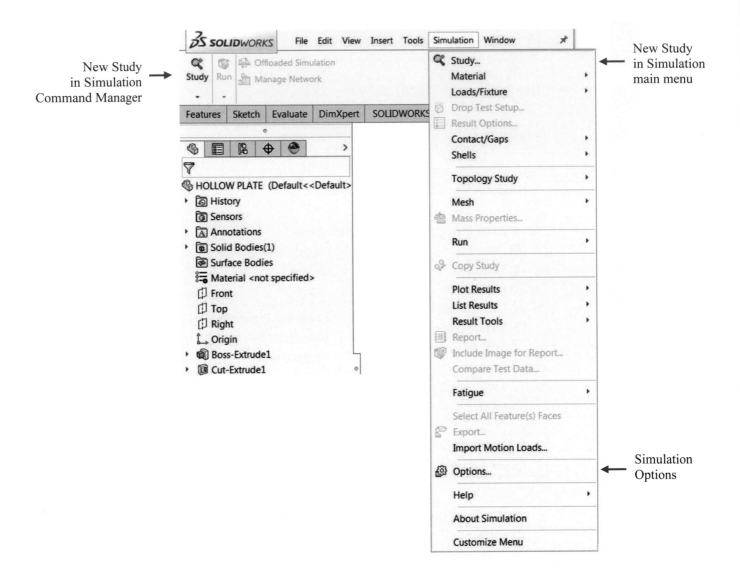

Figure 2-4: Simulation main menu.

Like the Simulation Command Manager shown in Figure 2-3, only the New Study icon is available. Notice that some commands are available both in the Command Manager and in the Simulation menu.

Simulation studies can be executed entirely from the **Simulation** drop-down menu shown in Figure 2-4. In this book we will use the **Simulation** main menu and/or Command Manager to create a new Study.

Click on the **Simulation Options** shown in Figure 2-4 to open the **Simulation Options** window (Figure 2-5); select **Default Options** tab and **Units** from the menu in **Default Options**.

Default Options

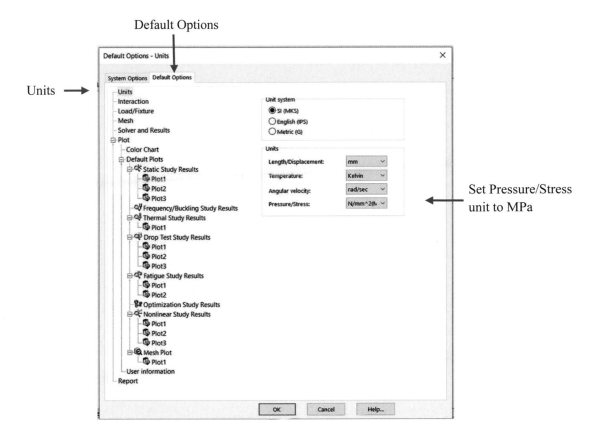

Units →

Set Pressure/Stress unit to MPa

Figure 2-5: Simulation Options window.

The Options window has two tabs: system Options and Default Options. In this illustration, Default Options tab is shown.

In the **Units** options, make the choices shown in Figure 2-5. In this book we will mostly use the SI system of units using MPa rather than Pa as a unit of stress and pressure. Occasionally we will switch to the IPS system.

Notice that **Default Plots** can be added, modified, deleted, or grouped into sub-folders which are created by right-clicking on the results folders, for example, **Static Study Results** folder, **Thermal Study Results** folder, etc.

In this chapter we will return to **Default Options** (Figure 2-5) twice to review choices in **Mesh** and in **Solver and Results**.

Creation of an FEA model starts with the definition of a study. To define a new study, select **New Study** in either the **Simulation** tab in the Command Manager (Figure 2.3) or **Simulation** main menu (Figure 2-4). This will open the **Study Property Manager**. Notice that the **New Study** icon in the **Simulation** Command Manager can also be used to open the **Study Advisor**. We won't be using the **Study Advisor** in this book. Name the study *tensile load 01* (Figure 2-6).

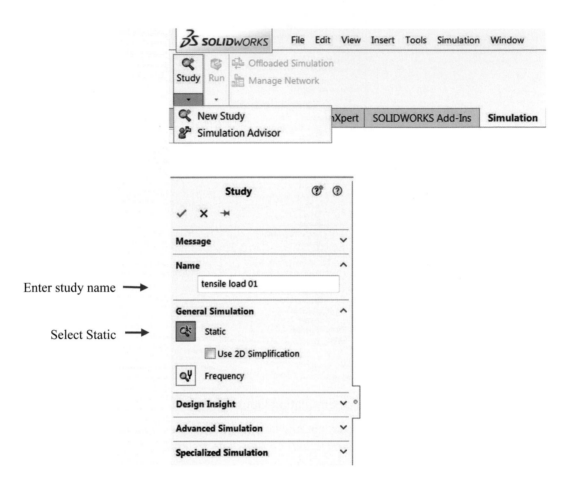

Figure 2-6: Creating a new study.

The study definition window offers choices of the type of study; here we select Static. Static study and Frequency study belong to the group called General Simulation.

Once a new study has been created, **Simulation** Commands can be invoked in three ways:

- From the Simulation Command Manager (Figure 2-4)
- From the Simulation main menu (Figure 2-4); in this book, we will most often use this method

When a study is defined, **Simulation** creates a study window located below the **Feature Manager Design Tree** and places several folders in it. It also adds a study tab located next to **Model** and **Motion Study** tabs. The tab provides access to the study (Figure 2-7). We won't be using the **Motion Studies** in this book.

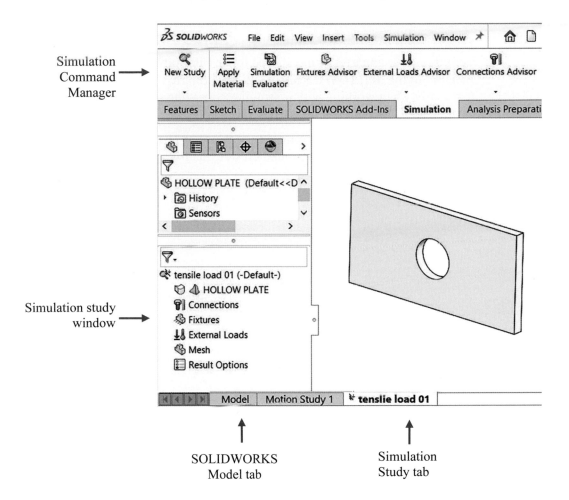

Figure 2-7: The Simulation window and Simulation tab.

You can switch between the SOLIDWORKS Model and Simulation Studies by selecting the appropriate tab.

We are now ready to define the analysis model. This process generally consists of the following steps:

- CAD geometry idealization and/or simplification in preparation for analysis. This is usually done in **SOLIDWORKS** by creating an analysis specific configuration and making your changes there
- Material properties assignment
- Restraints definition
- Loads definition

In this case, the geometry does not need any preparation because it is already very simple; we can start by assigning material properties.

Notice that if a material is defined for a **SOLIDWORKS** part model, the material definition is automatically transferred to the **Simulation** model. Assigning a material to the **SOLIDWORKS** model is a preferred modeling technique, especially when working with assemblies consisting of parts with different materials. We will do this in later exercises.

To apply material to the **Simulation** model, right-click the HOLLOW PLATE folder in the *tensile load 01* simulation study and select **Apply/Edit Material** from the pop-up menu (Figure 2-8).

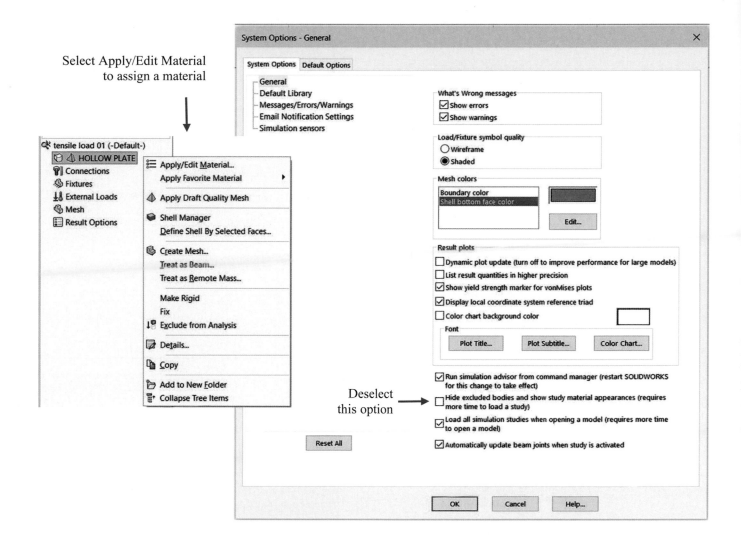

Figure 2-8: Assigning material properties.

The left window shows the first step in the process of applying material properties to the model. The right window shows modification to System Options; this needs to be done only once.

The action in Figure 2-8 opens the **Material** window shown in Figure 2-9.

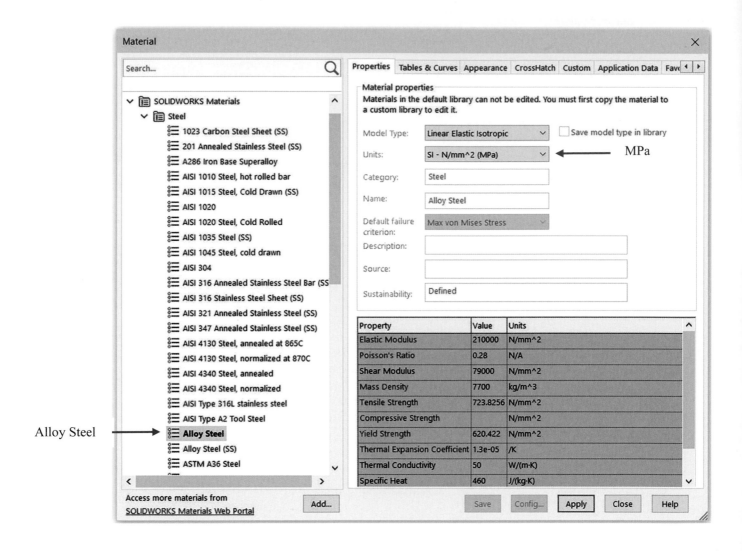

Figure 2-9: Material window.

Select Alloy Steel to be assigned to the model; select MPa as units.

Click Apply, and then click Close.

In the **Material** window, the properties are color coded to indicate the mandatory and optional properties. Red color (Elastic modulus, Poisson's ratio, Mass Density, Yield Strength) indicates a property that is mandatory based on the active study type and the material model. Blue color (Tensile strength, Compressive Strength, Thermal expansion coefficient) indicates optional properties. A black color description indicates properties calculated from the mandatory properties (Shear Modulus) or properties not applicable to the current study.

The analysis will not use volume loads; therefore, Mass Density will not be used.

In the **Material** window, open the **SOLIDWORKS Materials** menu, followed by the **Steel** menu. Select **Alloy Steel.** Select **SI** units under the **Properties** tab (other units could be used as well). Notice that the HOLLOW PLATE folder in the *tensile load 01* study now shows a check mark and the name of the selected material to indicate that a material has been assigned. If needed, you can define your own material by selecting **Custom Defined** material.

Material definition consists of two steps:

- Material selection (or material definition if a custom material is used)
- Material assignment (either to the entire model, to selected bodies of a multi-body part, or to selected components of an assembly)

Having assigned the material, we now move to defining the restraints. To display the pop-up menu that lists the options available for defining restraints, right-click the *Fixtures* folder in the *tensile load 01* study (Figure 2-10).

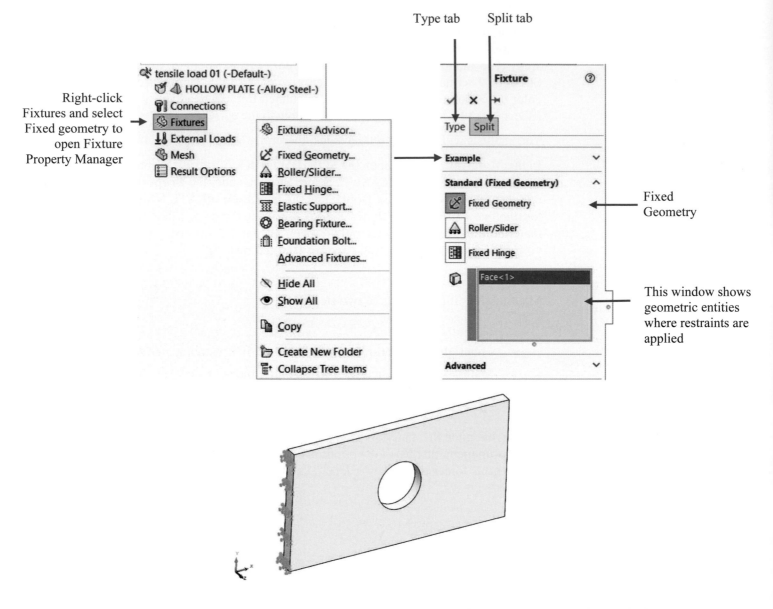

Figure 2-10: Pop-up menu for the Fixtures folder and Fixture definition window (Fixture Property Manager).

All restraints' definitions are done in the Type tab. The Split tab is used to define a split face where restraints are applied. The same can be done in SOLIDWORKS by defining a Split Face.

Once the **Fixtures** definition window is open, select the **Fixed Geometry** restraint type. Select the end-face entity where the restraint is to be applied. Click the green check mark in the Fixture Property manager window to complete the restraint definition.

Notice that in **SOLIDWORKS Simulation,** the term "Fixture" implies that the model is firmly "fixed" to the ground. However, aside from **Fixed Geometry**, which we have just used, all other types of fixtures restrain the model in certain directions while allowing movements in other directions. Therefore, the term "restraint" may better describe what happens when choices in the **Fixture** window are made. In this book we will switch between the terms "fixture" and "restraint" freely.

The existence of restraints is indicated by symbols shown in Figure 2-10. In the **Symbol** Settings of the Fixture window, the size of the symbol can be changed. Notice that symbols shown in Figure 2-10 are distributed over the highlighted face meaning the entire face has been restrained. Each symbol consists of three orthogonal arrows symbolizing directions where translations have been restrained. Each arrow has a disk symbolizing that rotations have also been restrained. The symbol implies that all six degrees of freedom (three translations and three rotations) have been restrained. However, the element type we will use to mesh this model (second order solid tetrahedral element) has only translational degrees of freedom. Rotational degrees of freedom can't be restrained because they don't exist in this type of element. Therefore, disks symbolizing restrained rotations are irrelevant in our model. Please see the following table for more explanations.

Before proceeding, explore other types of restraints accessible through the **Fixture** window. Restraints can be divided into two groups: **Standard** and **Advanced**. Review animated examples available in the **Fixture** window and review the following table. Some less frequently used types of restraints are not listed here.

Standard Fixtures

Fixed	Also called built-in or rigid support. All translational and all rotational degrees of freedom are restrained.
Immovable (No translations)	Only translational degrees of freedom are restrained, while rotational degrees of freedom remain unrestrained. If solid elements are used (like in this exercise), **Fixed** and **Immovable** restraints would have the same effect because solid elements do not have rotational degrees of freedom. Therefore, the **Immovable** restraint is not available if solid elements are used alone.
Roller/Slider	Specifies that a planar face can move freely on its plane but not in the direction normal to its plane. The face can shrink or expand under loading.
Fixed Hinge	Applies only to cylindrical faces and specifies that the cylindrical face can only rotate about its own axis. This condition is identical to selecting the **On cylindrical face** restraint type and setting the radial and axial components to zero.

Advanced Fixtures

Symmetry	Applies symmetry boundary conditions to a flat face. Translation in the direction normal to the face is restrained and rotations about the axes aligned with the face are restrained.
Cyclic symmetry	Allows analysis of a model with cyclic patterns around an axis by modeling a representative segment. The geometry, restraints, and loading conditions must be identical for all other segments making up the model. Turbine, fans, flywheels, and motor rotors can usually be analyzed using cyclic symmetry.
Use Reference Geometry	Restrains a face, edge, or vertex only in certain directions, while leaving the other directions free to move. You can specify the desired directions of restraint in relation to the selected reference plane or reference axis.
On Flat Faces	Provides restraints in selected directions, which are defined by the three directions of the flat face where restraints are being applied.
On Cylindrical Faces	This option is similar to **On flat face**, except that the three directions of a cylindrical face define the directions of restraints.
On Spherical Face	Similar to **On Flat Faces** and **On Cylindrical Faces**. The three directions of a spherical face define the directions of the applied restraints.

When a model is fully supported (as it is in our case), we say that the model does not have any rigid body motions (the term "rigid body modes" is also used), meaning it cannot move without experiencing deformation.

Notice that the presence of restraints in the model is manifested by both the restraint symbols (showing on the restrained face) and by the automatically created icon, **Fixture-1**, in the *Fixtures* folder. The display of the restraint symbols can be turned on and off by either:

- Right-click *Fixture*s folder and select **Hide All** or **Show All** in the pop-up menu shown in Figure 2-10

  or

- Right-click fixture icon and select **Hide** or **Show** from the pop-up menu.

Use the same method to control display of other **Simulation** symbols.

To define the load right-click the *External Loads* folder and select **Force** from the pop-up menu. This action opens the **Force** window as shown in Figure 2-11.

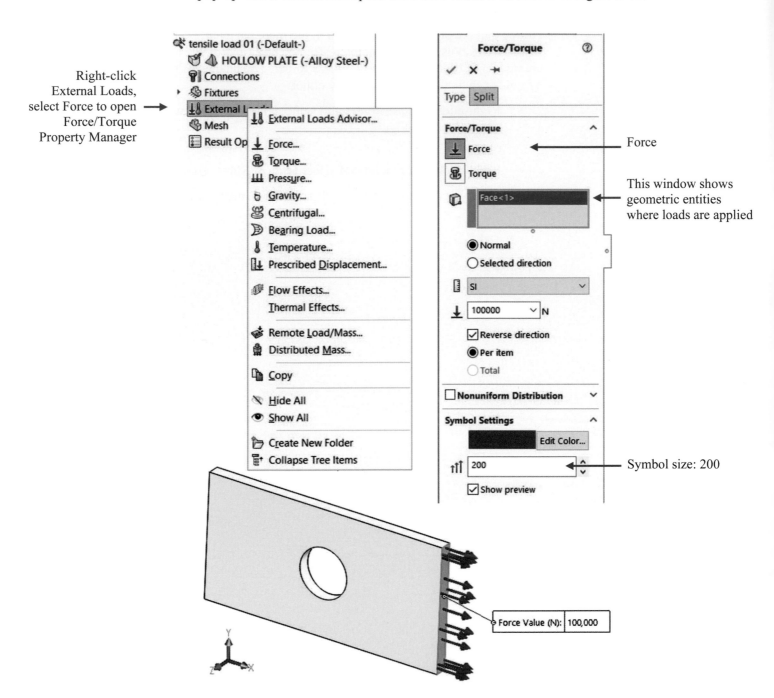

Right-click External Loads, select Force to open Force/Torque Property Manager

Force

This window shows geometric entities where loads are applied

Symbol size: 200

Force Value (N): 100,000

Figure 2-11: Pop-up menu for the External Loads folder and Force window.

The Force window displays the selected face where the tensile force is applied. If only one entity is selected, there is no distinction between Per Item and Total. In this illustration, load symbols have been enlarged by adjusting the Symbols Settings. Symbols of previously defined restraints have been hidden.

In the **Type** tab, select **Normal** in order to load the model with a 100000N tensile force uniformly distributed over the end face, as shown in Figure 2-11. Check the **Reverse direction** option to apply a tensile load.

Generally, forces can be applied to faces, edges, and vertices using different methods, which are reviewed below:

Force normal	Available for flat faces only, this option applies load in the direction normal to the selected face.
Force selected direction	This option applies a force or a moment to a face, edge, or vertex in the direction defined by the selected reference geometry. Moments can be applied only if shell elements are used. Shell elements have six degrees of freedom per node: three translations and three rotations. Nodes of shell element can take a moment load. Solid elements only have three degrees of freedom (translations) per node and, therefore, cannot take a moment load directly. If you need to apply moments to solid elements, they must be represented with appropriately applied forces.
Torque	This option applies torque (expressed by traction forces) about a reference axis using the right-hand rule.

Try using the click-inside technique to rename the **Fixture-1** and **Force/Torque-1** items. Notice that renaming using the click-inside technique works on all items in **SOLIDWORKS Simulation**.

The model is now ready for meshing. Before creating a mesh, let's make a few observations about defining the geometry, material properties, loads and restraints.

Geometry preparation is a well-defined step with few uncertainties. Geometry that is simplified for analysis can be compared with the original CAD model.

Material properties are most often selected from the material library and do not account for local defects, surface conditions, etc. Therefore, the definition of material properties usually has more uncertainties than geometry preparation.

The definition of loads is done in a few menu selections but involves many assumptions. Factors such as load magnitude and distribution are often only approximately known and must be assumed. Therefore, significant idealization errors can be made when defining loads.

Defining restraints is where severe errors are most often made. For example, it is easy enough to apply a fixed restraint without giving too much thought to the fact that a fixed restraint means a rigid support – a mathematical abstraction. A common error is over-constraining the model, which results in an overly stiff structure that underestimates displacements and stresses. The relative level of uncertainties in defining geometry, material, loads, and restraints is qualitatively shown in Figure 2-12.

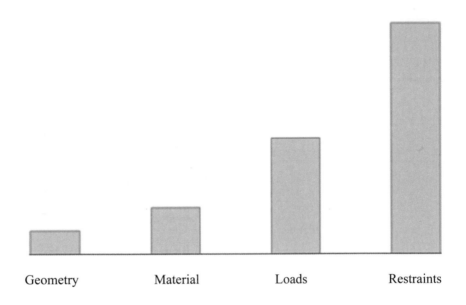

| Geometry | Material | Loads | Restraints |

Figure 2-12: Qualitative comparison of uncertainty in defining geometry, material properties, loads, and restraints.

The level of uncertainty (or the risk of error) has no relationship to the time required for each step, so the message in Figure 2-12 may be counterintuitive. In fact, preparing CAD geometry for FEA may take hours, while defining material properties and applying restraints and loads takes only a few clicks.

In all the examples presented in this book, we assume that definitions of material properties, loads, and restraints represent an acceptable idealization of real conditions. However, we need to point out that it is the responsibility of the FEA user to determine if all those idealized assumptions made during the creation of the mathematical model are indeed acceptable.

Before meshing the model, we need to verify under the **Default Options** tab, in the **Mesh** properties, that **High** quality, **Standard** mesh is selected (Figure 2-13). The **Options** window can be opened from the **SOLIDWORKS Simulation** menu as shown in Figure 2-4.

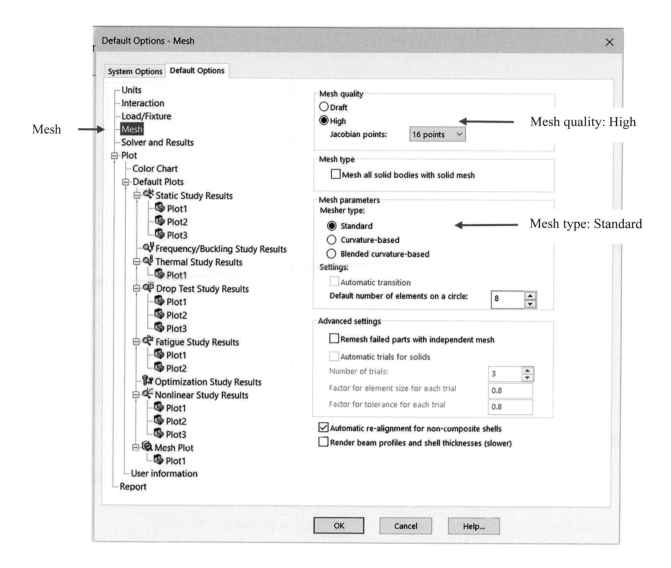

<u>Figure 2-13: Mesh settings in the Default Options window.</u>

High Mesh quality means that the second order elements are used. Standard mesh type is one of three types of meshed available in SOLIDWORKS Simulation. We will use curvature based meshes in later exercises.

The element order is visually indicated by a tetrahedron symbol next to Solid Body symbol in Study window. Tetrahedron with straight edges indicates the first order element. Tetrahedron with curvilinear edges indicates the second order element (Figure 2-14). If more than one body are present in a Simulation study, they can be meshed with different element order mesh. The current study has only one body called HOLLOW PLATE after the study name.

The difference between **High** and **Draft** mesh quality is:

- Draft quality mesh uses first order elements
- High quality mesh uses second order elements

Differences between first and second order elements were discussed in chapter 1.

The difference between **Curvature based** mesh and **Standard** mesh will be explained in chapter 3. Now, right-click the *Mesh* folder to display the pop-up menu (Figure 2-14).

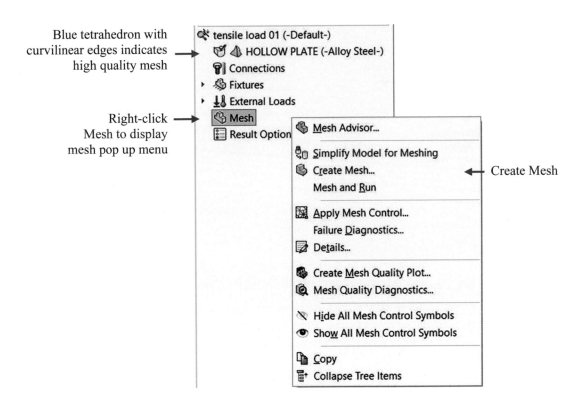

Figure 2-14: Mesh pop-up menu.

Select Create Mesh from the pop-up menu.

In the pop-up menu, select **Create Mesh**. This opens the **Mesh** window (Figure 2-15) which offers a choice of element size and element size tolerance.

This exercise demonstrates the impact of mesh size on results. Therefore, we will solve the same problem using three different meshes: coarse, medium (default), and fine. Figure 2-15 shows the respective selection of meshing parameters to create the three meshes. All three studies use the High Quality (second order) elements.

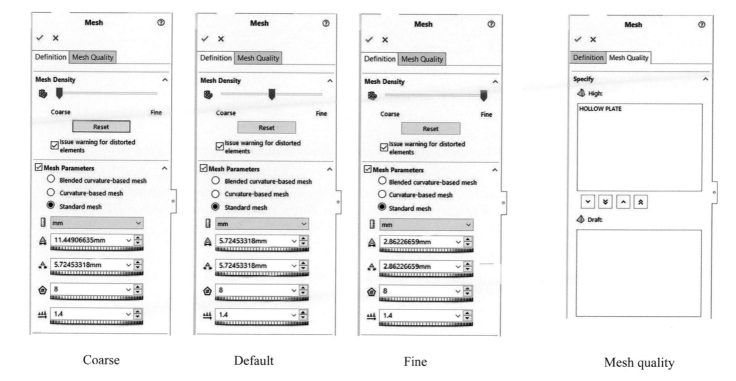

Coarse   Default   Fine   Mesh quality

Figure 2-15: Three mesh sizes: coarse, default, fine and mesh quality: high.

Show Mesh Parameters to see the element size. In all three cases use Standard mesh. Notice different slider positions in the three windows.

In all three cases use High Quality mesh (second order elements). Use Definition tab and Mesh Quality tab to switch between the above windows.

The medium mesh density (size), shown in the middle window in Figure 2-15, is the default that **SOLIDWORKS Simulation** proposes for meshing our model. The element size of 5.72 mm is established automatically based on the geometric features of the model. The 5.72 mm size is the characteristic element size in the mesh, as explained in Figure 2-16.

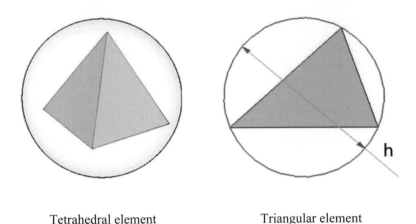

Tetrahedral element   Triangular element

Figure 2-16: Characteristic element size of a tetrahedral element and a triangular element.

The characteristic element size of a tetrahedral element can be defined as diameter h of a circumscribed sphere. This is easier to illustrate with a 2D analogy of a circle circumscribed on a triangle.

Element size has a direct impact on the accuracy of results. The smaller the elements, the lower the discretization error, but the meshing and solving time both take longer.

Right-click the Mesh folder again and select **Create…** to open the **Mesh** window. With the **Mesh** window open, set the slider all the way to the left (as illustrated in Figure 2-15, left) to create a coarse mesh, and click the green checkmark button. The mesh will be displayed as shown in Figure 2-17.

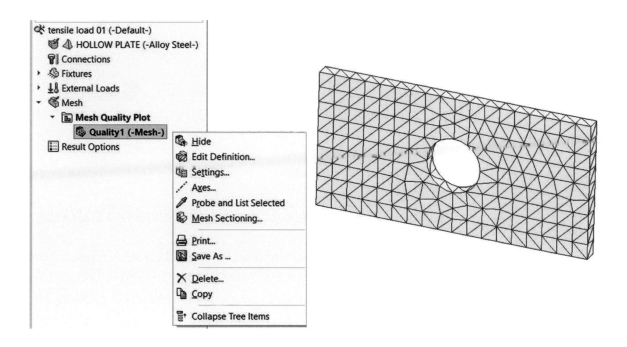

Figure 2-17: A coarse mesh created with second order, solid tetrahedral elements.

You can control the mesh display by selecting Hide Mesh or Show Mesh from pop-up menu in mesh plot. Here the plot is named Quality 1. Activate the pop-up menu by Right-click Quality 1 plot.

The presence of a mesh is reflected in the appearance of the solid folder and the mesh folder in a **Simulation** study (Figure 2-18).

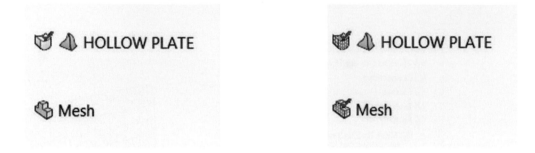

Before meshing   After meshing

Figure 2-18: Solid folder and mesh folder in a Simulation study before and after meshing.

After meshing, cross hatching is added to the Solid Body symbol. Assigning material properties adds a green check mark to the Solid Body symbol.

After meshing, cross hatching and a green check mark is added to Mesh folder.

When Solid folder contains more than one body, High Quality and Draft Quality meshes may be individually selected for each body. We do not use this option in this exercise.

To start the solution, right-click the *tensile load 01* study folder which displays a pop-up menu (Figure 2-19). Select **Run** to start the solution.

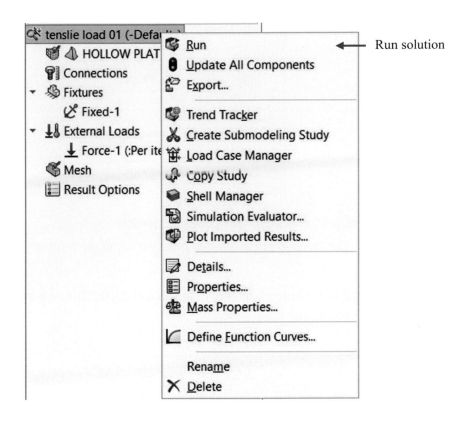

Figure 2-19: Pop-up menu for the *tensile load 01* folder.

Start the solution by right-clicking the tensile load 01 folder to display a pop-up menu. Select Run to start the solution.

The solution can be executed with different properties, which we will investigate in later chapters. You can monitor the solution progress while the solution is running (Figure 2-20).

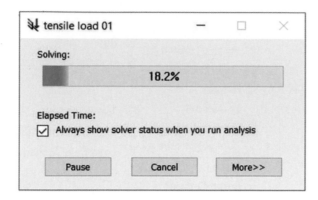

Figure 2-20: Solution Progress window.

The solver reports solution progress while the solution is running.

If the solution fails (not in this exercise), the failure is reported as shown in Figure 2-21.

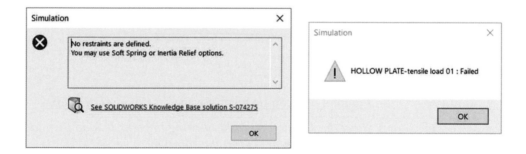

Figure 2-21: Failed solution warning windows.

Here, solution of a model with no restraints was attempted. Once the error message has been acknowledged (left), the solver displays the final outcome of the run (right).

When the solution completes successfully, **Simulation** creates a *Results* folder with result plots which are defined in **Simulation Default Options** as shown in Figure 2-5.

In a typical configuration three plots are created automatically in the *Static* study; make sure that the above plots are defined in **Simulation** options, if not, define them:

- *Stress1*   showing von Mises stresses
- *Displacement1*   showing resultant displacements
- *Strain1*   showing equivalent strain

Once the solution completes, you can add more plots to the *Results* folder. You can also create subfolders in the *Results* folder to group plots (Figure 2-22).

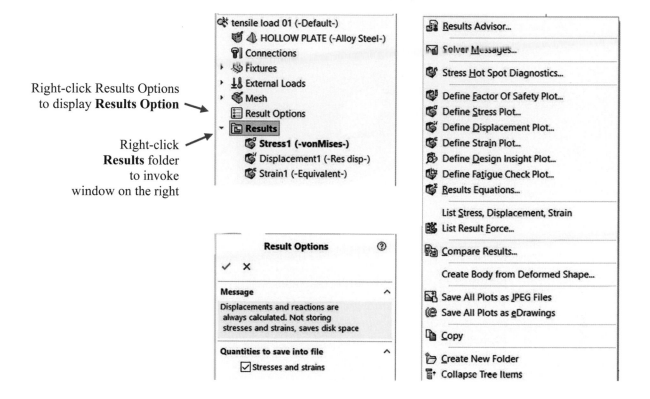

Figure 2-22: More plots and folders can be added to the *Results* folder.

Right-clicking on the Results folder activates this pop-up menu from which plots may be added. Three plots are automatically created if Results Options include these three plots: Stress, Displacement, Strain. You may define additional plots.

The **Result Options** window shown in Figure 2-22 has different choices depending on the type of study. In a **Static** study, deselecting **Stress and strain** disables calculation of stress and strain which reduces calculation time and the size of solution data base. This may be important in analysis of very large models.

To display stress results, double-click on the **Stress1** icon in the *Results* folder or right-click it and select **Show** from the pop-up menu. The stress plot is shown in Figure 2-23.

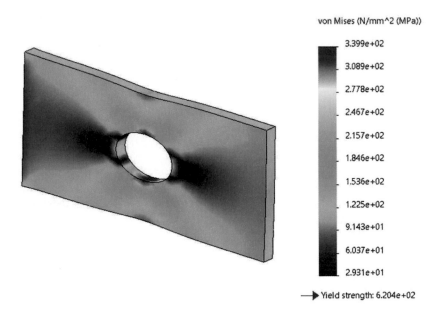

Figure 2-23: Stress plot displayed using default stress plot settings.

Von Mises stress results are shown by default in the stress plot window. Notice that results are shown in [MPa] as was set in the Default Options tab shown in Figure 2-5. The highest stress 345 MPa is below the material yield strength, 620 MPa.

Once the stress plot is showing, right-click the stress plot icon to display the pop-up menu featuring different plot options (Figure 2-24).

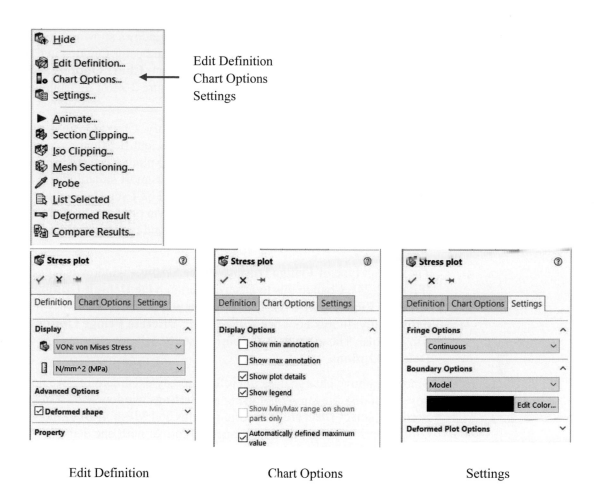

Edit Definition   Chart Options   Settings

Figure 2-24: Plot display options.

Plots can be modified using selections from the pop-up menu (top). The same menus may by invoked by right-clicking the color legend of result plots.

Explore all selections offered by these three windows shown in Figure 2-24. Select scientific, floating and general format and color Options in **Chart Options**. Notice that you may switch between **Definition**, **Chart Options** and **Settings** either by using the pop-up menu shown in Figure 2-24 or by using tabs at the top of these windows.

Use **Edit Definition** to change units if necessary. **Chart Options** offers control over the format of numerical results, such as scientific, floating, and general, and offers a different number of decimal places. Explore these choices. In this book, results will be presented using different choices most suitable for the desired plot. In many plots in this book **Color Options** will be **User Defined** to replace dark blue color with grey color which shows better in print. Figure 2-24 shows the **User Defined** menu. Position of the color legend may be modified for improved layout.

The default type of **Fringe Options** in the **Settings** window is **Continuous** (Figure 2-24). Change this to **Discrete** through the **Default Options** window, by selecting **Plot** (Figure 2-5). This way you will not have to modify the future plots individually. In this book we will be using **Discrete Fringe Options** to display fringe plots. The plots from the current study will not change after changing the **Default Options**.

Since the above change does not affect already existing plots, we now examine how to modify the stress plot using the **Settings** window shown in Figure 2-25. In **Settings,** select **Discrete** in **Fringe options** and **Mesh** in **Boundary options**. In **Chart Options** select floating numerical format with one decimal place.

The edited stress plot shown in Figure 2-25.

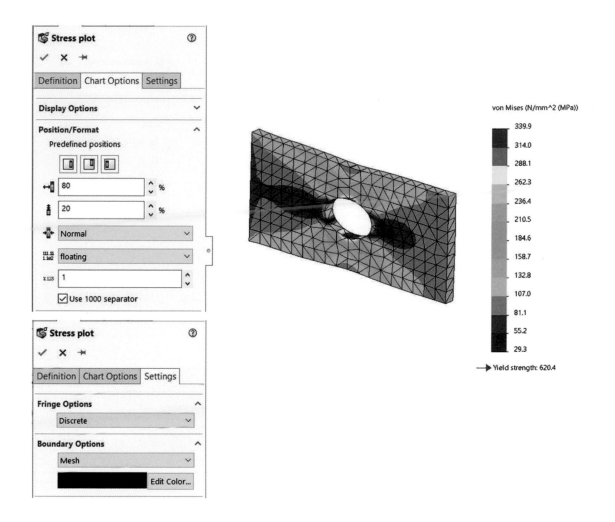

<u>Figure 2-25: The modified stress plot is shown with floating numerical format with one decimal place as selected in Chart Options and with discrete fringes and the mesh superimposed on the stress plot as selected in Settings.</u>

The stress plot in Figure 2-25 shows node values, also called averaged stresses. Element values (or non-averaged stresses) can be displayed by proper selection in the **Stress Plot** window in **Advanced Options**. Node values are most often used to present stress results. See chapter 3 and the glossary of terms in chapter 23 for more information on node values and element values of stress results.

Before you proceed, investigate stress plot with other selections available in the windows shown in Figure 2-24.

We now review the displacement and strain results. All these plots are created and modified in the same way. Sample results are shown in Figure 2-26 (displacement) and Figure 2-27 (strain).

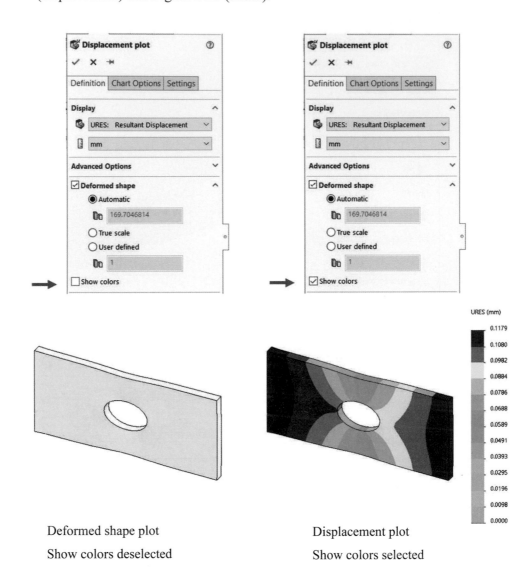

Deformed shape plot

Show colors deselected

Displacement plot

Show colors selected

Figure 2-26: Deformed shape plot (left) and Displacement plot (right).

A Displacement plot can be turned into a Deformation plot by deselecting Show Colors in the Displacement Plot Definition window. The same window has the option of showing the model with an exaggerated scale of deformation as shown above. Both plots show the deformed shape; this option may also be deselected. The blue color in the Displacement plot has been replaced with a gray color for better appearance in this illustration. That was done in Chart Options, Color Options, User defined. Numerical format specifies four decimal places.

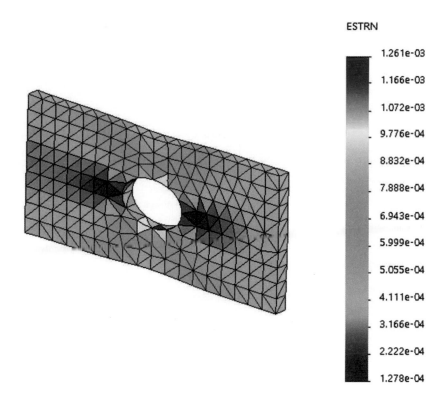

Figure 2-27: Strain plot.

Strain plot is shown using Element values. The mesh is also shown.

The plots in Figures 2-23, 2-25, 2-26, 2-27 all show the deformed shape in an exaggerated scale. You can change the display from deformed to undeformed or modify the scale of deformation in the **Displacement Plot**, **Stress Plot**, and **Strain Plot** windows, activated by right-clicking the plot icon, then selecting **Edit Definition**.

Now, construct a **Factor of Safety** plot using the menu shown in Figure 2-22. The definition of the **Factor of Safety** plot requires three steps. Follow steps 1 through 3 using the selection shown in Figure 2-28. Refer to chapter 1 and review Help to learn about failure criteria and their applicability to different materials.

Click right arrow to move
through windows

Help

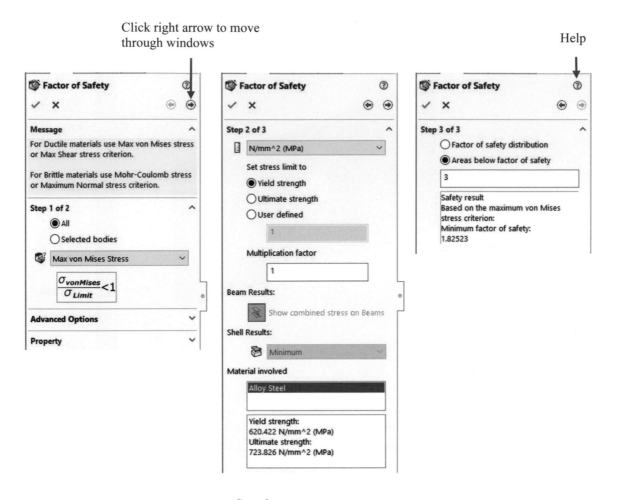

Step 1
Use von Mises Stress criterion.
Review Property options to
insert text and/or use specific
views for the plot

Step 2
Select stress limit and
multiplication factor

Step 3
Select "Areas below factor of safety"
enter 3

Figure 2-28: Three windows show the three steps in the Factor of Safety plot
definition. Select the Max von Mises Stress criterion in the first window.

*To move through steps, click on the right and left arrows located at the top of the
Factor of Safety dialog.*

Step 1 *selects the failure criterion;* ***Step 2*** *selects display units, sets the stress limit
and sets multiplication factor;* ***Step 3*** *selects what will be displayed in the plot.
Here we select areas below the factor of safety 3. Click green check mark to
display the plot.*

Review Help to learn more about failure criteria.

The factor of safety plot in Figure 2-29 shows the area where the factor of safety is below the specified.

Figure 2-29: The red color displays the areas where the factor of safety is below 3.

Irregular shape of red areas is caused by the coarse mesh used in this study.

We have completed the analysis with a coarse mesh and now wish to see how a change in mesh density will affect the results. Therefore, we will repeat the analysis two more times using medium and fine density meshes respectively. We will use the settings shown in Figure 2-15. All three meshes used in this exercise (coarse, default, and fine) are shown in Figure 2-30.

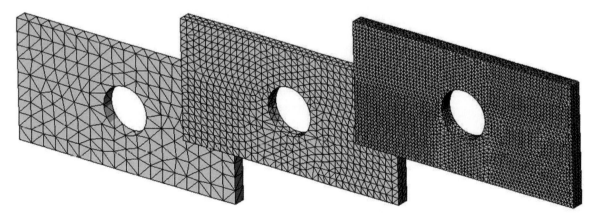

Figure 2-30: Coarse, medium, and fine meshes.

Three meshes used to study the effects of element size on results.

To compare the results produced by different meshes, we need more information than is available in the plots. Along with the maximum displacement and the maximum von Mises stress, for each study we
need to know:

- The number of nodes in the mesh
- The number of elements in the mesh
- The number of degrees of freedom in the model

The information on the number of nodes and number of elements can be found in **Mesh Details** accessible from the menu in Figure 2-14. The mesh Details window is shown in Figure 2-31.

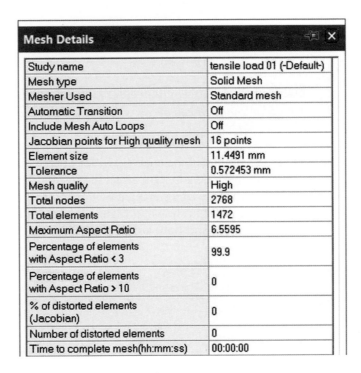

Mesh Details	
Study name	tensile load 01 (-Default-)
Mesh type	Solid Mesh
Mesher Used	Standard mesh
Automatic Transition	Off
Include Mesh Auto Loops	Off
Jacobian points for High quality mesh	16 points
Element size	11.4491 mm
Tolerance	0.572453 mm
Mesh quality	High
Total nodes	2768
Total elements	1472
Maximum Aspect Ratio	6.5595
Percentage of elements with Aspect Ratio < 3	99.9
Percentage of elements with Aspect Ratio > 10	0
% of distorted elements (Jacobian)	0
Number of distorted elements	0
Time to complete mesh(hh:mm:ss)	00:00:00

Figure 2-31: Mesh details window.

Right-click the Mesh folder and select Details from the pop-up menu to display the Mesh Details window. Information on the number of degrees of freedom is not available here.

Another way to find the number of nodes and elements and the only way to find the number of degrees of freedom is to use the pop-up menu shown in Figure 2-32. Right-click the *Results* folder and select **Solver Messages** to display the window shown in Figure 2-32.

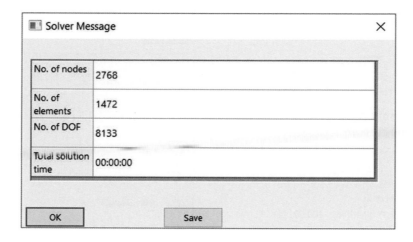

No. of nodes	2768
No. of elements	1472
No. of DOF	8133
Total solution time	00:00:00

Figure 2-32: The Solver Message window lists information pertaining to the solved study.

Total solution time was below 0.01s. Click the Save button to save this solver message.

Now create and run two more studies: *tensile load 02* with the default element size (medium), and *tensile load 03* with a fine element size, as shown in Figure 2-15 and Figure 2-30. To create a new study, we could just repeat the same steps as before, but an easier way is to copy the original study. To copy a study, follow the steps in Figure 2-33.

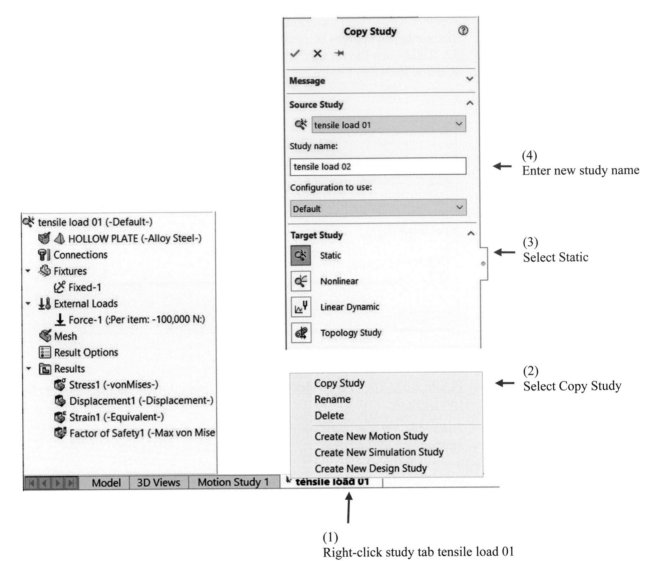

Figure 2-33: A study can be copied into another study in three steps as shown.

All definitions in a study (material, restraints, loads, mesh) can also be copied individually from one study to another by dragging and dropping them into a new study tab.

A study is copied complete with results and plot definitions. Before remeshing with default element size, you must acknowledge the message shown in Figure 2-34.

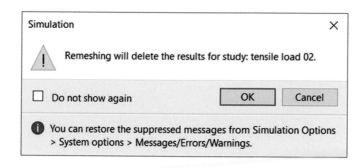

Figure 2-34: Remeshing deletes existing results in the study.

The summary of results produced by the three studies is shown in Figure 2-35.

Study	Element size	Number of nodes	Number of elements	Number of DOF	Max. resultant displacement	Max. von Mises stress
	mm				mm	MPa
tensile load 01	11.45	2768	1472	8133	0.117863	339.9
tensile load 02	5.72	12222	7040	36111	0.118068	371.3
tensile load 03	2.86	84112	54969	250851	0.118074	377.0

Figure 2-35: Summary of results produced by the three meshes.

Notice that these results are based on the same problem. Differences in the results arise from the different mesh densities used in studies tensile load 01, tensile load 02, and tensile load 03.

The actual numbers in this table may vary slightly depending on the type of solver and release of the software used for solution.

Figure 2-36 shows the maximum resultant displacement and the maximum von Mises stress as a function of the number of degrees of freedom. The number of degrees of freedom is in turn a function of mesh density.

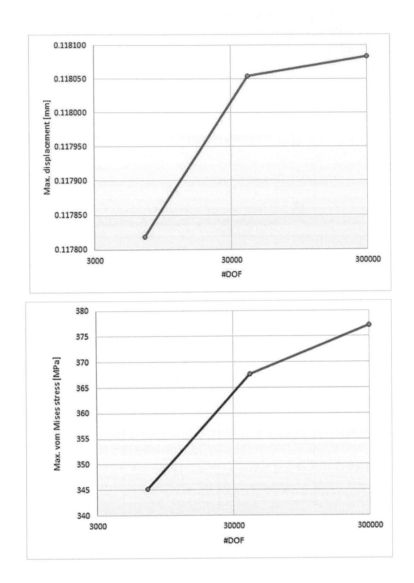

Figure 2-36: Maximum resultant displacement (top) and maximum von Mises stress (bottom).

Both are plotted as a function of the number of degrees of freedom in the model. The three points on the curves correspond to the three models solved. Straight lines connect the three points only to visually enhance the graphs.

Having noticed that the maximum displacement increases with mesh refinement, we can conclude that the model becomes "softer" when smaller elements are used. With mesh refinement, a larger number of elements allows for better approximation of the real displacement and stress field. Therefore, we can say that the artificial restraints imposed by element definition become less imposing with mesh refinement.

Displacements are the primary unknowns in structural FEA, and stresses are calculated based on displacement results. Therefore, stresses also increase with mesh refinement. If we continued with mesh refinement, we would see that both the displacement and stress results converge to a finite value which is the solution of the mathematical model. Differences between the solution of the FEA model and the mathematical model are due to discretization errors, which diminish with mesh refinement.

We will now repeat our analysis of the HOLLOW PLATE by using prescribed displacements in place of a load. Rather than loading it with a 100000N force that caused a 0.118 mm displacement of the loaded face, we will apply a prescribed displacement of 0.118 mm to this face to see what stresses this causes. For this exercise, we will use only one mesh with the default (medium) mesh density.

Define a fourth study, called *prescribed displ*. The easiest way to do this is to copy one of the already completed studies, for example study *tensile load 02*. The definition of material properties, the fixed restraint to the left-side end-face and mesh are all identical to the previous design study. We need to delete the current **External Loads** (right-click the load icon and select **Delete**) and apply in its place a prescribed displacement.

To apply the prescribed displacement to the right-side end-face, right-click the **Fixtures** folder and select **Advanced Fixtures** from the pop-up menu. This opens the **Fixture** definition window. Select **On Flat Face** from the **Advanced** menu and define the displacement as shown in Figure 2-37. Check **Reverse direction** to obtain displacement in the tensile direction. Notice that the direction of a prescribed displacement is indicated by a restraint symbol.

Prescribed displacement may also be defined from the **External Loads** folder by selecting **Prescribed Displacement** from the pop-up menu. This opens the same **Fixture** definition window. Figure 2-37 illustrates both methods of defining prescribed displacement.

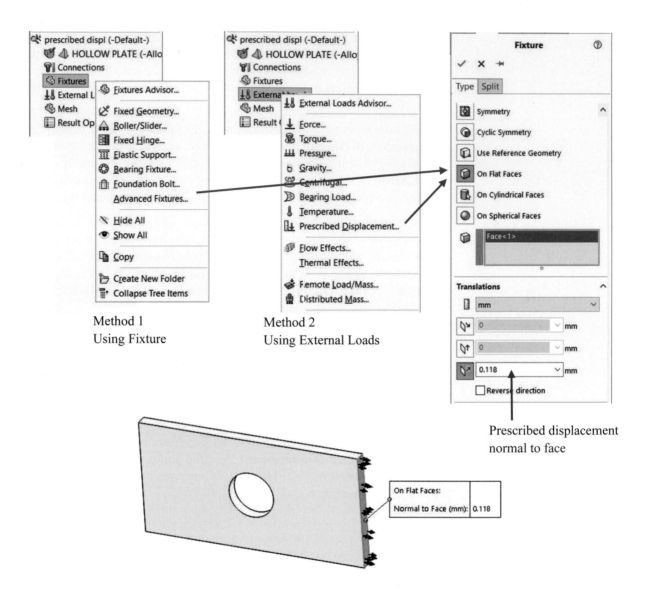

Method 1
Using Fixture

Method 2
Using External Loads

Prescribed displacement
normal to face

Figure 2-37: Two methods of defining prescribed displacement.

The prescribed displacement of 0.118 mm is applied to the same face where the tensile load of 100000N was applied in previous studies.

Verify that the arrows (shown here in black color) are pointed away from the selected face. Fixed restraints symbols on the other side are not shown.

Once again, notice that the visibility of all loads and restraints symbols is controlled by right-clicking the symbol and making the desired choice (**Hide/Show**). All load symbols and all restraint (fixture) symbols may also be turned on/off all at once by right-clicking the **Fixtures** or **External loads** folders and selecting **Hide all/ Show all** from the pop-up menu.

Once a prescribed displacement is defined to the end face, it overrides any previously applied loads to the same end face. While it is better to delete the load in order to keep the model clean, a load has no effect if a prescribed displacement is applied to the same entity in the same direction.

Figure 2-38 compares stress results for the model loaded with force to the model loaded with prescribed displacement.

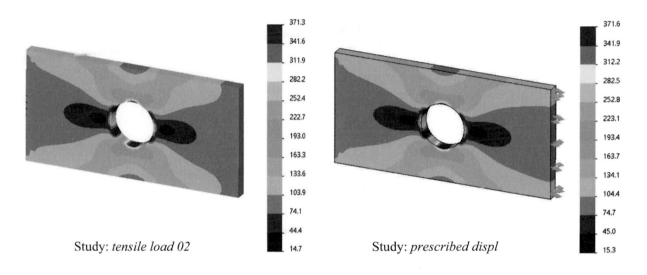

Study: *tensile load 02*   Study: *prescribed displ*

Figure 2-38: Comparison of von Mises stress results.

Von Mises stress results with a force load (left) and with a prescribed displacement (right).

Results produced by applying a force load and by applying a prescribed displacement load are very similar but not identical. The reason for this discrepancy is that in the model loaded by force, the loaded face does not remain flat. In the prescribed displacement model, this face remains flat, even though it experiences displacement as a whole. Also, while the prescribed displacement of 0.118 mm applies to the entire face in the prescribed displacement model, it is only seen as a maximum displacement in one point in the force load model. You may plot the displacement along the edge of the end face in study *tensile load 02* by following the steps in Figure 2-39.

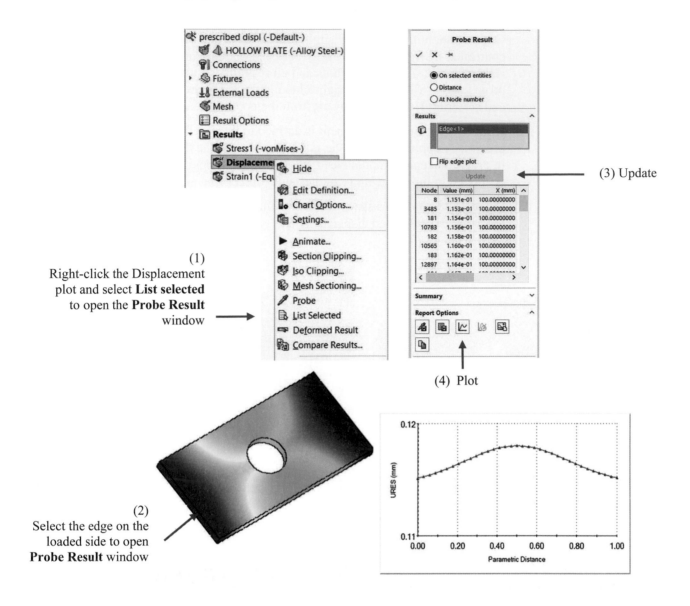

Figure 2-39: Plotting displacement along the edge of the force loaded face in study *tensile load 02.*

Right click the resultant displacement plot in the tensile load 02 study to invoke a pop-up menu shown in the top left corner. Follow steps 1 through 4 to produce a graph of displacements along the loaded edge. Repeat this exercise for a model loaded with prescribed displacement to verify that the displacement is constant along the edge.

We conclude the analysis of the HOLLOW PLATE by examining the reaction forces using the results of study *tensile load 02*. In the study *tensile load 02*, right-click the *Results* folder. From the pop-up menu, select **List Result Force** to open the **Result Force** window. Select the face where the fixed restraint is applied and click the **Update** button. Information on reaction forces will be displayed as shown in Figure 2-40.

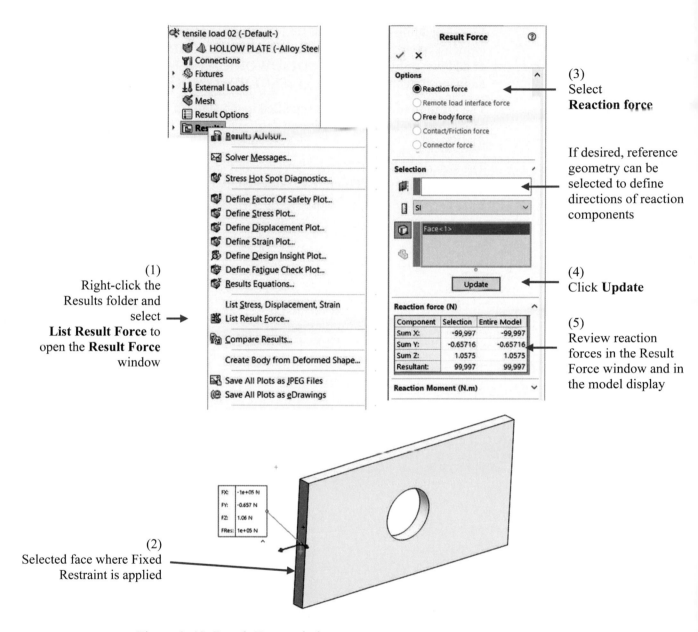

Figure 2-40: Result Force window.

Right-click the Results folder and follow steps 1 through 4 to analyze and display reaction forces. Reaction forces can also be displayed in components other than those defined by the global reference system. To do this, reference geometry such as a plane or axis must be selected.

A note on where **Simulation** results are stored: By default, all study files are saved in the same folder with the **SOLIDWORKS** part or assembly model. Mesh data and results of each study are stored separately in *.CWR files. For example, the mesh and results of study tensile load 02 have been stored in the file: HOLLOW PLATE-tensile load 02.CWR

When the study is opened, the CWR file is extracted into a number of different files depending on the type of study. Upon exiting **SOLIDWORKS Simulation** (which is done by means of deselecting **SOLIDWORKS Simulation** from the list of add-ins, or by closing the **SOLIDWORKS** model), all files are compressed allowing for convenient backup of **SOLIDWORKS Simulation** results.

The location of CWR files is specified in the **Default Options** window (Figure 2-5). For easy reference, the **Default Options** window is shown again in Figure 2-41.

The size of the CWR file may be reduced if stresses and strains are deselected in the Results Options (Figure 2-22). We will not use this option.

Default Options

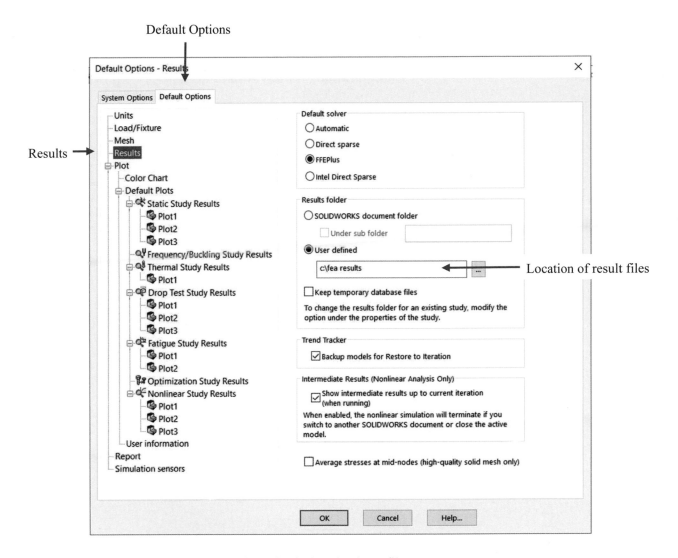

Results

Location of result files

Figure 2-41: Location of solution database files.

You may use the SOLIDWORKS document folder, or a user defined folder as shown above.

Using the settings shown in Figure 2-41, the solution data base files are located in folder c:\fea results. The default location is the SOLIDWORKS document folder.

Models in this chapter

Model	Configuration	Study Name	Study Type
HOLLOW PLATE.sldprt	*Default*	*tensile load 01*	Static
		tensile load 02	Static
		tensile load 03	Static
		prescribed displ	Static

3: Static analysis of an L-bracket

Topics covered

- ❑ Stress singularities
- ❑ Differences between modeling errors and discretization errors
- ❑ Using mesh controls
- ❑ Analysis in different **SOLIDWORKS** configurations
- ❑ Nodal stresses, element stresses

Project description

An L-shaped bracket (part L BRACKET) is supported and loaded as shown in Figure 3-1. We would like to find the displacements and stresses caused by a 1000N bending load. We are interested in stresses along the edge where the 2 mm fillet is located. Since the radius of the fillet is small compared to the overall size of the model, we decide to suppress it. As will be proven, suppressing the fillet is a bad mistake.

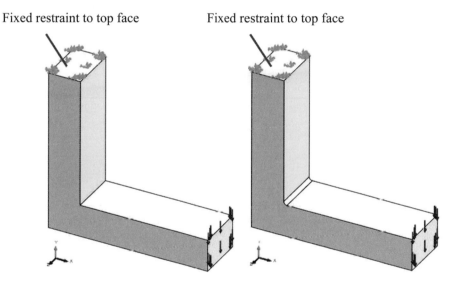

Fixed restraint to top face   Fixed restraint to top face

Configuration: *01 sharp edge*   Configuration: *02 round edge*

Figure 3-1: Loads and supports applied to the L BRACKET model.

1000N force in –y direction is uniformly distributed over the end face.

Loads and restraints are the same in both configurations.

The L BRACKET model has two configurations: *01 sharp edge* and *02 round edge*. The material definition (Alloy Steel) is applied to the **SOLIDWORKS** model and is automatically transferred to **Simulation**.

Procedure

Make sure the model is in configuration *01 sharp edge*. Following the same steps as those described in chapter 2, create a **Static** study called *mesh 1* and define a **Fixed** restraint to the top face shown in Figure 3-1. Define the load using the **Force/Torque** window choices shown in Figure 3-2.

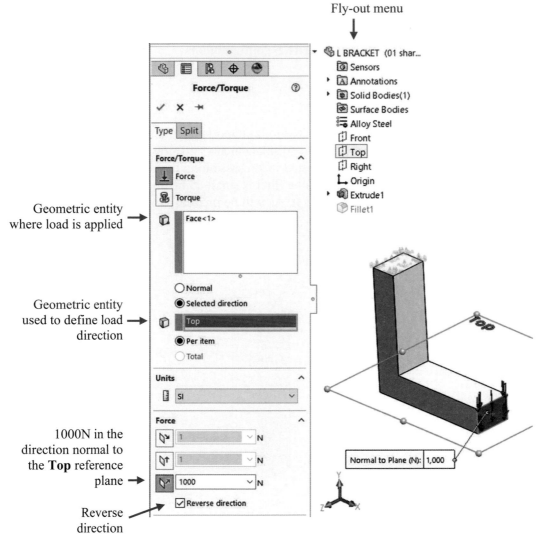

Figure 3-2: The force definition window specifies force in a selected direction. The direction is specified as normal to the Top reference plane.

The Top reference plane is used as a reference to determine the force direction. The reference plane can be conveniently selected from the fly out SOLIDWORKS menu to the right of the Force/Torque window.

Next, make sure the mesh setting is **Standard mesh** and mesh the model with second order tetrahedral elements, accepting the default element size. The finite element mesh is shown in Figure 3-3.

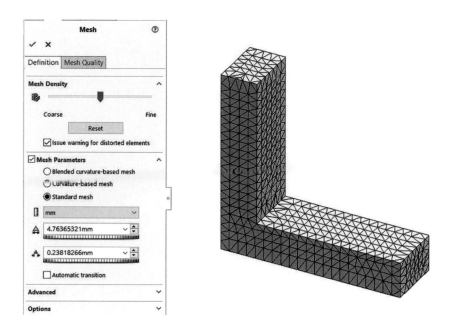

Figure 3-3: The finite element mesh created with the default settings of the mesher.

To invoke the Mesh window right-click the mesh folder in study main menu. In this mesh, the global element size is 4.76 mm.

The displacement and stress results obtained in the *mesh 1* study are shown in Figure 3-4.

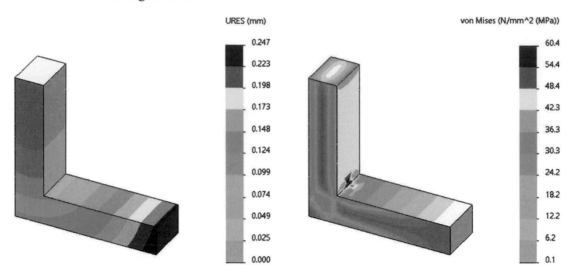

Figure 3-4: Displacement plot (left) and von Mises stress plot (right) produced using study *mesh 1*.

The maximum displacement is 0.247mm and the maximum von Mises stress is 60MPa. As it will be explained later, these stress results are meaningless.

Custom colors are in these fringe plots. Custom colors are defined in plot Chart Options.

Now we will investigate how using smaller elements affects the results. In chapter 2, we did this by refining the mesh uniformly so that the entire model was meshed with elements of a smaller size. Here we will use a different technique. Having noticed that the stress concentration is located near the sharp re-entrant edge, we will refine the mesh locally in that area by applying mesh controls. The element size everywhere else will remain the same as defined previously: 4.76mm.

Copy the *mesh 1* study into a new study, naming it *mesh 2*. Select the edge where mesh controls will be applied, then right-click the *Mesh* folder in the *mesh 2* study dialog (this folder is currently empty) to display the pop-up menu shown in Figure 3-5.

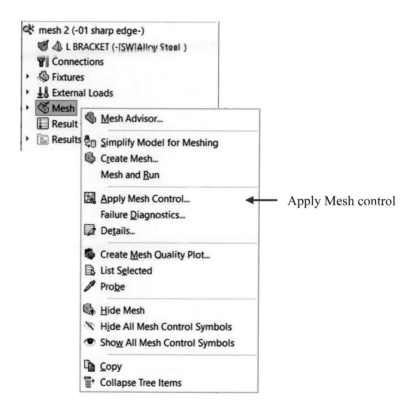

Figure 3-5: Mesh pop-up menu.

Select Apply Mesh control.

Select **Apply Mesh Control...**, which opens the **Mesh Control** window (Figure 3-6). It is also possible to open the **Mesh Control** window first and then select the desired entity or entities (here the re-entrant edge) where mesh controls are being applied.

Green check mark →

Create Mesh →

Element size along the selected entity →

Relative element size in adjacent layers of elements →

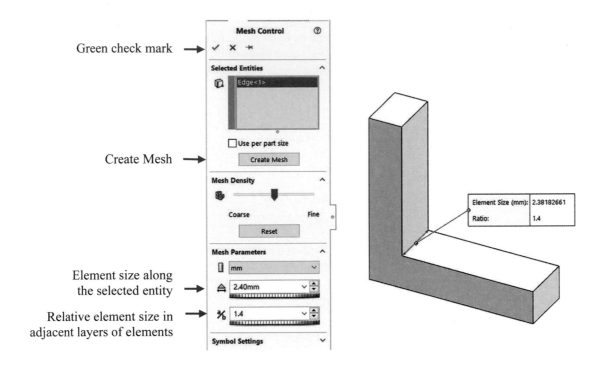

Figure 3-6: Mesh Control window in study *Mesh 2*.

Mesh controls allow for the definition of a local element size on selected entities. Accept the default settings of the Mesh Control window.

Relative element size in adjacent layers of elements refers to elements in the transition zone between smaller elements along the edge and larger elements everywhere else.

Using **Mesh Control** window in Figure 3-6 you may click **Create Mesh** or green check mark to exit without creating a mesh.

The element size along the selected edge is now controlled independently of the global element size. **Mesh Control** can also be applied to vertices, faces and to entire components of assemblies. Having defined a **Mesh Control**, create a mesh with the same global element size as before (4.76 mm), while making elements along the specified edge to be 2.4 mm. The added mesh controls display as the **Control-1** icon in the *Mesh* folder and can be edited using the pop-up menu displayed by right-clicking the mesh control icon (Figure 3-7).

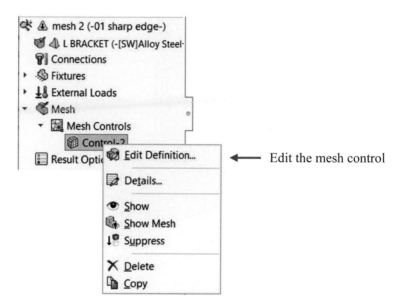

Figure 3-7: Pop-up menu activated by right-clicking the mesh control icon.

If desired, right-click Mesh Control and select Edit Definition to open the Mesh Control window and edit the mesh control.

Warning signs by the study name, mesh and results indicate that results (copied from study mesh 1) are no longer valid because mesh has been changed.

The mesh with applied control (also called mesh bias) is shown in Figure 3-8.

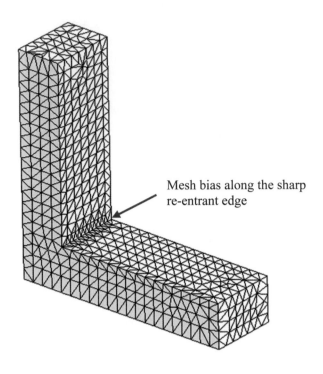

Mesh bias along the sharp re-entrant edge

Figure 3-8: Mesh with applied controls (mesh bias).

The mesh in study mesh 2 is refined along the sharp re-entrant edge.

The maximum displacements and stress results obtained in study *mesh 2* are 0.2478 mm and 68.7MPa, respectively. The number of digits shown in a result plot is controlled using **Chart Options** (right-click on a plot and select **Chart Options)**.

Now repeat the same exercise three more times using progressively smaller elements along the sharp re-entrant edge. Create studies *mesh 3, mesh 4, mesh 5* with an element size along the sharp reentrant edge as shown in Figure 3-9. Use **Standard mesh**.

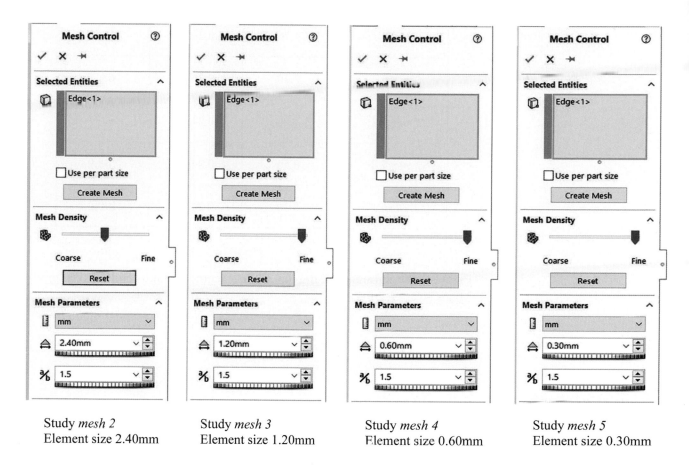

Study *mesh 2*   Study *mesh 3*   Study *mesh 4*   Study *mesh 5*
Element size 2.40mm   Element size 1.20mm   Element size 0.60mm   Element size 0.30mm

Figure 3-9: Mesh Control windows in studies *mesh 2, mesh 3, mesh 4, and mesh 5.*

a/b ratio remains 1.5 in all the above settings.

The summary of results of all five studies is shown in Figures 3-10 and 3-11.

Study	Element size along the edge	Max. resultant displacement	Max. von Mises stress
	mm	mm	MPa
mesh 1	4.76	0.2473	60.4
mesh 2	2.40	0.2478	68.7
mesh 3	1.20	0.2481	92.8
mesh 4	0.60	0.2483	138.9
mesh 5	0.30	0.2485	198.2

Figure 3-10: Summary of maximum displacement results and maximum von Mises stress results.

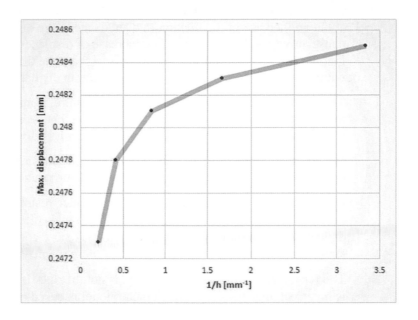

Max. resultant displacement

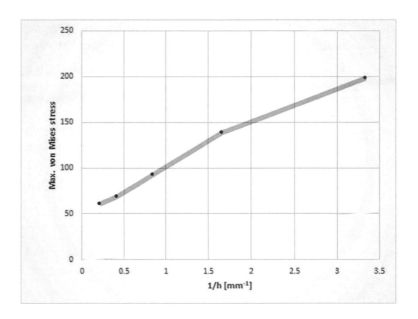

Max. von Mises stress

Figure 3-11: Max. resultant displacement (top) and max. von Mises stress (bottom) as a function of 1/h, where h is the element size (Figure 2-16) along the sharp re-entrant edge.

Stress results show no sign of convergence.

Upon examining Figure 3-11, we notice that while each mesh refinement brings about an increase in the maximum displacement, the difference between consecutive results decreases. The increase of displacement in results is so minute that the results need four decimal places to show the difference. The first study, without any mesh refinement, provides accurate displacement results.

The stress behaves very differently. Each mesh refinement brings about an increase in the maximum stress. The difference between consecutive results increases, proving that the maximum stress result is divergent.

We could continue with this exercise by refining mesh locally near the sharp re-entrant, or globally, by reducing the global element size, as we did in chapter 2.

Given enough time and patience, we can produce results showing any stress magnitude we want. All that is necessary is to make the element size small enough! We should avoid a temptation to make any conclusions based on the stress graph in Figure 3-11 because all of these results are meaningless.

The reason for divergent stress results is not that the finite element model is incorrect, but that the finite element model is based on the wrong mathematical model.

According to the theory of elasticity, stress in a sharp re-entrant corner is infinite. A mathematician would say that stress in a sharp re-entrant edge is singular. Stress results along sharp re-entrant edges are completely dependent on mesh size: the smaller the element, the higher the stress.

We used convergence process to detect stress singularity. **Simulation** offers an automated detection of stress singularities called **Stress Hot Spot Diagnostics**; it will be discussed in Chapter 21.

In this exercise we'll eliminate stress singularity by un-suppressing the fillet. This is done by changing from configuration *01 sharp edge* to *02 round edge* in the **SOLIDWORKS Configuration Manager**.

Notice that after we return from the **SOLIDWORKS Configuration Manager** window to the **SOLIDWORKS Simulation** window, all studies pertaining to the model in configuration *01 sharp edge* are not accessible. They can be accessed only if the model configuration is changed back to *01 sharp edge* (Figure 3-12).

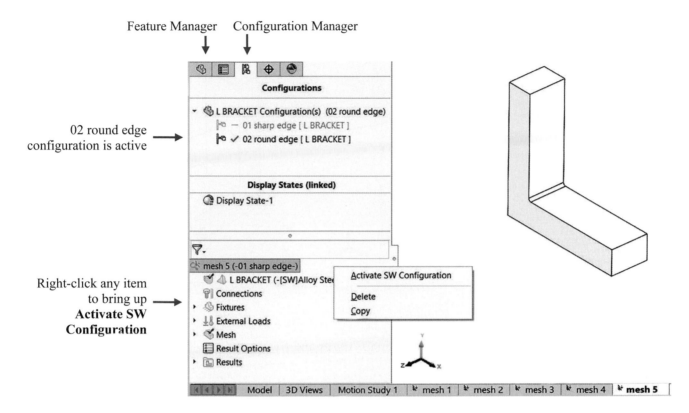

Figure 3-12: Studies become inaccessible when the model configuration is changed to a configuration other than that corresponding to the now grayed-out studies.

Studies mesh 1, mesh 2, mesh 3, mesh 4, mesh 5 have all been created in configuration 01 sharp edge and are now inaccessible.

The configuration of the SOLIDWORKS model can be changed to a configuration corresponding to a given study by right-clicking any item in the study window and selecting Activate SW configuration.

To analyze the bracket with the round fillet, define a study named *round edge*. Copy the *Fixtures* and *External Loads* folders from any inactive studies to the *round edge* study by clicking them and dragging them to the new study tab (Figure 3-13).

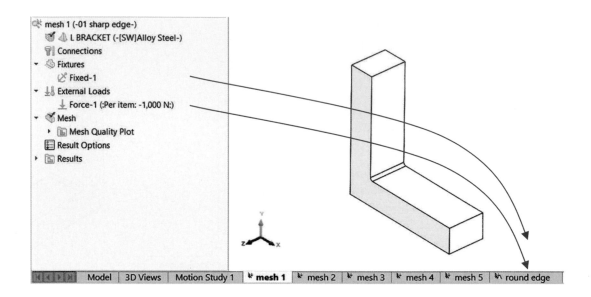

Figure 3-13: Copying *Fixtures* and *External Loads* from study *mesh 2* to *study round edge.*

Entities can be copied between studies by dragging them and dropping them into a study tab as shown, even if the source information is from an inaccessible study.

In some cases, the entire Fixtures or External Loads folders cannot be copied but you may still copy the contents of these folders individually.

Meshing with the default element size and **Standard Mesh** properties produces elements with an excessive turn angle in the area where it is particularly important to have a correct mesh (Figure 3-14).

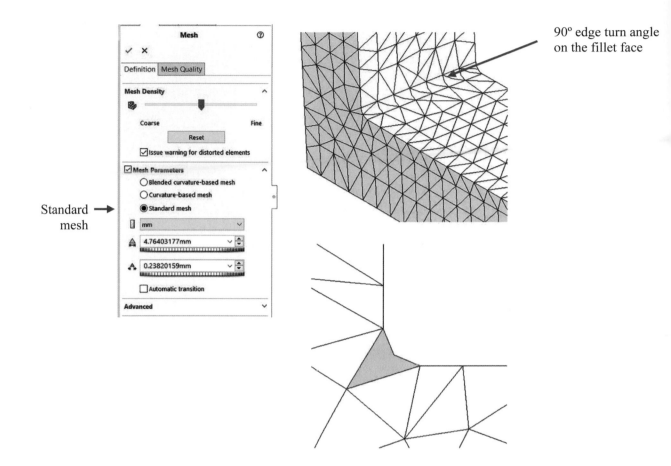

Figure 3-14: A mesh created as a Standard Mesh with a default element size is not acceptable because of the high turn angle.

The turn angle of the element edge in the fillet is 90°; one element edge covers 90° angle. A section through the middle of the bracket thickness (lower illustration) shows a "kink" in the element edge. This is a result of simplified graphics, not the actual element edge shape which is circular.

To eliminate excessive turn angles, we use the **Curvature based mesh** option and enter values shown in Figure 3-15.

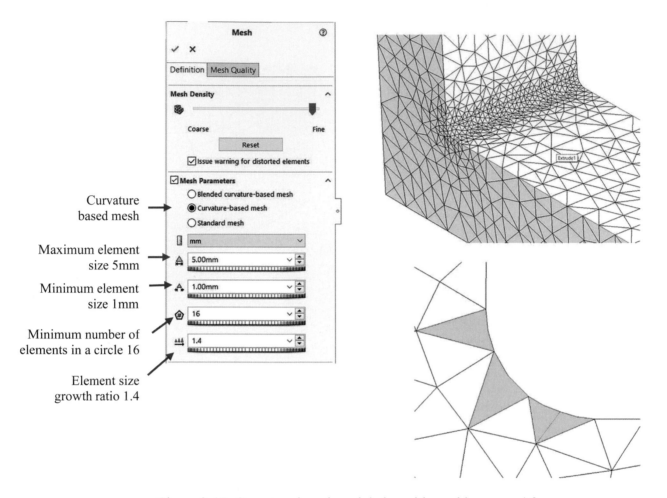

Figure 3-15: Curvature based mesh helps with meshing curved faces.

Using the above settings, a minimum of 4 elements are created on the 90° fillet.

With the minimum number of elements in a circle (360°) set to 16, an element turn angle is no more than 360°/16=22.5°. Elements with turn angle 22.5° are highlighted in the lower illustration.

The element size growth ratio controls the transition between the refined mesh on curved faces and the coarser mesh on flat faces.

It is generally recommended that the turn angle does not exceed 30° in locations where stresses must be correctly modeled.

The L-BRACKET example is a good place to review the different ways of displaying stress results. Stresses can be presented either as **Node Values** or **Element Values**. To select either node values or element values, right-click the plot icon and select **Edit Definition**. This will open the **Stress Plot** window. Figure 3-16 shows the node values of von Mises stress results produced in the study *02 round edge*.

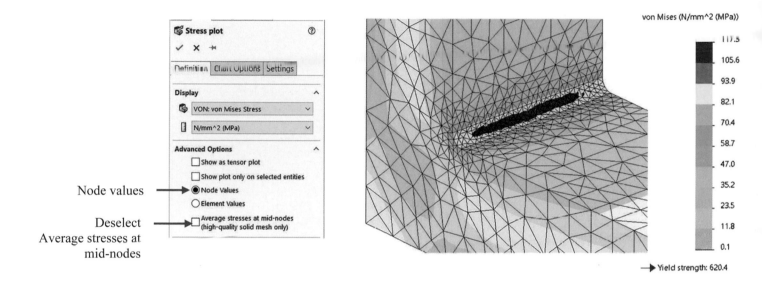

Node values →

Deselect
Average stresses at
mid-nodes →

Figure 3-16: Von Mises stresses displayed as node values.

The irregularities in the shape of discrete fringes showing nodal stress results may be used to decide if more mesh refinement is needed around a stress concentration. Here, regular shapes indicate sufficient mesh refinement.

The maximum stress (117MPa) is now bounded. In the convergence process it will converge to a finite value, close to the one shown in Figure 3-16. We cannot compare this result to the maximum stress results produced in the studies using the sharp re-entrant edge because those results are all meaningless.

As was explained in chapter 1, nodal displacements are computed first, from which strains and then stresses are calculated. Stresses are first calculated inside the element at certain locations, called Gauss points. Next, stress results are extrapolated to all the elements' nodes. If one node belongs to more than one element (which is always the case unless it is a vertex node), then the stress results from all the elements sharing a given node are averaged and one stress value, called a node value, is reported for each node. This stress value is called a nodal stress.

There are two ways of calculating nodal stress results. Stress plot in Figure 3-16 has been prepared using default settings. Another way to average stresses is **Stress Averaging at Mid-side Nodes** discussed in Chapter 21.

An alternate procedure to stress averaging is not to average stresses between different elements. Using this approach, stresses are calculated at Gauss points, and then averaged in between themselves and one stress value is calculated for the element. This stress value is called an element stress. Figure 3-17 shows the element values of von Mises stress results produced in the study *02 round edge*.

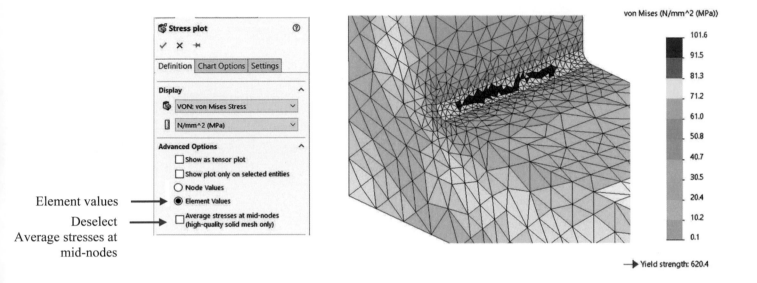

Element values —→

Deselect
Average stresses at
mid-nodes

Figure 3-17 Von Mises stresses displayed as element values.

Element values are not averaged across different elements. A single stress value is assigned to each element.

Nodal stresses are used more often because they offer smoothed out, continuous stress results. However, examination of element stresses provides important feedback on the quality of the results. If element stresses in two adjacent elements differ too much, it indicates that the element size at this location is too large to properly model the stress gradient. By examining the element stresses, we can locate mesh deficiencies without running a convergence analysis.

To decide how much is "too much" of a difference requires some experience. As a general guideline, we can say that if the element values of stress in adjacent elements are apart by several colors on the default color chart (12 colors), then a more refined mesh should be used. You are encouraged to perform a convergence analysis using a **Curvature Based mesh**.

Models in this chapter

Model	Configuration	Study Name	Study Type
L BRACKET.sldprt	01 sharp edge	mesh 1	Static
		mesh 2	Static
		mesh 3	Static
		mesh 4	Static
		mesh 5	Static
	02 round edge	round edge	Static

Notes:

4: Static and frequency analyses of a pipe support

Topics covered

- Use of shell elements
- Frequency analysis
- Bearing load

Project description

We will analyze a support bracket (Figure 4-1) with the objective of finding stresses and the first few modes of vibration. This will require running both static and frequency analyses. Open the PIPE SUPPORT model with assigned material properties of Galvanized Steel. Open a new **Static** study and name it *01 static*.

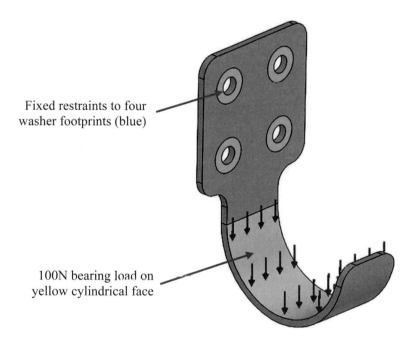

Fixed restraints to four washer footprints (blue)

100N bearing load on yellow cylindrical face

Figure 4-1: PIPE SUPPORT model.

The model has been designed in SOLIDWORKS as a sheet metal.

Procedure

Before defining the study, consider that thin wall geometry would be difficult to mesh with solid elements. Generally, it is recommended that two layers of second order tetrahedral elements be used across the thickness of a wall undergoing bending. Therefore, a large number of solid elements would be required to mesh this thin model.

A sheet metal model has inherently thin walls. Therefore, when a sheet metal model is presented to **SOLIDWORKS Simulation**, and Default Option **Mesh all solid bodies with solid mesh** is deselected (Figure 4-4), the sheet metal model is designated for meshing with shell elements. Create static study *01 static* and notice that the familiar *Solid* body icon is replaced with a *Shell* body (Figure 4-2).

Before proceeding with PIPE SUPPORT model, we will review different types of bodies and explain what elements are used to mesh them. Solid bodies and surface bodies are denoted by different icons in the study menu. Additionally, a tetrahedron symbol indicates what element order will be used to mesh the body. Tetrahedron with straight edges indicates the first order elements; tetrahedron with curvilinear edges indicates the second order elements (Figure 4-2). We will not use the first order elements in this book.

	First order elements	Second order elements
Solid body used in exercises in chapter 2 and chapter 3. Solid body is meshed with solid elements.		
Surface body used in this exercise. This surface body symbol indicates surface derived from sheet metal geometry. The surface body is meshed with shell elements.		
Surface body (not used in this exercise). This symbol indicates surface created from faces of solid bodies or surfaces created without using solid body geometry. The surface body is meshed with shell elements.		

Figure 4-2: The solid body, shell body derived from sheet metal geometry and surface body created from faces of solid bodies or surfaces created without using any solid geometry.

Body icons may carry a green check mark which indicates that material has been assigned to them as shown in this illustration.

All icons are shown here together with tetrahedrons indicating the element order used to mesh the body.

If body symbols show cross-hatching this means that mesh has been already created (not in this illustration).

If solid elements are preferred after all, the designation **Shell** can be changed to **Solid** as shown in Figure 4-3.

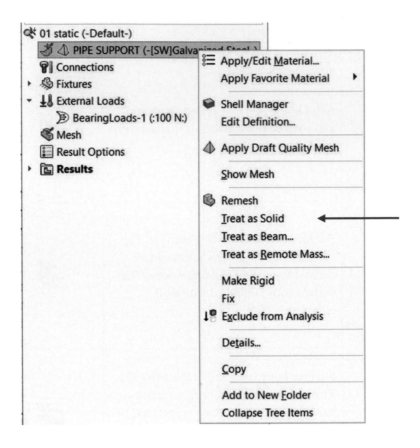

Figure 4-3: Changing from Shell elements to Solid elements.

If you prefer to mesh the sheet metal model with solid elements, right click the Shell folder and select Treat as Solid. This exercise uses shell elements; the illustration is for information only.

Another way to change the designation **Shell** to **Solid** is changing Simulation **Default Option** (Figure 4-4).

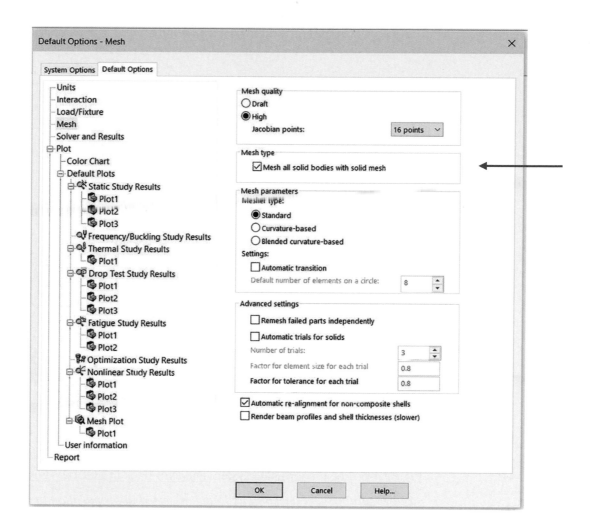

Figure 4-4: Changing from Shell elements to Solid elements using Default Options; select Mesh all solid bodies with solid mesh.

This illustration is for information only. Do not select this option.

Apply a fixed restraint to the four washer footprints, as shown in Figure 4-1. Review the split lines in the **SOLIDWORKS** model that define split faces where restraints are applied. Split faces are commonly used in preparation of CAD models for analysis with **SOLIDWORKS Simulation**.

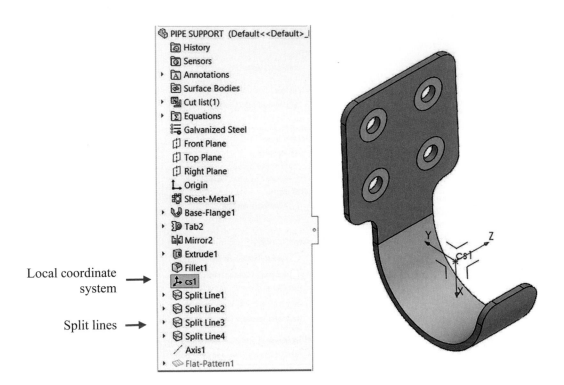

Figure 4-5: Local coordinate system and split lines in SOLIDWORKS model.

Adding split lines is a technique frequently used in preparation of a CAD model for analysis with FEA. In this model, split lines define faces where restraints are applied.

Apply the **Fixed** restraints as shown in Figure 4-1.

The total load carried by the hanger is 100N in the *x* direction of the local coordinate system **cs1**. We approximate the load of a pipe onto the hanger by applying a **Bearing Load**. Right-click the *External Loads* folder and select **Bearing Load** from the pop-up menu. Select the cylindrical face where the load is to be applied, and then select **cs1** as the reference coordinate system using the *fly-out* menu (shown previously in Figure 3-2) and apply 100N in the *x* direction of that coordinate system (Figure 4-6).

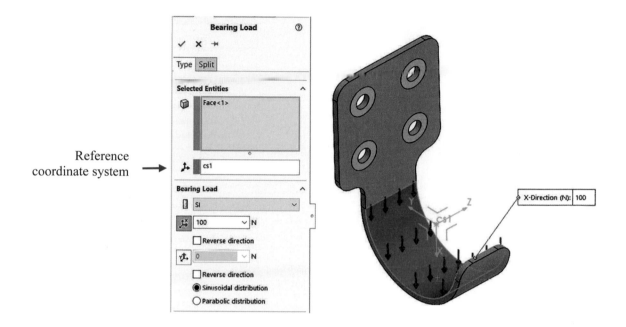

Figure 4-6: Bearing Load definition.

A bearing load can only be applied to a cylindrical face. It is not uniformly distributed over the face but follows a sinusoidal or a parabolic distribution.

A typical application of a Bearing Load is modeling interactions between a shaft and housing.

The model is now ready for meshing (right-click the *Mesh* folder and **Create Mesh**). Use the default element size. The shell element mesh is shown in Figure 4-7.

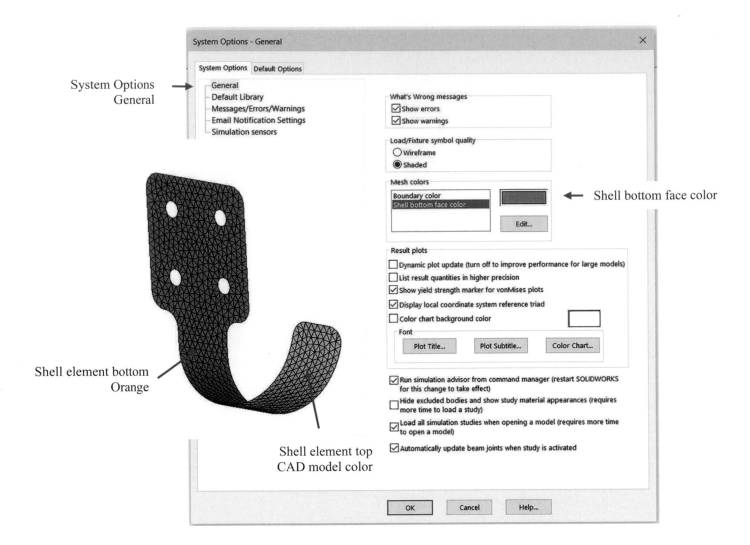

Figure 4-7: A shell element mesh. Mesh elements have been placed on a mid-surface defined between the outer faces defining the thin wall.

Different colors distinguish between the top and bottom of the shell elements. The bottom face color is specified in the System Options window as shown above. The top face color is the same as the color of the SOLIDWORKS model.

In this model the side opposite to where the load is applied is meshed with shell element tops. The side where the load is applied is meshed with the bottoms of shell elements. The side where the load is applied appears with a color specified in **Shell bottom face color** in the **Simulation Options** window.

Review the mesh colors to ensure that the shell elements are aligned. Try reversing the shell element orientation: select the face where you want to reverse the orientation, right-click the Mesh folder to display a pop-up menu and select **Flip shell elements** (Figure 4-8).

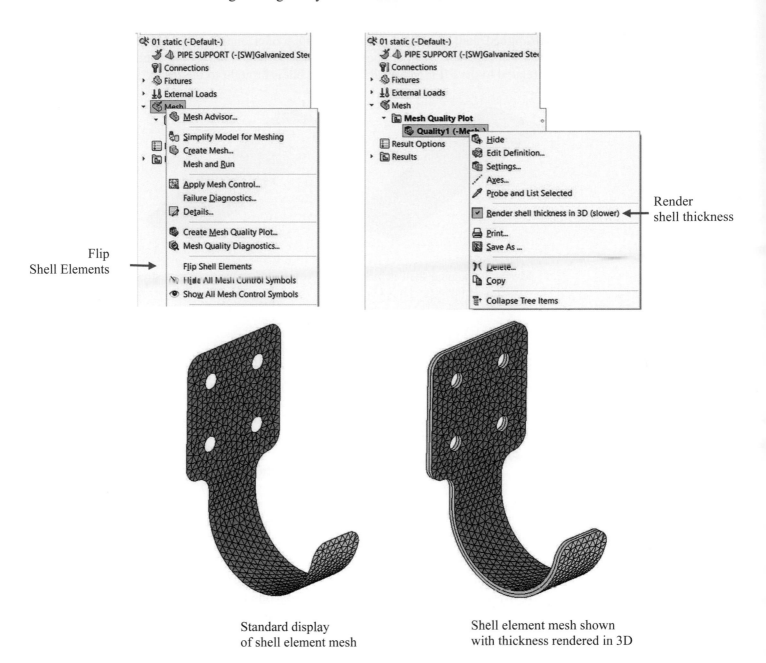

Standard display
of shell element mesh

Shell element mesh shown
with thickness rendered in 3D

Figure 4-8: The pop-up menu may be used for modifying shell element orientation and for modifying shell element mesh display.

If desired, you may reverse shell element orientation with this menu choice. In this exercise reversing shell element orientation is not required.

You may also use this menu to display shell element mesh shown with 3D thickness as shown in this illustration.

Misaligned shell elements lead to the creation of erroneous plots like the one shown in Figure 4-9, which shows a rectangular plate undergoing bending; this is unrelated to the PIPE SUPPORT exercise but is brought to the reader's attention.

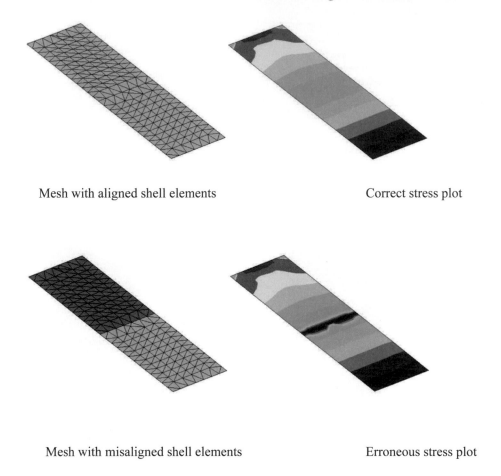

Mesh with aligned shell elements   Correct stress plot

Mesh with misaligned shell elements   Erroneous stress plot

Figure 4-9: Aligned shell element mesh produces correct von Mises stress plot (top). Misaligned shell elements produce erroneous von Mises Stress plot (bottom). Review model MISALIGNMENT which comes with studies defined.

The erroneous plot is the result of stress averaging between top-top and bottom-bottom location, which in the case of misaligned mesh is situated on the opposite side of the model.

Return to model PIPE SUPPORT and obtain the solution and display displacement results. Select the **Superimpose model on the deformed shape** option. This option is available in plot settings (Figure 4-10).

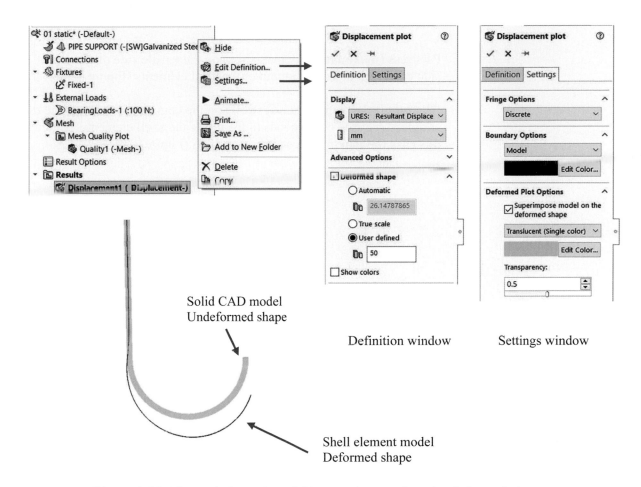

Definition window   Settings window

Solid CAD model
Undeformed shape

Shell element model
Deformed shape

Figure 4-10: The undeformed model is superimposed on the deformed shape.

This plot shows that the shell element mesh has been placed in the mid-plane of the solid geometry. The plot may show displacements or deformations by selecting or deselecting Show colors option in Definition window. You may Superimpose model on the deformed shape by selecting this in Settings window.

Explore all options in Definition, Chart Options (not shown here) and Setting tabs.

Shell elements differentiate between stress results on the top and bottom of the element. In the case of bending, one side will show tensile stress, the other compressive stress. For correct interpretation of results, we must know on which side of the element results are presented. To illustrate this, we prepare two stress plots:

P1 stress (maximum principal stress) on the tensile side of the model. The tensile side corresponds to the bottom of the shell elements (Figure 4-11).

P3 stress (minimum principal stress) on the compressive side of the model. The compressive side corresponds to the top of the shell elements. Notice that P3 is in this case the maximum compressive stress (Figure 4-11).

Maximum principal stress P1 on the bottom faces of the shell elements. This is the visible side.

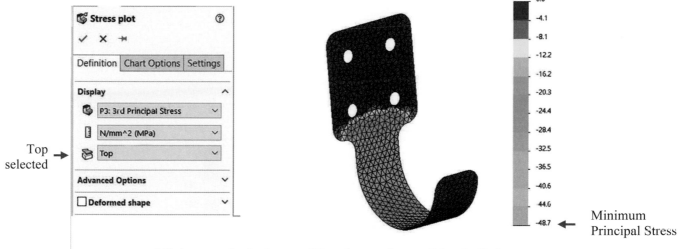

Minimum principal stress P3 on the top faces of the shell elements. This is the back side, invisible in this illustration.

Figure 4-11: Maximum and minimum principal stress plots in the shell element model.

Information on the element side is shown in Plot details. Plot details can be turned off in Chart Options and are not shown in this illustration.

As demonstrated in Figure 4-11, what stress is visible (top or bottom) does not depend on view direction (which side is visible) but only on the selection made in the **Stress Plot** window.

Stress plots showing results produced by shell element models may be confusing because the model behaves as if it was transparent; results pertaining to the top side of elements can be seen from the bottom side and vice versa. This confusion may be avoided if shell thickness is rendered on the stress plot. This option is available in **Advanced Options** of the stress plot. Figure 4-12 demonstrates this using the von Mises stress plot as an example.

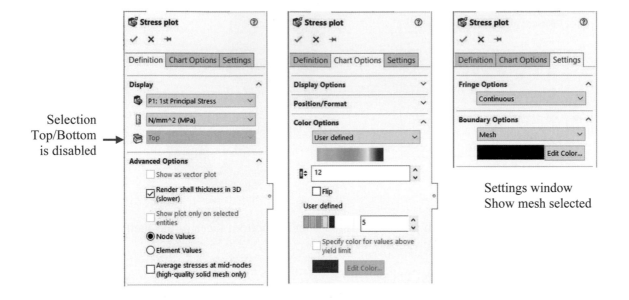

Selection
Top/Bottom
is disabled

Definition window
Render shell thickness selected

Chart Options window
Use defined colors selected

Settings window
Show mesh selected

Figure 4-12: Maximum principal stress (P1) shown with shell thickness rendered.

Mesh plot is extruded making it look as if the mesh was made of prism elements.

Rotate the stress plot to see that results are now shown differently on the inside and the outside of the model. Custom colors are used in this and many other illustrations in this book.

Figure 4-12 now shows P1 stress results with a rendered thickness of the shell element mesh. Notice that the elements look like prisms because the mesh shown in this illustration has an appearance of a 3D mesh. This is the result of 3D rendering on the shell element mesh.

Stress distribution through that rendered thickness shown in Figure 4-12 does not show a distribution of bending stress, it shows the distribution of P1 stress. To see how bending stress changes across the thickness requires a plot of directional stresses such as SX, SY, SZ. These directional stresses are aligned by default with the global coordinate system. To show the distribution of bending stresses in the curved portion of the model we need to align them with an axis that defines the local cylindrical coordinate system. After the alignment, SX becomes radial stress, SY becomes circumferential stress and SZ becomes axial stress (Figure 4-13).

Default stress component	Stress component when aligned with cylindrical coordinate system defined by the axis
SX	Radial
SY	Circumferential
SZ	Axial

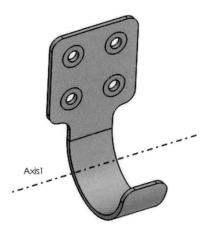

Figure 4-13: Alignment of directional stresses with an axis defining a local cylindrical coordinate system.

Directional stresses may also be aligned with a coordinate system or a reference plane in which case stress components are defined in a local Cartesian coordinate system.

To show the distribution of bending stress in the curved portion of the model, use SY stress and align the plot with Axis1 as shown in Figure 4-14.

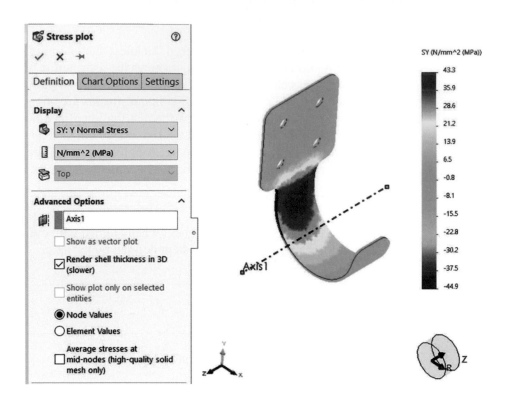

Figure 4-14: Stress plot SY aligned with Axis1, shell thickness is rendered. Circumferential stress can be seen as changing from compressive -45MPa on the back side of the model to tensile 43MPa on the front side of the model.

Notice the symbol in the lower right corner; it indicates that stresses are aligned with a local cylindrical coordinate system.

The distribution of bending stress across the shell element thickness may also be shown without rendering shell element thickness, but that requires two plots as shown in Figure 4-15 and Figure 4-16. These plots use vectors to show the line and direction of stress. Remember that as opposed to von Mises stress, which is a scalar entity, directional stresses are vectors.

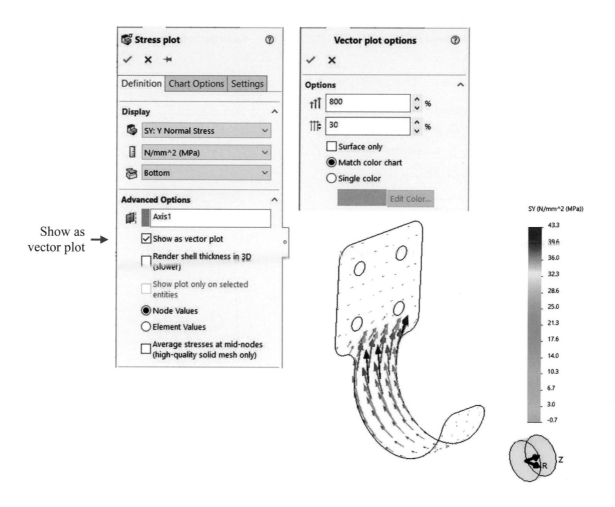

Figure 4-15: Stress plot SY aligned with Axis1, shown on the front side of the model. This is the tensile side, so stresses are positive.

Vector display may be modified using the Vector plot option window. The vector plot option window is called using Vector Plot Options in the pop-up menu.

Vector Plot Options becomes available in the pop-up menu when Show as vector plot is selected in the Stress Plot definition.

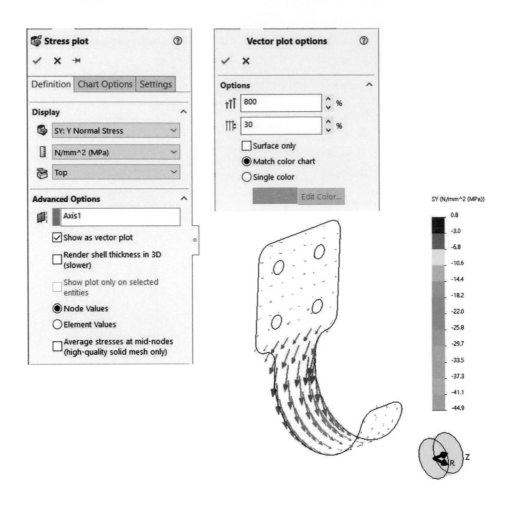

Figure 4-16: Stress plot SY aligned with Axis1, shown on the back side of the model. This is the compressive side where stresses are shown as negative.

The structural analysis of the support bracket has been completed. We now proceed to find modes of vibration of the same bracket. This requires **Frequency** study. Create a new study and name it *02 modal* (Figure 4-17).

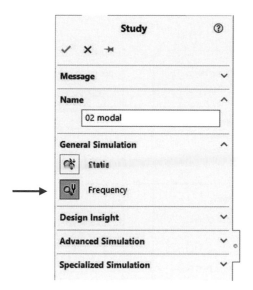

Figure 4-17: Frequency study definition.

This type of analysis is also known as Modal analysis.

You can copy restraints and mesh definitions from the static study to the frequency study by dropping them into the frequency study tab. No loads are defined in this **Frequency** study. We assume that loads, if any are present, will not significantly affect the natural frequencies.

In the properties of the **Frequency** study, verify that five modes will be calculated (Figure 4-18). In **Simulation Default Options**, verify that displacement plots are automatically created in the *Results* folder for all calculated modes (Figure 4-19).

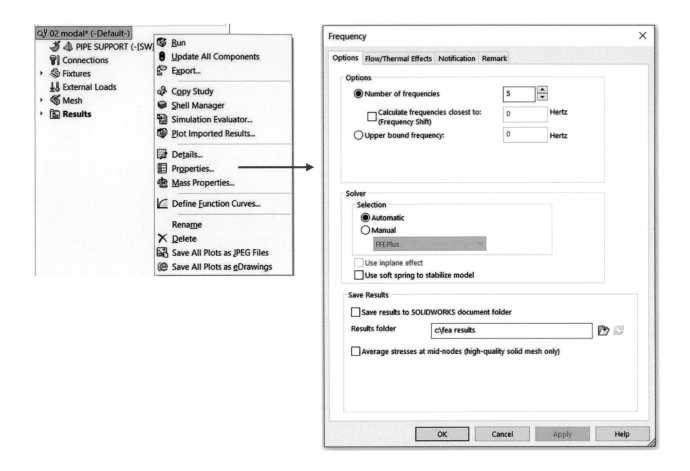

Figure 4-18: Properties of a frequency study.

Five frequencies are calculated by default, meaning that five frequencies and five corresponding shapes of vibration will be calculated. Frequency and the corresponding shape are jointly called a mode of vibration.

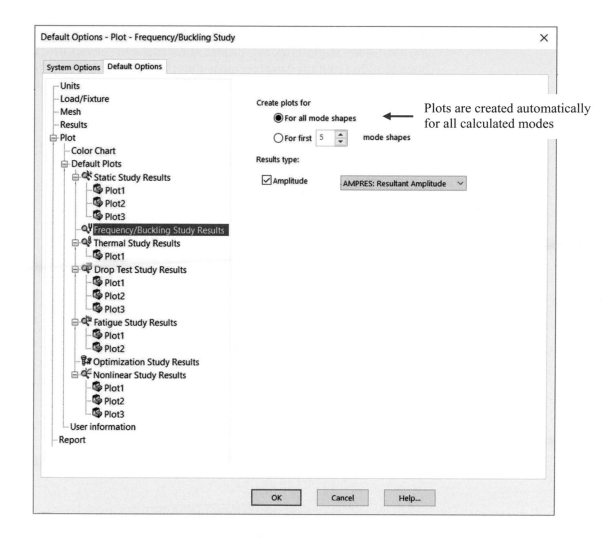

Figure 4-19: Using the above settings, Amplitude plots are automatically created for all the modes of vibration specified in the options of the Frequency study window (Figure 4-18).

Five plots will be created because five plots were selected in Properties of Frequency study as shown in Figure 4-18.

Run the solution and verify that five amplitude plots have been placed in the *Results* folder. Right-click any displacement plot and select **Edit Definition** to open the **Mode Shape/Displacement Plot** panel. Try displaying the mode shape plot with and without colors by making the appropriate selection shown in Figure 4-20.

Read this important message →

Colors may be deselected →

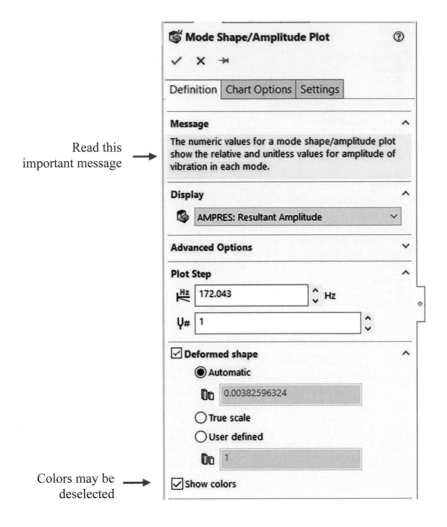

Figure 4-20: Mode Shape/Amplitude Plot definition window.

You can switch between Mode Shape and Amplitude plot by showing or hiding colors. However, you must remember that absolute displacements values are meaningless. Only ratios between displacements in the same mode are valid.

Figure 4-21 shows the first mode of vibration presented as a mode shape and as an amplitude plot.

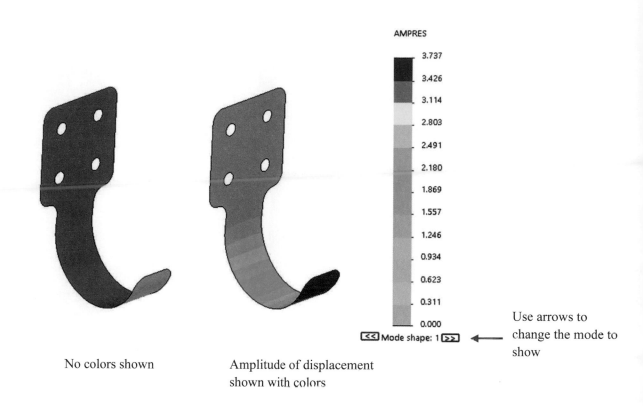

No colors shown   Amplitude of displacement
                         shown with colors

Use arrows to change the mode to show

Figure 4-21: Mode shape plot and Displacement plot.

Numerical results shown in the color legend are qualitative only. They may be used only for comparison of displacements in different locations within the same mode. No units of displacements are shown.

Even though the displacement plot does show displacement magnitude, the absolute displacement results are meaningless. Displacement results are purely qualitative and can be used only for a comparison of displacements within the same mode of vibration. Relative comparisons of displacements between different modes are invalid.

One of the worst errors an FEA user can make is to take displacement results from a modal analysis for their face value!

Examine deformation plots of higher modes (Figure 4-22) to notice that higher modes are associated with more deformation. The best way to analyze the results of a frequency analysis is by examining the animated deformation plots.

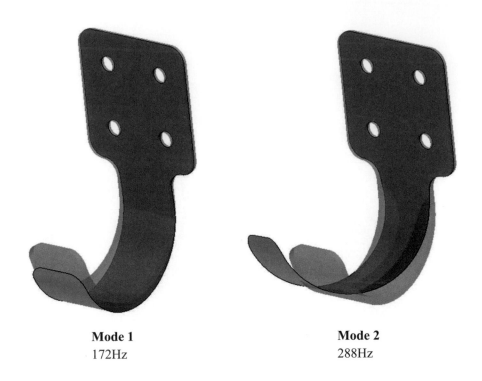

Mode 1
172Hz

Mode 2
288Hz

<u>Figure 4-22: Deformation plots showing the shape of deformation (mode shape) of the first two modes. Undeformed model is superimposed on the deformed shape.</u>

Adjust color and transparency of the superimposed model for the best effect.

To animate any plot, right-click an active plot icon to display an associated pop-up menu, and then select **Animate**.

To **List Resonance Frequencies**, right-click the *Results* folder and make the selection as shown in Figure 4-23.

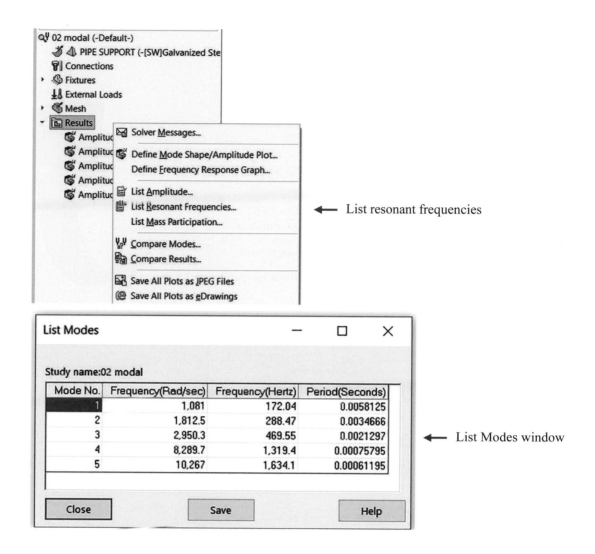

Figure 4-23: The summary of frequency results includes the list of all calculated modal (resonant) frequencies.

Select List Resonant Frequencies to open List Modes window.

If desired, displacement result may be normalized to 1 (Figure 4-24).

Normalize
Model Shape →

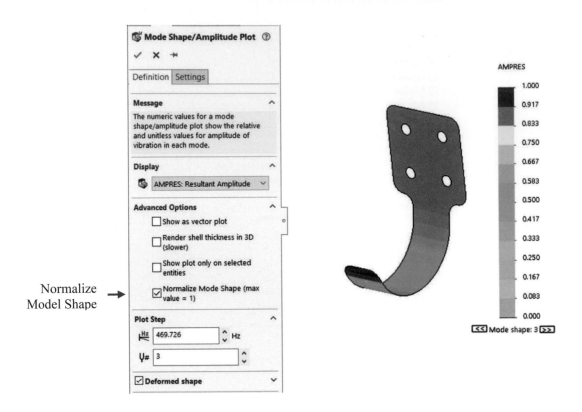

Figure 4-24: Normalized displacement amplitude results.

Displacement plot for mode 3 is shown in this illustration.

List Modes window in Figure 4-23 presents frequency results in three different units:

- ω - circular (angular) frequency [rad/s]
- f - frequency (cycles per second) [Hz]
- T - vibration period [s]

Circular frequency, frequency and period are related as follows:

$$\omega = 2\pi f \qquad T = \frac{1}{f}$$

Models in this chapter

Model	Configuration	Study Name	Study Type
PIPE SUPPORT.sldprt	*Default*	*01 static*	Static
		02 modal	Frequency
MISALIGNMENT.sldprt*	*Default*	*01 misaligned*	Static
		02 aligned	Static

* This model comes with studies defined.

Notes:

5: Static analysis of a link

Topics covered

- Symmetry boundary conditions
- Preventing rigid body motions
- Limitations of the small displacement theory

Project description

We need to calculate displacements and stresses of the link shown in Figure 5-1. The link is supported by tight-fitting pins in the two end holes and is loaded by a loose fitting pin at the central hole with a force of 30000N.

Open part file LINK. It has been assigned the material properties of Chrome Stainless Steel.

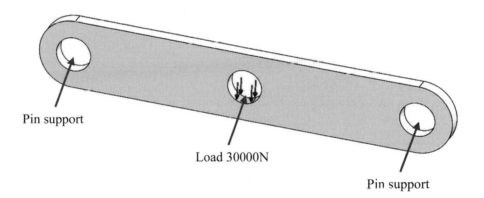

Figure 5-1: LINK model with load and supports.

Notice that the supporting pins and the loaded pin are not present in the model. Load is applied to the blue face on the face in the middle hole.

Procedure

One way to conduct this analysis would be to model both the link and all three pins, and then conduct an analysis of the assembly. However, we are not interested in the contact stresses that will develop between the pins and the link. Our focus is on displacements and stresses that will develop in the link. Therefore, the analysis can be simplified; instead of modeling the pins, we can simulate their effects by properly defining restraints and loads. Notice that the link geometry, restraints, and loads are all symmetrical. We can take advantage of this symmetry and analyze only half of the model, replacing the other half with symmetry boundary conditions.

To work with half of the model, switch to the *02 half model* configuration using the **Configuration Manager**. This also suppresses all the small chamfers in the model. The chamfers have negligible structural effect and would unnecessarily complicate the mesh. Removing geometric details deemed unnecessary for analysis is called defeaturing.

Finally, notice a split face in the middle hole that defines the area where the load will be applied. The geometry in FEA-ready form is shown in Figure 5-2. Figure 5-2 also explains how the restraints should be applied.

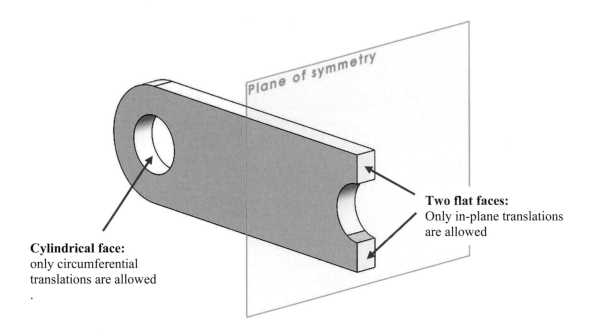

Figure 5-2: Half of the link with restraints explained.

The applied restraints: a hole where the pin support is simulated and two faces in the plane of symmetry where symmetry boundary conditions are defined: translations in the plane of symmetry are allowed; translation in the direction perpendicular to the plane of symmetry is not allowed.

The model is ready for the definition of restraints – the highlight of this exercise. Move to **SOLIDWORKS Simulation** and define a static study.

Right-click *Fixtures* folder and select **Fixed Hinge** from the pop-up menu. This opens the **Fixture** window with the **Fixed Hinge** button already selected (Figure 5-3). This restraint simulates hinge support. An identical restraint can be obtained using either **Hinge** or **On Cylindrical Face** (with suppressed radial and axial directions) (Figure 5-3).

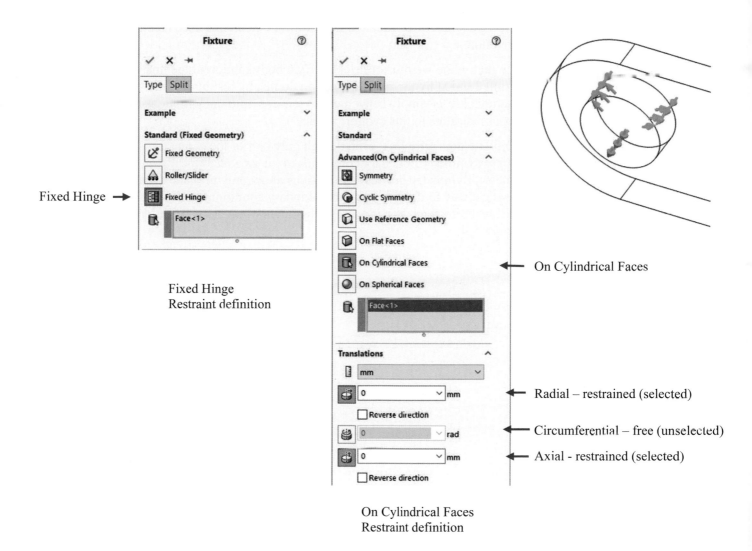

Fixed Hinge

Fixed Hinge
Restraint definition

On Cylindrical Faces

Radial – restrained (selected)

Circumferential – free (unselected)

Axial - restrained (selected)

On Cylindrical Faces
Restraint definition

Figure 5-3: Two alternative ways of defining hinge support.

Symbols of restraints on cylindrical face are shown.

Once the Fixture definition window has been opened, you can move between different types of restraints. You are not committed to the default selection made in the pop-up menu choice.

When the **Fixture** definition window specifies the restraint type as **Hinge** or **On cylindrical face**, the restraint directions are associated with the directions of the cylindrical face (radial, circumferential, and axial), rather than with global directions x, y, z.

To simulate a pin support that allows the link to rotate about the pin axis, radial displacements need to be restrained and circumferential displacements allowed. Furthermore, displacements in the axial direction need to be restrained in order to avoid any rigid body motions of the entire link along this direction.

Using the **Hinge** support is easier but does not describe how the restraint functions, so the **On cylindrical face** restraint method is used here instead for clarification.

Notice that while we must restrict the rigid body motion of the link in the direction defined by the pin axis, we can do this by restraining any point of the model. It is simply convenient to remove rigid body motions by applying the axial restraints to this cylindrical face.

To simulate the entire link, even though only half of the geometry is present, we apply symmetry boundary conditions to the two faces located in the plane of symmetry. Symmetry boundary conditions allow only in-plane displacements. The easiest way to define symmetry boundary conditions is to use **Symmetry** as a type of restraint. The definition of the symmetry boundary conditions is illustrated in Figure 5-4.

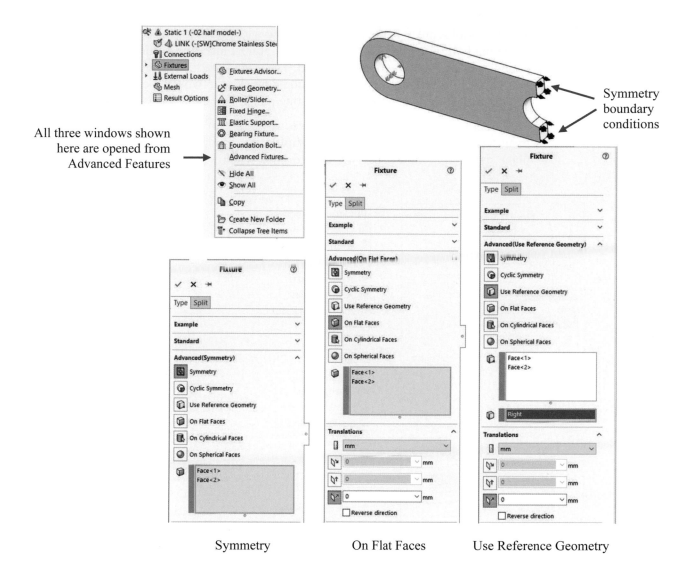

Figure 5-4: Definition of symmetry boundary conditions.

Three ways of defining the same symmetry condition are shown; all produce the same result:

(1) Symmetry restraint.

(2) On Flat Face restraint where in-plane translations are allowed, but translation in the direction normal to the face is set to zero.

(3) Use Reference Geometry where free and restrained directions are defined with reference to selected reference geometry (here, the Right reference plane). The reference plane can be selected from the fly-out menu which is not shown here.

Recall from Figure 5-1 that the link is loaded with 30000N. Since we are modeling half of the link, we must apply a 15000N load to a portion of the cylindrical face, as shown in Figure 5-5. The size of the load application area is arbitrarily created with a split line. It should be close to what we expect the contact area to be between the loose fitting pin and the link.

When defining the load (Figure 5-5), take advantage of **SOLIDWORKS'** fly-out menu visible in the **SOLIDWORKS Simulation** window to select the reference plane required in the **Force/Torque** definition window.

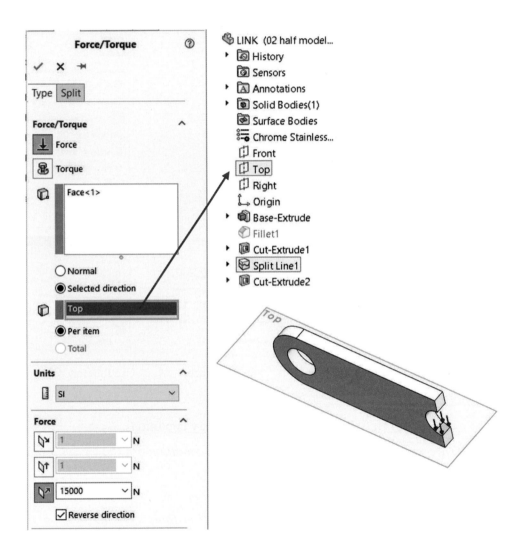

Figure 5-5: 15000N force applied to the central hole.

Force is applied in the selected direction. The Top reference plane is used to determine the load direction. Notice that the load is distributed uniformly. We are not trying to simulate a contact stress problem.

The final task of model preparation is meshing. Right-click the *Mesh* folder to display the related pop-up menu, and then select **Create Mesh...** Verify that the mesh preferences are set on high quality (meaning that second order elements will be created) and mesh the geometry using the default element size. For more information on the created mesh, you may wish to review the **Mesh Details** (Figure 5-6).

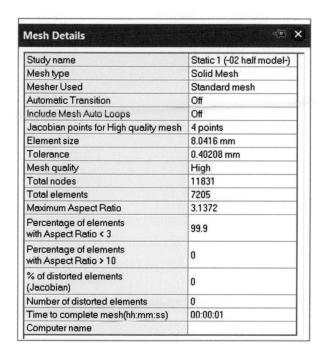

Mesh Details	
Study name	Static 1 (-02 half model-)
Mesh type	Solid Mesh
Mesher Used	Standard mesh
Automatic Transition	Off
Include Mesh Auto Loops	Off
Jacobian points for High quality mesh	4 points
Element size	8.0416 mm
Tolerance	0.40208 mm
Mesh quality	High
Total nodes	11831
Total elements	7205
Maximum Aspect Ratio	3.1372
Percentage of elements with Aspect Ratio < 3	99.9
Percentage of elements with Aspect Ratio > 10	0
% of distorted elements (Jacobian)	0
Number of distorted elements	0
Time to complete mesh(hh:mm:ss)	00:00:01
Computer name	

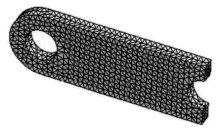

Figure 5-6: The meshed model shown together with the Mesh Details window.

After solving the model, we first need to check if the pin support and symmetry boundary conditions have been applied properly. This includes checking whether the link can rotate around the pin and whether it behaves as a half of the whole link. This is best done by examining the animated displacements, preferably with the undeformed shape visible (Figure 5-7).

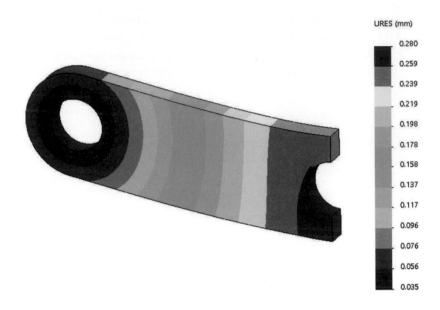

URES (mm)

0.280
0.259
0.239
0.219
0.198
0.178
0.158
0.137
0.117
0.096
0.076
0.056
0.035

Figure 5-7: Deformed displacements plot.

Animate this plot and observe that the link rotates around the imaginary pin while faces in the plane of symmetry remain flat and perform only in-plane translation.

To conclude this exercise, review the stress results. Examine different stress components, including the maximum principal stresses, minimum principal stresses, etc.

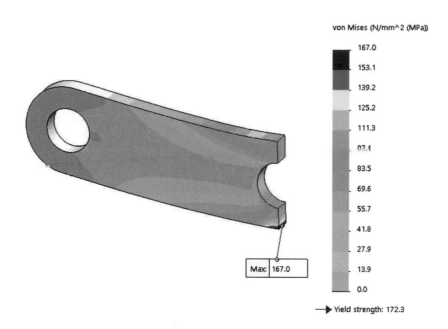

Figure 5-8: Sample of stress results: von Mises stress.

In this plot, the location of the maximum stress is shown, as requested in the Chart Options window.

Repeat this exercise using the full model to perform an analysis of the complete model without using symmetry boundary conditions.

Before finishing the analysis of LINK, we should notice that the link supported by two pins as modeled in this exercise corresponds to the configuration shown in Figure 14-25, where one of the hinges is supported by rollers and is free to move horizontally.

Since linear analysis does not account for changes in model stiffness during the deformation process (nor does it account for material yielding), linear analysis is unable to model stresses that would have developed if both pins were in a fixed position. If both pins were fixed, a nonlinear geometry analysis would be required to analyze the model. Refer to chapter 14 for more information on non-linear analyses.

Models in this chapter

Model	Configuration	Study Name	Study Type
LINK.sldprt	*01 full model*		
	02 half model	*Static 1*	Static

6: Frequency analysis of a tuning fork and a plastic part

Topics covered

- Frequency analysis with and without supports
- Rigid body modes
- The role of supports in frequency analysis
- Symmetric and anti-symmetric modes

Project description

Structures have preferred frequencies of vibration, called resonant frequencies. A mode of vibration is the shape in which a structure will vibrate at a given natural frequency. The only factor controlling the amplitude of vibration in resonance is damping. While any structure has an infinite number of resonant frequencies and associated modes of vibration, only a few of the lowest modes are important when analyzing response to dynamic loading. A frequency analysis is used to calculate these resonant frequencies and their associated modes of vibration.

Open the part file called TUNING FORK. It has material properties already assigned (Chrome Stainless Steel). The model is shown in Figure 6-1.

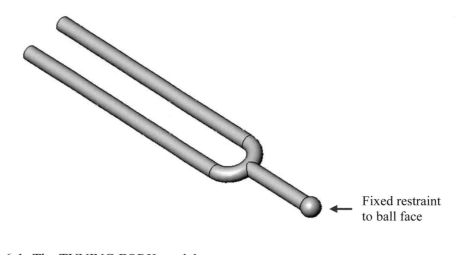

Fixed restraint to ball face

Figure 6-1: The TUNING FORK model.

A fixed restraint is applied to the surface of the ball. Do not apply an "On spherical face" restraint.

A quick inspection of the CAD geometry reveals a sharp re-entrant edge. This condition renders the geometry unsuitable for stress analysis but is acceptable for frequency analysis unless the omitted fillet significantly changes the model stiffness.

Procedure

Define a **Frequency** study called *Frequency 1*. Once the study has been created right-click it to open a pop-up menu and select Properties (Figure 6-2).

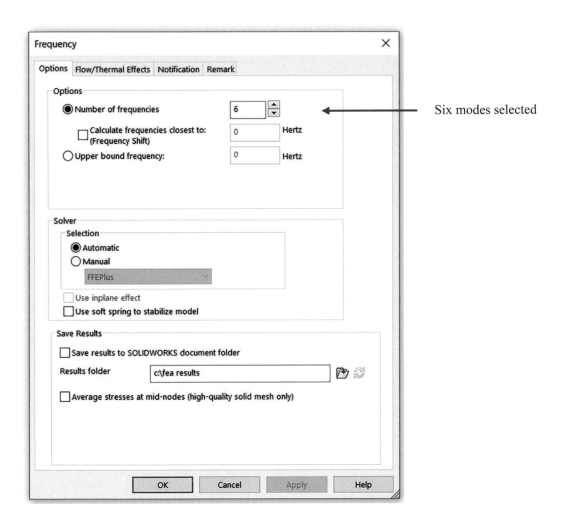

Figure 6-2: Frequency study properties.

Six modes of vibration will be calculated.

Next, define fixed restraints to the ball surface, as shown in Figure 6-1. This approximates the situation when the TUNING FORK is held with two fingers.

Finally, mesh the model with the default element size. The meshed model is shown in Figure 6-3. The mesher selects the element size to satisfy the requirements of a stress analysis. A frequency analysis is less demanding on the mesh, so generally, a less refined mesh is acceptable. Nevertheless, since this is a very simple model, we accept the mesh without making any attempt to simplify it.

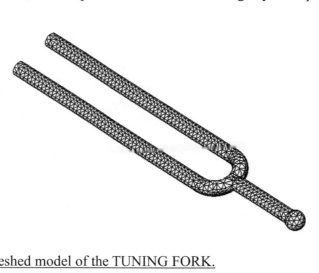

Figure 6-3: Meshed model of the TUNING FORK.

Sharp re-entrant edges are present in the model which is acceptable for frequency analysis if their omission does not significantly change the model's stiffness. Sharp reentrant edges will not cause a singularity in the solution for modes of vibration.

After the solution is complete, **SOLIDWORKS Simulation** automatically creates displacement plots (Figure 6-4). For reasons already explained in chapter 4, we ignore numerical values in amplitude results. Using the **Mode Shape/ Amplitude** window we deselect colors, turning the amplitude plot into a deformation plot; we call this a **Mode Shape** plot.

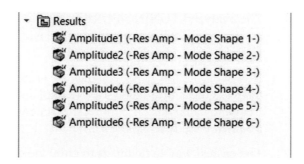

Figure 6-4: Six amplitude plots are created automatically.

Plot Amplitude1 shows Mode Shape 1, etc.

The **Mode Shape** plot in the modal analysis shows the associated modal shape of vibration and lists the corresponding natural frequency. The first four modes of vibration are presented in Figure 6-5.

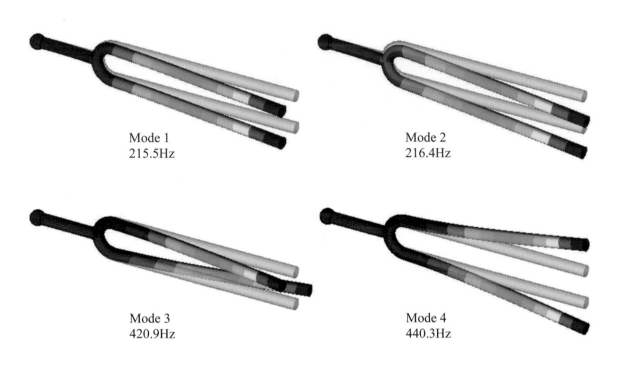

Mode 1
215.5Hz

Mode 2
216.4Hz

Mode 3
420.9Hz

Mode 4
440.3Hz

<u>Figure 6-5: The first four modes of vibration and their associated frequencies.</u>

The undeformed model is superimposed on the mode shape plots.

The analyzed TUNING FORK is the most common type of tuning fork and as any musician will tell us, should produce a lower A sound, with a frequency of 440 Hz.

However, the lower A frequency of 440 Hz, which is expected to be the first mode, is the fourth mode. Before explaining why this occurs, let us run the frequency analysis once more, this time without any restraints using a new frequency study *Frequency 2*.

The easiest way to do this is to copy the existing *Frequency 1* study and either delete or suppress the restraint (right-click the restraint icon and make the proper selection).

In the absence of restraints, the model has six **Rigid Body Modes**. To calculate six modes of vibration, which for consistency should now be called **Elastic Modes**, we must specify a total of 12 modes in the properties of the frequency study.

After the solution has been completed, right-click the *Results* folder and select **List Resonant Frequencies** (Figure 6-6).

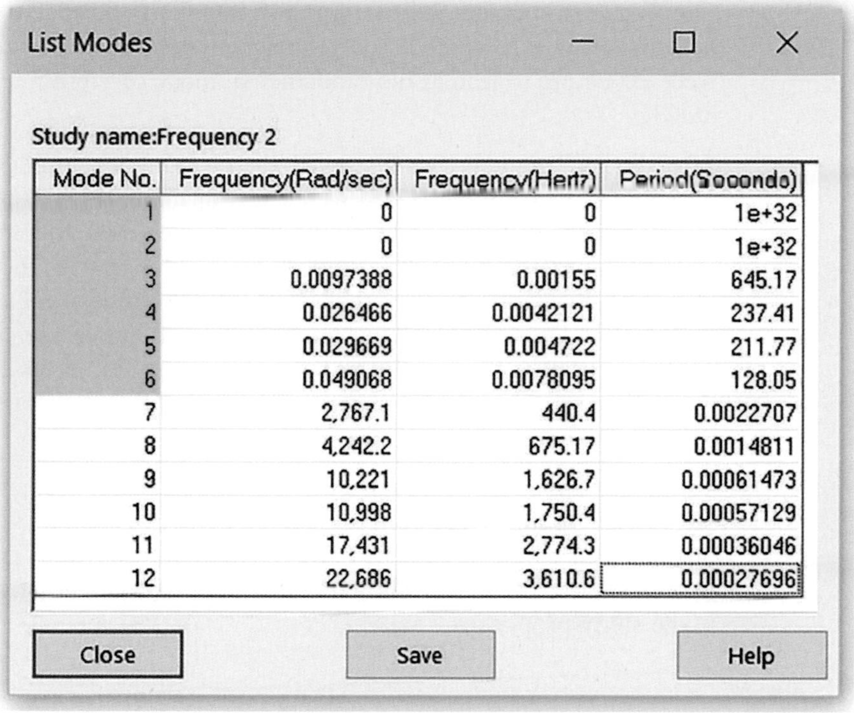

List Modes			— ☐ ✕
Study name:Frequency 2			
Mode No.	Frequency(Rad/sec)	Frequency(Hertz)	Period(Seconds)
1	0	0	1e+32
2	0	0	1e+32
3	0.0097388	0.00155	645.17
4	0.026466	0.0042121	237.41
5	0.029669	0.004722	211.77
6	0.049068	0.0078095	128.05
7	2,767.1	440.4	0.0022707
8	4,242.2	675.17	0.0014811
9	10,221	1,626.7	0.00061473
10	10,998	1,750.4	0.00057129
11	17,431	2,774.3	0.00036046
12	22,686	3,610.6	0.00027696

Close   Save   Help

Figure 6-6: The **List Modes** window in the analysis of TUNING FORK.

Modes 1-6, highlighted in green, are rigid body modes with frequency equal to, or very close to 0Hz. Numerical error is the reason why not all rigid body modes have a frequency exactly equal to 0Hz. Mode 7 is the first elastic mode of vibration.

Notice that if you re-run the solution with more modes requested (here 11) but you use the model where five modes have been previously calculated, then only five plots will be automatically created. You will have to define the remaining plots manually. Alternatively, you may delete all plots before re-running the solution and all twelve plots will be created automatically.

We notice that the first six modes have the associated frequency 0Hz or very close to 0Hz. Why? The first six modes of vibration correspond to rigid body modes. Because the TUNING FORK is not supported, it has six degrees of freedom as a rigid body: three translational and three rotational.

SOLIDWORKS Simulation detects these rigid body modes and assigns them with a frequency of zero (0Hz). Modes 3, 4, 5, and 6 do not have a frequency of exactly zero due to numerical error.

The first elastic mode of vibration, meaning the first mode requiring the fork to deform is mode 7, which has a frequency of 440.4 Hz. This is close to what we were expecting to find as the fundamental mode of vibration for the TUNING FORK.

Why did the frequency analysis with the restraint not produce the first mode with a frequency near to 440 Hz? If we closely examine the first three modes of vibration of the supported TUNING FORK, we notice that they all need the support in order to exist. The support is needed to sustain these modes, but while this support makes these first three modes possible, in reality it also provides damping. After modes 1, 2, and 3 have been damped out, the TUNING FORK vibrates the way it was designed to: in mode 4 (as calculated in the analysis with supports) or mode 7 (as calculated in the analysis without supports). These two modes are identical.

To learn more about modes of vibration of an unsupported elastic model (all models in linear FEA are considered elastic), calculate six elastic modes of the unsupported model PLASTIC PART. **Mode Shape** results corresponding to the first six elastic modes are shown in Figure 6-7. You will need to request 12 modes in the study properties.

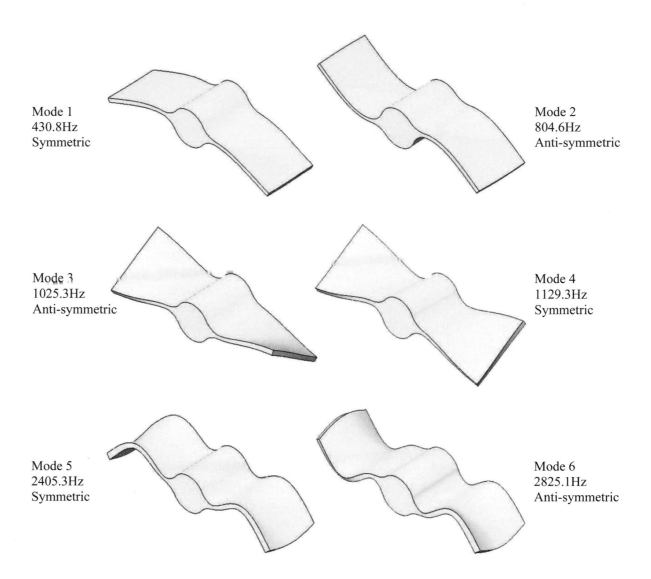

Mode 1
430.8Hz
Symmetric

Mode 2
804.6Hz
Anti-symmetric

Mode 3
1025.3Hz
Anti-symmetric

Mode 4
1129.3Hz
Symmetric

Mode 5
2405.3Hz
Symmetric

Mode 6
2825.1Hz
Anti-symmetric

Figure 6-7: The first six elastic modes of vibration of the unsupported model
PLASTIC PART.

Based on the deformation results of PIPE SUPPORT, TUNING FORK and
PLASTIC PART, we can make an interesting observation about the nature of the
modes of vibration. If a model is symmetric and has symmetric restraints (or no
restraints at all), then its modal shapes are either symmetric or anti-symmetric.

All 12 modes: six rigid body modes and six elastic modes, are listed in Figure 6-8.

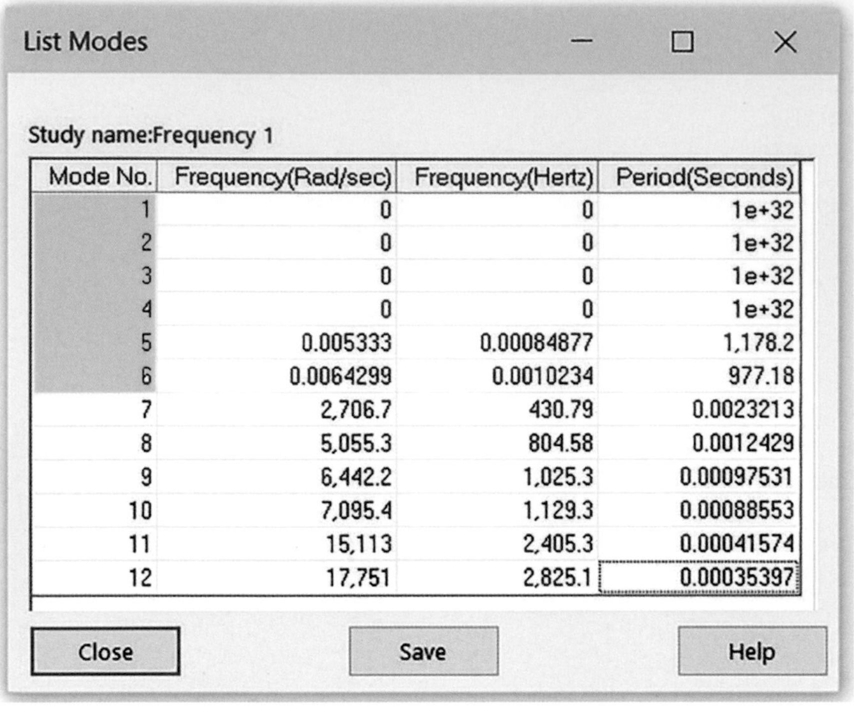

Mode No.	Frequency(Rad/sec)	Frequency(Hertz)	Period(Seconds)
1	0	0	1e+32
2	0	0	1e+32
3	0	0	1e+32
4	0	0	1e+32
5	0.005333	0.00084877	1,178.2
6	0.0064299	0.0010234	977.18
7	2,706.7	430.79	0.0023213
8	5,055.3	804.58	0.0012429
9	6,442.2	1,025.3	0.00097531
10	7,095.4	1,129.3	0.00088553
11	15,113	2,405.3	0.00041574
12	17,751	2,825.1	0.00035397

Study name:Frequency 1

List Modes — □ ✕

Close   Save   Help

Figure 6-8: The List Modes window in the analysis of PLASTIC PART.

Modes 1-6, highlighted in green, are rigid body modes with frequency equal or very close to 0Hz. Modes 7-12 are elastic modes shown in Figure 6-7 where they are numbered from 1 to 6.

Models in this chapter

Model	Configuration	Study Name	Study Type
TUNING FORK.sldprt	*Default*	*Frequency 1*	Frequency
		Frequency 2	Frequency
PLASTIC PART.sldprt	*Default*	*Frequency 1*	Frequency

Notes:

7: Thermal analysis of a pipe connector and a heater

Topics covered

- Analogies between structural and thermal analysis
- Steady state thermal analysis
- Analysis of temperature distribution and heat flux
- Thermal boundary conditions
- Thermal stresses
- Vector plots

Project description

So far, we have performed static analyses and frequency analyses, which both belong to the class of structural analyses. Static analysis provides results in the form of displacements, strains, and stresses, while frequency analysis provides results in the form of natural frequencies and associated modes of vibration. We will now examine a thermal analysis example. Numerous analogies exist between thermal and structural analyses. The most direct analogies are summarized in Figure 7-1.

Structural Analysis	Thermal Analysis
Displacement [m]	Temperature [K]
Strain [dimensionless]	Temperature gradient [K/m]
Stress [N/m^2]	Heat flux [W/m^2]
Load [N] [N/m] [N/m^2] [N/m^3]	Heat source [W] [W/m] [W/m^2] [W/m^3]
Prescribed displacement [m]	Prescribed temperature [K]

Figure 7-1: Selected analogies between structural and thermal analysis with corresponding units in the SI system.

Procedure

Open part model PIPE CONNECTOR in *01 full* configuration. Our objective is to find the steady state temperature and heat flux in the part when prescribed temperatures are applied to the end faces as shown in Figure 7-2. As indicated in Figure 7-1, prescribed temperatures in thermal analysis are analogous to prescribed displacements in structural analysis.

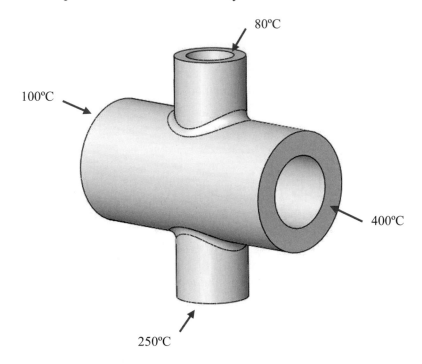

Figure 7-2: PIPE CONNECTOR model.

The prescribed temperatures are applied to the end faces as boundary conditions. Values are given are in degrees Celsius; degrees Fahrenheit or degrees Kelvin could also be used.

Since no convection coefficients are defined on any surfaces, heat can enter and leave the model only through the end faces with the prescribed temperatures assigned. Even though the problem has little relevance to real heat transfer problems, it helps us understand the basics of thermal analysis.

The first step in thermal analysis is the study definition. Call this study *01 thermal* and define it as shown in Figure 7-3.

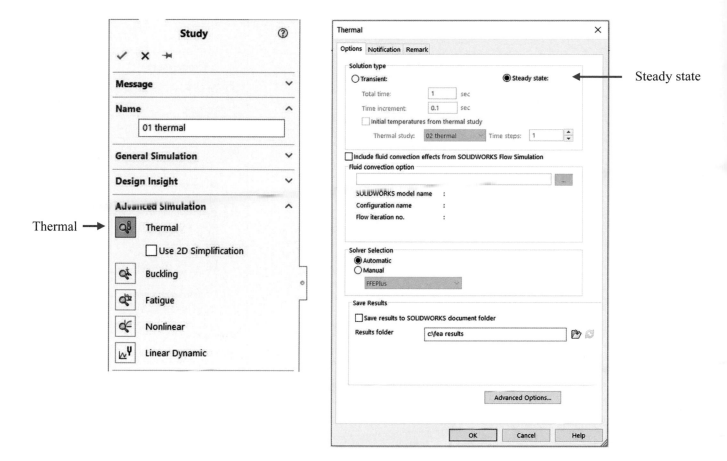

Figure 7-3: Definition of *01 thermal* study and the Thermal study properties window.

Steady state is the default option in Thermal studies.

In a **Steady state** thermal analysis, it is assumed that enough time has passed since the thermal conditions have been applied, and therefore all parameters characterizing heat flow no longer vary with time.

To define the prescribed temperature, right-click the *Thermal Loads* folder and select **Temperature** to open the **Temperature** definition window. Define prescribed temperatures to the four faces in four separate steps to define prescribed temperatures shown in Figure 7-2.

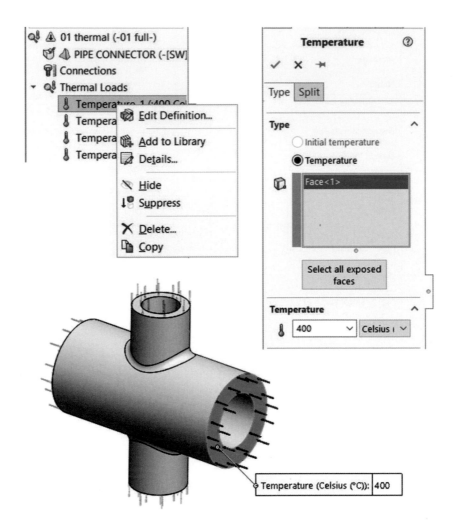

Figure 7-4: Defining prescribed temperature on the end face with temperature 400°C. Prescribed temperatures on the remaining faces have to be defined as shown in Figure 7-2.

Right-click the Thermal Loads folder and select Temperature from the pop-up menu. Define the prescribed temperatures in the Temperature definition window. The bottom illustration shows Temperature symbols on the selected face.

Mesh the model using the settings shown in Figure 7-5. To control the number of elements on the fillets, **Blended curvature-based mesh** is used in this exercise.

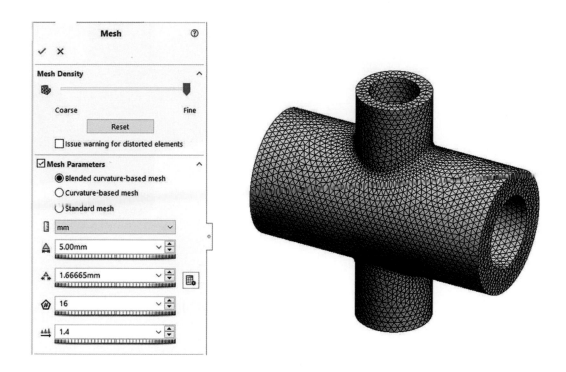

Figure 7-5: Curvature based mesh is used to assure correct meshing of fillets.

5mm is the maximum element size. Sixteen elements in a circle are specified for a low turn angle.

After solving the model, notice that only one result folder called *Thermal1* is present. By default, it shows the temperature distribution (Figure 7-6).

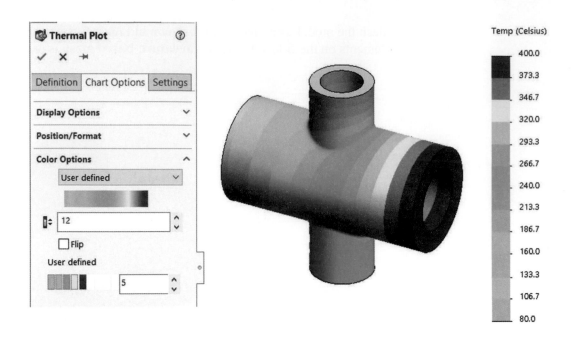

Figure 7-6: Thermal Plot window defining temperature distribution and corresponding plot.

As a reminder, notice that the dark blue color has been replaced by a grey color to improve appearance of this plot. As explained in chapter 2, this method is used in many illustrations throughout this book.

Create a plot showing resultant heat flux (Figure 7-7).

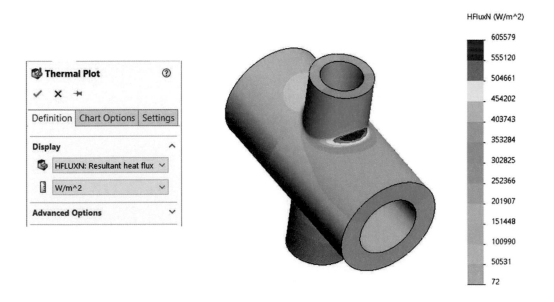

Figure 7-7: Thermal Plot window defining heat flux and corresponding plot.

Since heat flux is a vector quantity, it may be presented as a vector plot. Follow the steps in Figure 7-8 to redefine Figure 7-7 into a vector plot and edit it appropriately. To create a vector plot, right-click the existing heat flux plot and select **Edit definition** to open the **Thermal Plot** window. In **Advanced Options**, select **Show as vector plot** (1). Once the vector plot is showing, right-click its icon again and select **Vector Plot Options** to open the **Vector plot options window** (2). Adjust the settings to have a clear plot. Select **Surface only** in **Vector plot options**.

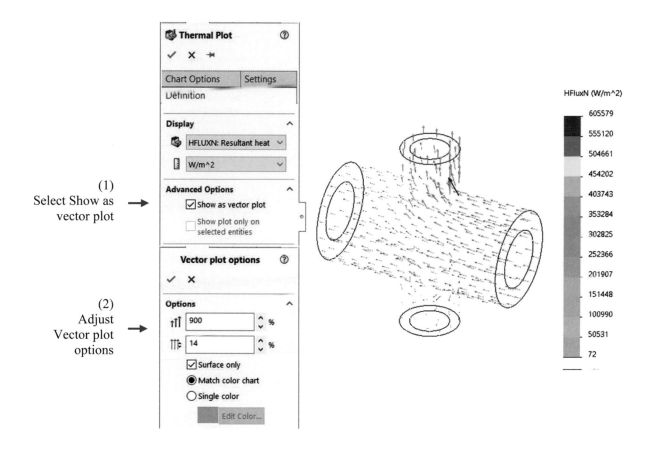

Figure 7-8: Heat flux result presented as a vector plot.

Experiment with different Vector plot options.

The non-uniform temperature field that establishes itself in the model (Figure 7-6) produces thermal stress due to non-uniform thermal expansion of different portions of the model. We will now analyze those thermal stresses. Define a static study *01 thermal stress*, with the properties shown in Figure 7-9.

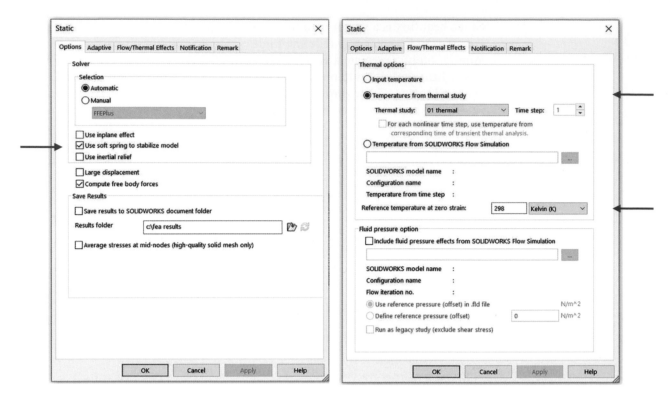

Options tab:
Select Use soft springs to stabilize the model

Flow/Thermal Effects tab:
Select Temperatures from thermal study
Use Reference temperature at zero strain
298K or 24.85°C

Figure 7-9: Properties of the study intended for analysis of thermal stresses are defined under two tabs.

Proceed as explained below to define Options and Flow/Thermal Effects.

In the **Options** tab select **Use soft springs to stabilize model**. This is because in a static analysis, the model will not be subjected to any structural loads or restraints. This way the pure effect of temperature is shown. The model is under internally balanced loads, but due to numerical errors, would experience rigid body movement. Soft springs eliminate those rigid body movements by attaching springs of very low stiffness to all nodes.

In the **Flow/Thermal Effects** tab specify **Temperature from thermal study**, *01 thermal*. This will import temperature results from the completed thermal study.

A static study with temperatures imported from a thermal study must use a mesh that is identical to the one used in the thermal study. To make sure the mesh is identical, copy mesh from the thermal study *01 thermal* to the static study *01 thermal stresses*. To copy the mesh, click and drag it from *01 thermal* study onto *01 thermal stresses* study tab.

Run the thermal stresses study and display a von Mises stress plot (Figure 7-10). Do not use "Average stresses at mid-nodes" option.

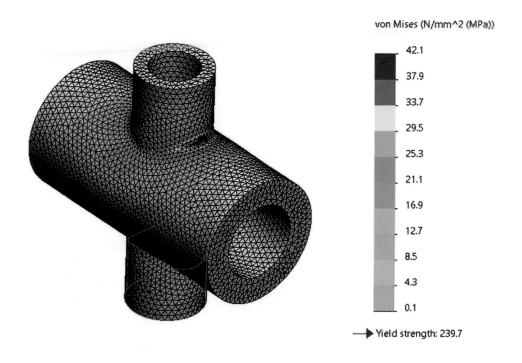

Figure 7-10: Thermal stresses develop due to non-uniform temperature distribution.

Notice an irregular shape of fringes where stress concentrations are located.

Analysis of the results in Figure 7-10 reveals irregularly shaped fringes. This is due to a coarse mesh. The mesh was adequate for the analysis of temperature and heat flux but is not sufficiently refined for an analysis of thermal stresses.

Repeat this exercise with a more refined mesh by using 32 elements mesh in a circle (Figure 7-11). This will require two more studies: thermal study *02 thermal* and static study *02 thermal stress*.

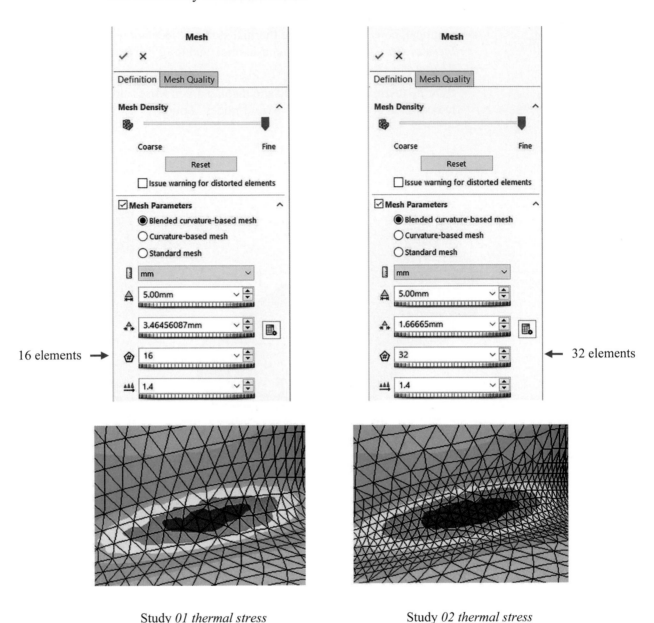

Study *01 thermal stress*   Study *02 thermal stress*

Figure 7-11: Thermal stress results produced studies *01 thermal stress* and *02 thermal stress.*

The original mesh uses 16 elements in a circle. The refined mesh uses 32 elements in a circle; both use 1.4 ratio. The "regularity" of fringes may be used to decide if a refinement is needed.

You may append this exercise as follows:

Review the animated displacement results and notice that in the absence of restraints the model expands about its center of mass.

Repeat this exercise using *02 half* configuration. Do not define any thermal conditions on the faces in the plane of symmetry; this way the symmetry of heat flow will be satisfied.

We will now conduct a thermal analysis of a pipe with cooling fins. The objective of this analysis is to find how much heat is dissipated by a 50mm long section (Figure 7-12). Open part model HEATER in *01 full* configuration and create a **Thermal** study *Thermal 1*.

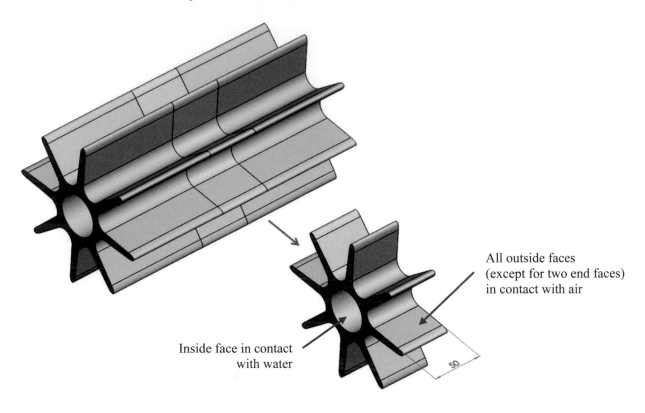

All outside faces (except for two end faces) in contact with air

Inside face in contact with water

Figure 7-12: Analysis of a pipe with cooling fins is conducted on a section 50mm long.

There are no thermal boundary conditions defined on two end faces meaning that these faces are insulated: no heat flows in or out.

Hot water at 100°C (373K) flows inside the heater. The coefficient of thermal convection between the water and the inside heater face is 1000W/m^{2}K, meaning that each 1m^2 of the inside face exchanges (gains or losses) 1000J of heat per second if the temperature difference between the face and water is 1K. The coefficient of thermal convection between the outside faces and the air is 20 W/(m^{2}K) meaning that each 1m^2 of the outside face exchanges (gains or losses) 20J of heat per second if the temperature difference between the face and air is 1K. The ambient air temperature is 27°C (300K).

Notice that we are using somewhat arbitrary values of convection coefficients. Finding convection coefficients that correctly describe the problem is often the most difficult part of a thermal analysis. For more information refer to "Thermal Analysis with SOLIDWORKS Simulation" published by SDC Publications.

Define convection coefficients and bulk temperatures as shown in Figure 7-13 and Figure 7-14.

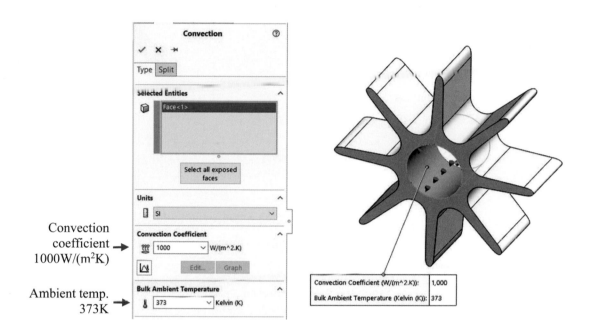

Figure 7-13: Convection coefficient and bulk temperature on the water side (inside the tube).

Convection symbols are shown.

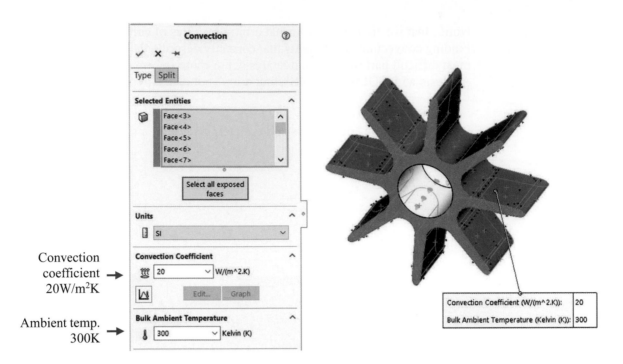

Convection coefficient 20W/m²K

Ambient temp. 300K

Figure 7-14: Convection coefficient and bulk temperature on the air side (outside).

Select all outside faces. Convection symbols are shown in small size.

Since the model represents a section of a longer pipe, we assume that there is no heat exchange through the end faces. Therefore, we do not define any convection coefficients on the end faces, treating them as insulated.

Use a **Curvature based mesh**. Specify a 2.5mm global element size to create two elements across the fin thickness. Use 8 for the minimum number of elements on a circle to produce a low element turn angle (Figure 7-15).

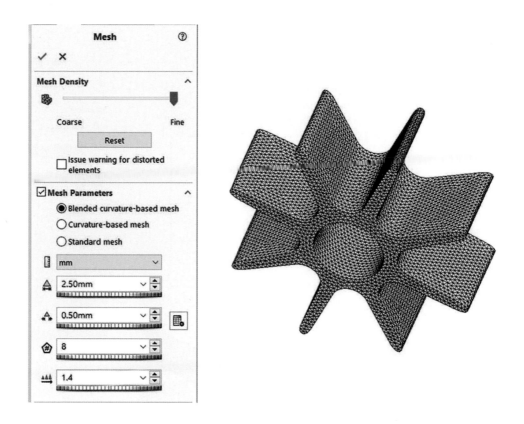

Figure 7-15: Blended curvature-based mesh is used to control the element turn angle in fillets.

Use element size 2.5mm and 8 elements in a circle.

Solve the study then Right-click the results folder and select **List Heat Power** (1) from the pop-up menu to open the **Heat Power** window. In the model window select the inside face (water side) (2), then click the **Update** button (3) in the **Probe Result** window. The total heat entering the model through the selected face is shown in the lower portion of the **Probe Result** window (4) as shown in Figure 7-16.

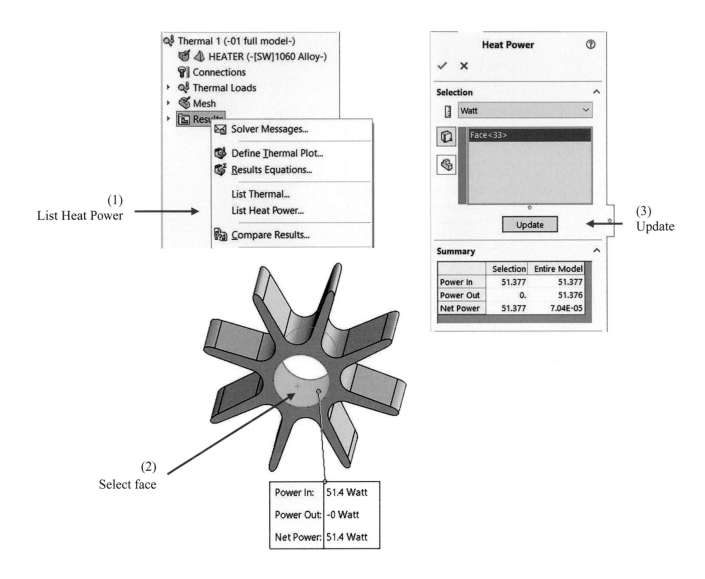

Figure 7-16: Total heat exchanged between the water and the model is 51.4W.

The positive sign of net power indicates that the model gains heat from the water. This can be visualized by constructing a vector plot of resultant heat flux.

Alternatively, we can obtain the same results by selecting all faces on the air side (Figure 7-17).

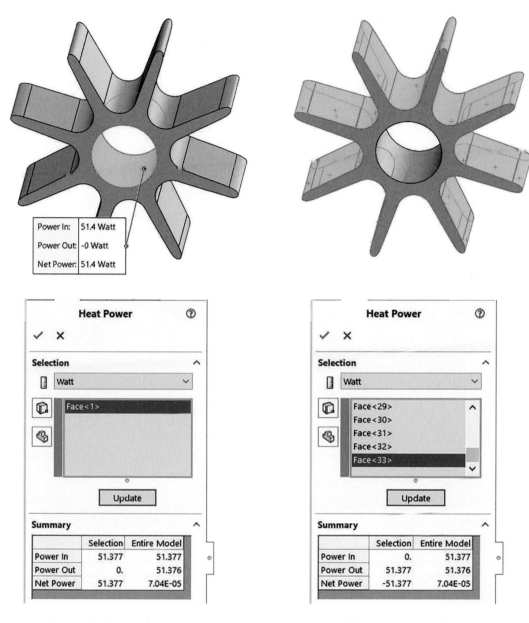

Heat gained on water side   Heat lost on air side

Figure 7-17: Total heat exchanged between water and the model (left), and between air and the model (right).

The heat gained from the water is +51.4W. The heat lost by the model to the air is -51.4W. The left window is a repetition from Figure 7-16.

To avoid clutter, callouts on the air side are not shown.

Since this is a steady state thermal analysis, the amount of heat entering the model and the amount of heat dissipated by the model must be equal.

The analysis can be significantly simplified by noticing that heat flow through the model has the property of circular symmetry. Therefore, instead of analyzing the entire model, we can analyze one section (Figure 7-18).

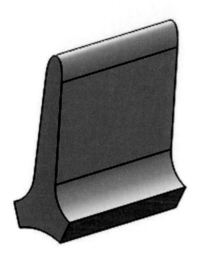

Figure 7-18: The model may be simplified to one 45° section.

This geometry is saved in configuration 02 section.

Repeat the analysis using configuration *02 section,* Standard mesh with default settings will suffice. Do not apply any convection conditions on the faces created by radial cuts. Multiply the total heat by 8 (Figure 7-19).

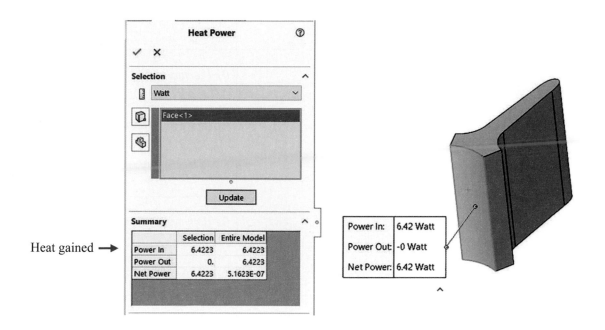

Figure 7-19: Total heat exchanged between water and 30° section of the heater.

Total heat gained from the water by the model is 6.42 x 8 = 51.4W.

Models in this chapter

Model	Configuration	Study Name	Study Type
PIPE CONNECTOR.sldprt	01 full	01 thermal	Thermal
		01 thermal stresses	Static
		02 thermal	Thermal
		02 thermal stresses	Static
	02 half		
HEATER.sldprt	01 full	Thermal 1	Thermal
	02 section	Thermal 2	Thermal

8: Thermal analysis of a heat sink

Topics covered

- Analysis of an assembly
- Global and local Contact conditions
- Steady state thermal analysis
- Transient thermal analysis
- Thermal resistance layer
- Use of section views in result plots

Project description

In this exercise, we continue with thermal analysis. However, this time we will analyze an assembly rather than a single part. Open the assembly HEAT SINK (Figure 8-1).

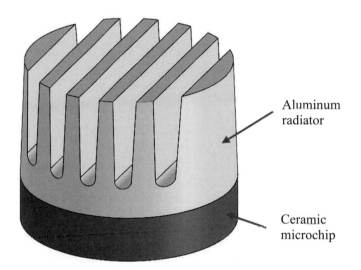

Aluminum radiator

Ceramic microchip

Figure 8-1: HEAT SINK assembly.

HEAT SINK assembly consists of two parts: a ceramic microchip and an aluminum radiator.

Analysis of an assembly allows assignment of different material properties to each assembly component. Notice that the *Solids* folder contains two icons corresponding to the two assembly components with material properties already assigned. This is because material has been assigned to each part that the assembly consists of: Ceramic Porcelain material to the microchip and 1060Alloy to the radiator.

The ceramic microchip generates a heat power of 25W and the aluminum radiator dissipates this heat. The ambient temperature is 27°C (300K). Heat is dissipated to the environment by convection through all exposed faces of the assembly. The convection coefficient (also called the film coefficient) is assumed to be 25 W/(m^2/K) on all exposed faces. This value of the convection coefficient corresponds to natural convection taking place without a cooling fan.

Heat flowing from the microchip to the radiator encounters thermal resistance on the boundary between the microchip and radiator. Therefore, a thermal resistance layer must be defined on the interface between these two components.

Our first objective is to determine the temperature and heat flux of the assembly after enough time has passed for temperatures to stabilize. This will require steady state thermal analysis.

The second objective is to study the temperature in the assembly as a function of time in a transient process when the assembly is initially at room temperature and the power is turned on at time t = 0. This will require transient thermal analysis.

Procedure

Create a thermal study called *01 steady state*. Before proceeding, we need to investigate the folder called *Connectors* which is found in the study window. By default, the **Global Contact** between parts in an assembly is **Bonded**. As the name implies, all parts in assembly behave as one. We need to change it by defining a **Local Interaction**.

Right-click the *Connections* folder and select **Local Interaction** to open the **Local Interaction** window shown in Figure 8-2. Select **Thermal Resistance** with **Node to surface** as the contact type. This **Local Interaction** condition overrides the global **Bonded** condition. Select the contacting faces and enter **Distributed Thermal Resistance** as 0.001Km^2/W. This value is quite high; we use it to clearly demonstrate the effect of a thermal resistance layer. The magnitude of thermal resistance is usually obtained by testing. Notice that the units of thermal resistance are the reciprocals of units of thermal convection.

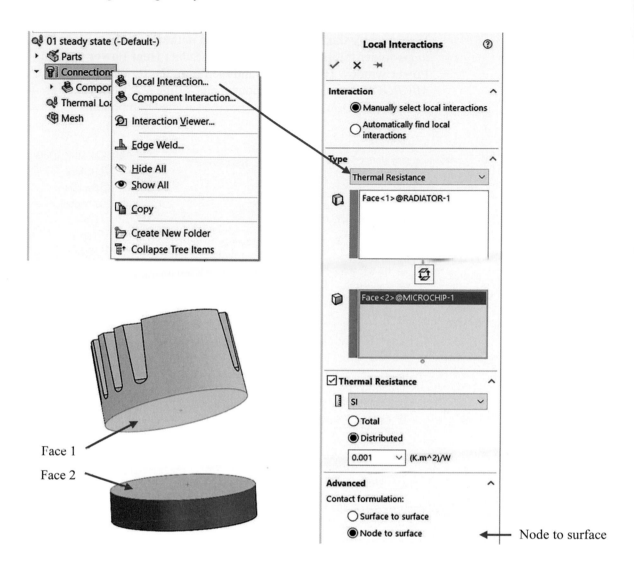

Node to surface

Figure 8-2: Definition of Thermal Resistance Contact Set.

We need to define a local Contact Set for contacting faces to introduce a thermal resistance layer between the contacting faces. This can only be done as a Local Interaction using Node to surface option in Advanced settings. Use Exploded View 1 to select two touching faces.

Next, specify the heat power generated in the microchip. Right-click the *Thermal loads* folder to open the pop-up menu. Select **Heat Power** to open the **Heat Power** window (Figure 8-3), and then from the fly-out menu select the *MICROCHIP* assembly component and define a 25W heat power. This applies heat power to the entire volume of the selected component.

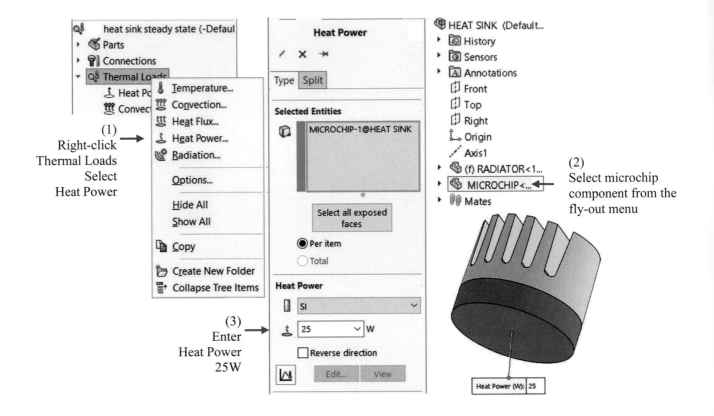

Figure 8-3: The SOLIDWORKS fly-out menu is used to make a selection of the part (here a microchip) necessary to define Heat Power.

Notice that MICROCHIP-1 appears in the Selected Entities window.

So far, we have assigned material properties to each component and also defined a heat source and a thermal resistance layer. For heat to flow, we must also establish a mechanism for heat to escape the model. This is accomplished by defining convective boundary conditions.

Right-click the *Thermal Loads* folder to open a pop-up menu and select **Convection...** to open the **Convection** window (Figure 8-4). Select all faces of the RADIATOR except the one touching the MICROCHIP. Do not select any face of the MICROCHIP. Enter 25W/m²/K as the value of the convection coefficient for all selected faces and enter the **Bulk Ambient Temperature** as 300K.

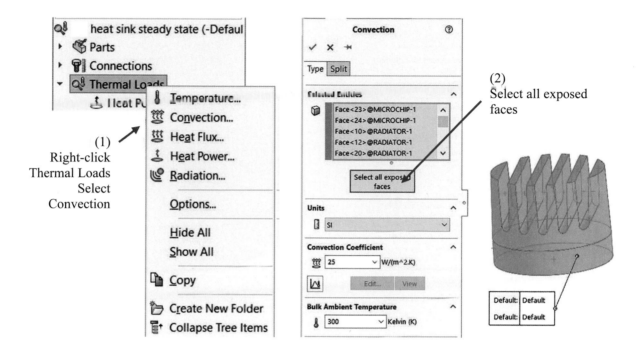

Figure 8-4: Definition of convective boundary conditions on all exposed faces of the assembly.

Definition of convective boundary conditions consists of definition of Convection Coefficient and Bulk Ambient Temperature.

The last step before solving is creating the mesh. Use **Blended curvature-based mesh** with settings shown in Figure 8-5.

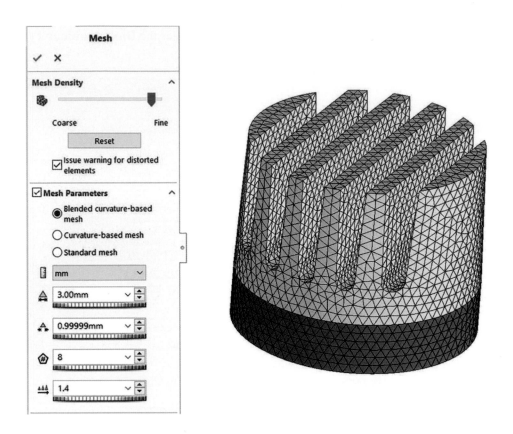

Figure 8-5: Blended curvature-based mesh of microchip assembly.

The use of Blended curvature-based mesh allows for control of mesh quality on round faces without the use of mesh controls.

Once the solution is ready, examine the two plots: temperature and resultant heat flux. The temperature plot is created automatically in the *Results* folder. The required choices for both plots are shown in Figure 8-6.

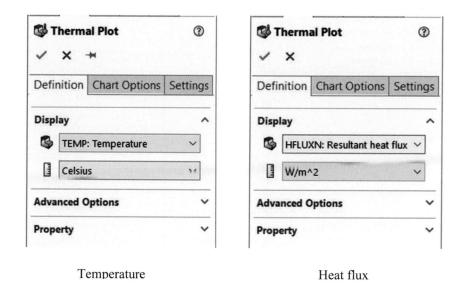

Temperature   Heat flux

Figure 8-6: Temperature and Heat Flux plots definition windows.

Both temperature and heat flux result plots are more informative if presented using section views. SOLIDWORKS Simulation offers a multitude of options for sectioned plots which are easier to practice than to read about. Here we describe the procedure of creating a flat section result plot using one of the default reference planes in SOLIDWORKS, but any reference plane can be used.

To show the section view of the temperature distribution plot, right-click the plot icon and select Section Clipping from the pop-up menu to open the Section window. By default, the cutting surface is aligned with the Front reference plane. To select another cutting plane select it from the SOLIDWORKS fly-out menu.

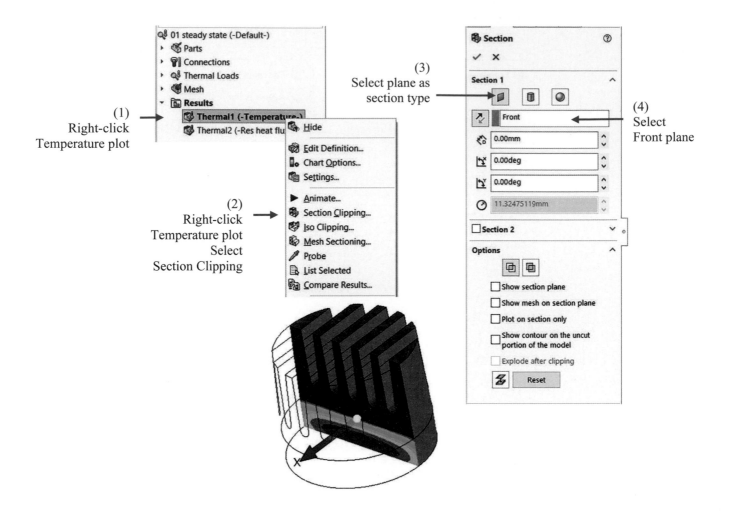

Figure 8-7: A section plot of the temperature distribution in the assembly using an exploded view. Color legend is not shown.

The cutting plane is aligned with the Right reference plane. The position of the cutting plane can be modified like SOLIDWORKS section view. Experiment with Options in the Section window.

Now, construct a plot of heat flux shown as a vector plot. Follow the steps described in chapter 7 to create a heat flux vector plot. In addition to using a vector plot, use the exploded view to produce a heat flux plot like the one shown in Figure 8-8.

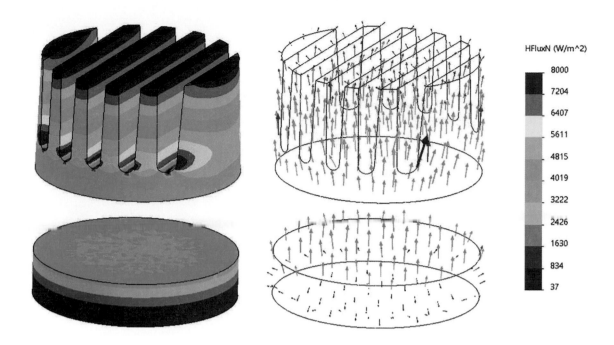

Figure 8-8: Fringe plot and vector plot of heat flux in the assembly.

Notice that arrows "coming out" of the microchip visually represent where heat leaves the model.

Some irregularities in vector display are attributed to irregularities in the mesh. The maximum heat flux magnitude has been capped at 8000W/m² to mask these irregularities. Exploded View 2 is used. Experiment with Vector plot options.

Try repeating the analysis with a more refined mesh.

This completes the steady state thermal analysis of the HEAT SINK assembly. We now proceed with a transient thermal analysis. Copy the study *01 steady state* into a new study named *02 transient*. Right-click the study *02 transient* folder and select **Properties** to open the window shown in Figure 8-9.

Select **Transient** analysis. Our objective is to monitor temperature changes every 360 seconds during the first 3600 seconds. Enter 3600 as the **Total time** and 360 as the **Time increment**.

Total time: 3600s

Time increment: 360s

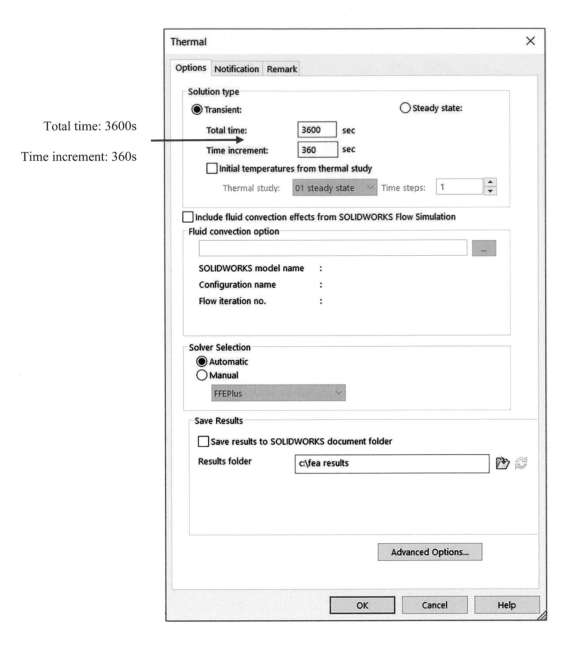

Figure 8-9: The Transient thermal analysis is specified in the thermal study Options tab.

Analysis will be carried on for 3600 seconds in 10 steps. Results will be reported every 360 seconds.

Transient thermal analysis requires that the initial temperature of the model be defined in addition to the already defined **Heat Power** and **Convection** coefficients which have been copied from the *heat sink steady state* study together with **Contact conditions** and the **Mesh**.

We assume that both components have the same initial temperature of 300K. Right-click the *Thermal Loads* folder in the *heat sink transient* study and select **Temperature** to open the window shown in Figure 8-10. Select **Initial Temperature** and enter 300K. From the fly-out menu select both assembly components.

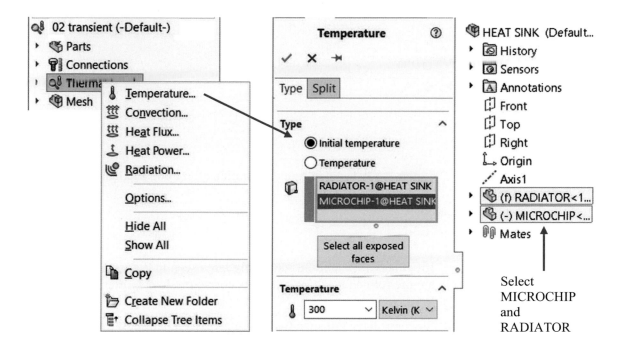

Figure 8-10: Initial Temperature specified for both assembly components.

Assembly components can be selected from the SOLIDWORKS fly-out menu.

Now run the analysis and display the temperature plot for step 10 (the last step) by right clicking the plot icon, selecting **Edit Definition** and setting the **Plot Step** to 10 (Figure 8-11).

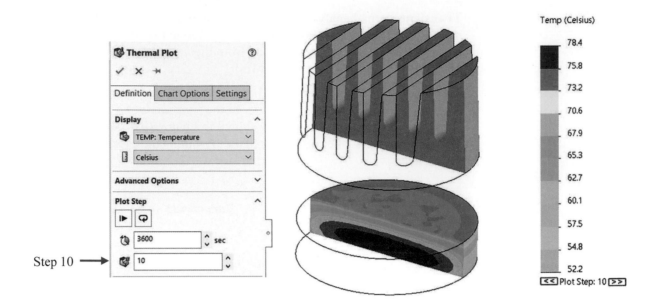

Step 10 →

Figure 8-11: Temperature distribution after 3600 seconds (step 10) from when the microchip was turned on.

Temperature results shown in 10th time step.

Irregular fringes on the contacting faces seen in Figure 8-11 indicate the need for a more refined mesh. We use this coarse mesh to shorten the solution time of the transient analysis.

Since we have not specified heat power as function of time, it is assumed that the full power is turned on at time t=0, when the assembly is at an initial temperature of 300K. Figure 8-11 shows the temperature distribution after 3600 seconds. This result is very close to the result of steady state thermal analysis, meaning that after 3600 seconds, the temperature of assembly has almost stabilized.

To see the temperature time history at selected locations of the model, proceed as follows: make sure the temperature plot is displaying the full model, not a section as in Fig 8-11. This can be done by right clicking the temperature plot, selecting **Section Clipping...** , and turning off **Clipping** under the **Options** menu. Next, right-click the temperature plot icon, and select **Probe** to probe temperature in the location shown in Figure 8-12.

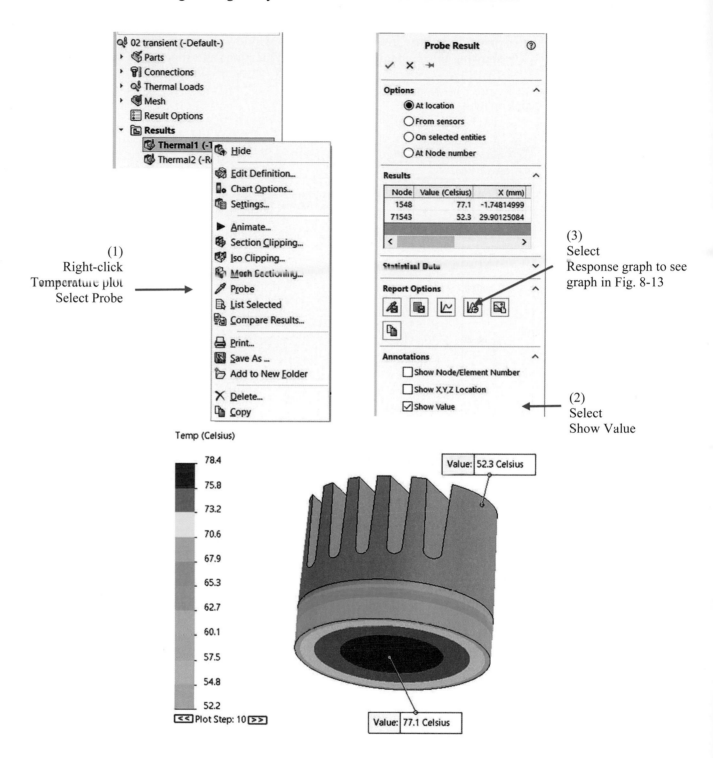

Figure 8-12: Temperature is probed in the indicated locations: top face of radiator and bottom face of microchip.

Only Show value is selected in the callout.

Select **Response** in the **Probe Results** window (Figure 8-12) to display a graph showing the temperature at the probed locations as a function of time (Figure 8-13).

Graph options

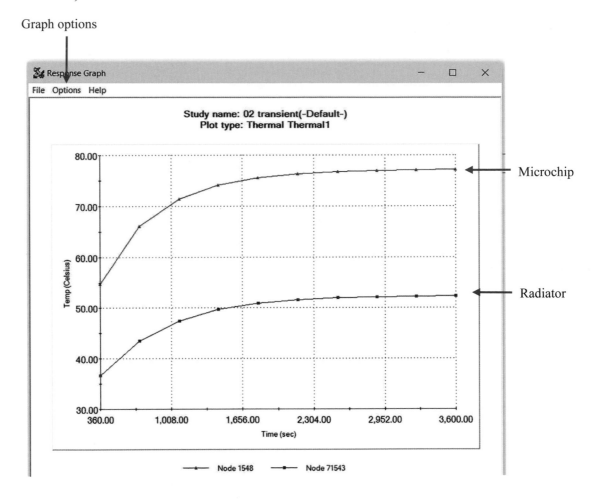

Figure 8-13: Temperatures as a function of time in the probed locations.

To produce a response graph (here temperature as a function of time), you may probe temperature from any of the 10 performed time steps. It does not have to be the last step. Experiment with different graph Options.

An examination of the **Response Graph** in Figure 8-13 proves that after 3600 seconds the probed locations have almost achieved the steady state temperature.

To conclude the thermal analysis of the HEAT SINK assembly, we will study the effect of the thermal resistance layer. Switch to the Exploded View 1 and show **Temperature** plot for step 10. Here we use the plot from the last step of the transient thermal analysis. Probe temperatures in corresponding locations on two contacting faces separated by the layer of thermal resistance and notice that a temperature gradient exists due to the presence of the thermal resistance layer (Figure 8-14). The temperature gradient is required to "push" heat through the layer of thermal resistance.

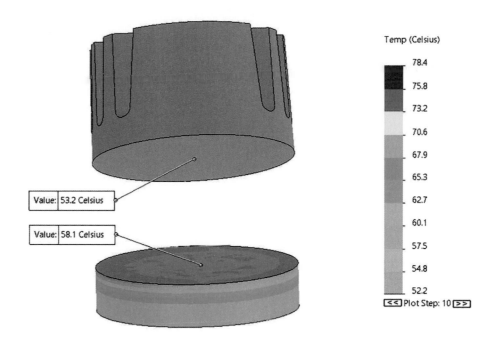

Figure 8-14: Temperature difference on two contacting faces separated by the layer of thermal resistance; microchip: 58.1°C, radiator 53.2°C.

Exploded View 1 is used in this illustration.

The presence of the thermal resistance layer creates a temperature gradient shown in the probed results and in the temperature time history graph. The same effect could be demonstrated using steady state study.

Models in this chapter

Model	Configuration	Study Name	Study Type
HEAT SINK.sldasm	Default	01 steady state	Thermal
		02 transient	Thermal

9: Static analysis of a hanger

Topics covered

- Global and Local Contact conditions
- Hierarchy of Contact conditions

Project description

In this exercise we review different options available for defining the interactions between assembly components in structural analysis.

Open the HANGER assembly and create a Static study *01 contact.* Apply restraint and load as shown in Figure 9-1.

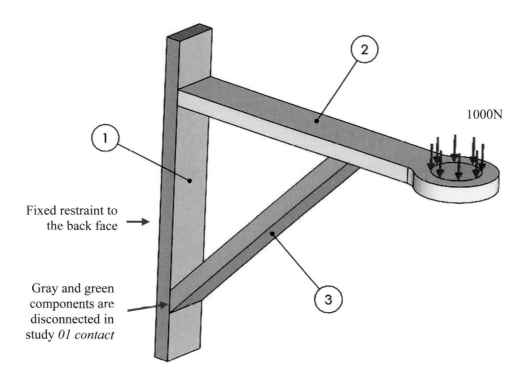

Figure 9-1: HANGER assembly consisting of three parts.

A uniformly distributed 1000N load is applied to blue face; fixed restraint is applied to the back face of the vertical component. Balloons indicate assembly components.

Procedure

In study *01 contact* we disconnect two components as indicated in Figure 9-1. Touching faces of assembly components are bonded by default. To disconnect selected faces, we need to modify local interactions between the selected faces.

Switch to exploded view, then right-click *Connections* folder and select *Local Interaction* from the pop-up menu to open **Local Interactions** window. Specify **Contact** as the type of **Local Interactions** and select two faces that we want to disconnect. **Local Interaction** is then added to **Local Interactions** folder (Figure 9-2).

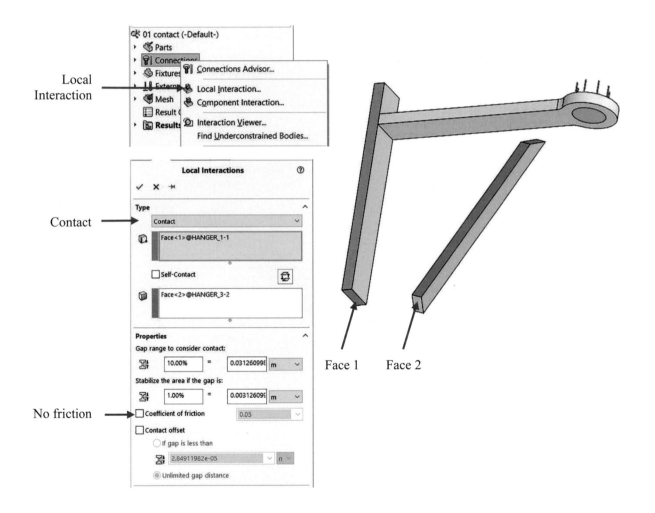

Figure 9-2: Definition of Contact type of Local Interactions.

Face 1 and Face 2 are disconnected. The faces can slide along each other without friction when the load is pointing down, or they can separate if the load is pointing up, but the grey and the green components cannot interfere.

The exploded view shown here includes translation and rotation of the green component to show Face 1 and Face 2 in the same view.

Mesh the assembly with Standard, Default mesh and solve study *01 contact*. Next, copy study *01 contact* into study *02 contact*. Reverse the load direction and solve the study. Displacement results of studies *01 contact* and *02 contact* are shown in Figure 9-3.

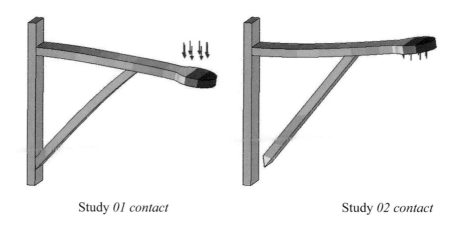

Study *01 contact*   Study *02 contact*

Figure 9-3: Displacement results in studies *01 contact* and *02 contact*.

Load pointing down causes sliding of faces; load pointing up produces separation.

Copy study *01 contact* into *03 bonded* and change **Local Interaction** to **Bonded**, obtain solution. Finally copy study *01 contact* into *04 free* and change **Local Interaction** to **Free**. Displacement results of studies *03 bonded* and *04 free* are shown in Figure 9-4.

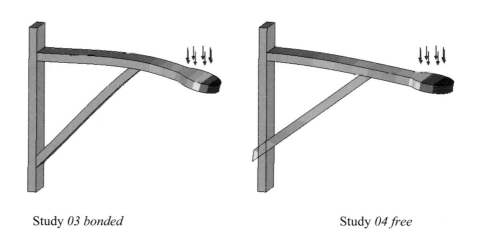

Study *03 bonded*   Study *04 free*

Figure 9-4: Displacement results in studies *03 bonded* and *04 free*.

If Local Interaction is defined as Bonded, components behave as one part. If Local Interaction is defined as Free, components may interfere or come apart with no consequences.

There are two more options in **Local Interaction**: **Virtual Wall** and **Shrink Fit**, but they are not applicable to HANGER assembly.

Local Interactions are defined between individual faces. A summary of different types of **Local Interactions** is given in the table below.

STRUCTURAL LOCAL INTERACTIONS	
Contact	Available for static, drop test and nonlinear studies. Contact prevents interference between source and target entities but allows gaps to form.
Bonded	The source and target entities are bonded. The entities may be touching or within a small distance from each other. The program gives a warning if the distance between bonded entities is larger than the average element size of the associated elements.
Free	The selected faces are disjoined and can freely go through each other with no interaction.
Shrink fit	Valid for faces from two components which show interference. This interference is eliminated after solution.
Virtual wall	This contact type defines contact between the source entities and a virtual wall defined by a target plane. The target plane may be rigid or flexible.
THERMAL LOCAL INTERACTIONS	
Thermal resistance	Specifies thermal resistance between source and target faces.
Bonded	Similar to Bonded option in structural studies.
Insulated	Similar to the Free option in structural studies. The program treats the source and target faces as disjointed and therefore prevents heat flow due to conduction through the source and target faces.

Interactions may also be defined between components or globally for the entire assembly as **Contact**, **Bonded** and **Free**. **Bonded Interaction** (between components or global) has different meshing options as shown in Figure 9-5.

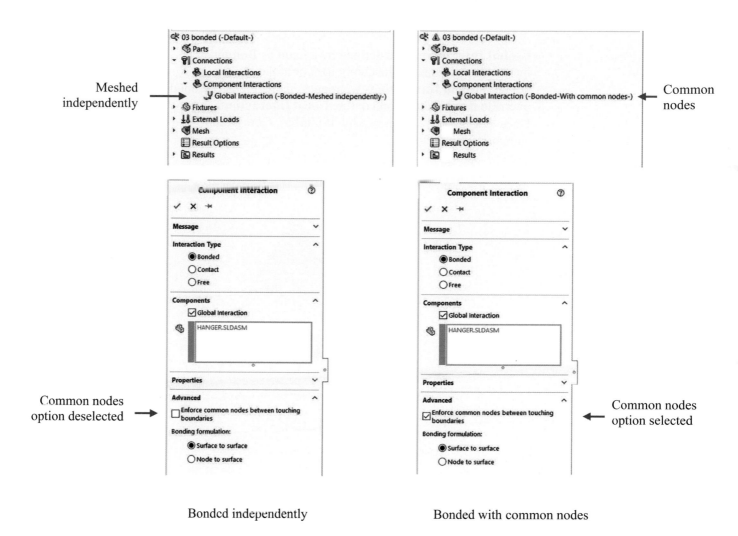

Bonded independently   Bonded with common nodes

Figure 9-5: Bonded contact may have common nodes of contacting faces or contacting faces may be meshed independently.

This illustration is not directly related to the analysis of HANGER assembly.

Interaction may be defined as **Node to Surface** or **Surface to Surface** condition selectable in Advanced Options. Faces in **Surface to Surface** contact conditions do not have to touch initially, but they are expected to come in contact under the load.

Global Interactions are defined by default as **Bonded**. This can be overridden by interactions defined at the component or at the local level. For example, using the default **Global Interactions** we request that all faces be bonded. Next, we locally override this condition and define **Local Interactions** for one or more sets of faces as **Contact**. The hierarchy of **Global**, **Component**, and **Local** interactions is shown in Figure 9-6.

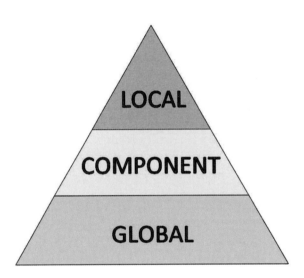

Figure 9-6: Hierarchy of Interactions.

Component Interactions override Global Interactions. Local Interactions override Component Interactions.

Notice that the study with **Contact Interaction** requires more time to run than study with **Bonded** or **Free** Interactions because the contact constraints must be resolved (Figure 9-7). **Contact Interactions** is a nonlinear problem and requires an iterative solution procedure. This takes longer to complete than the linear solutions of studies with **Bonded** or **Free** interactions.

Solving contact constraints

Figure 9-7: Iterative solution progress window in study *01 contact.*

The display of the solution status window can be toggled between more and less detail by selecting More >> and Less <<.

Models in this chapter

Model	Configuration	Study Name	Study Type
HANGER.sldasm	*Default*	*01 contact*	Static
		02 contact	Static
		03 bonded	Static
		04 free	Static

10: Thermal stress analysis of a bi-metal loop

Topics covered

- Thermal deformation and thermal stress analysis
- "Parasolid round trip"
- Saving model in a deformed shape

Project description

The temperature of the loop shown in Figure 10-1 increases uniformly from an initial 298K to 500K. We need to find the deformations and stresses induced by the increase in temperature.

Procedure

Open the LOOP assembly and notice that it consists of two parts of the same shape but different material (Figure 10-1).

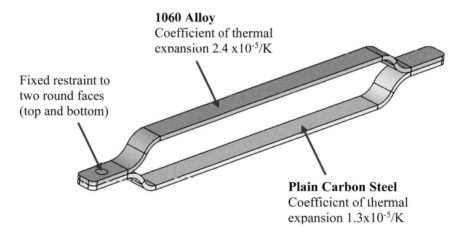

1060 Alloy
Coefficient of thermal expansion 2.4×10^{-5}/K

Fixed restraint to two round faces (top and bottom)

Plain Carbon Steel
Coefficient of thermal expansion 1.3×10^{-5}/K

Figure 10-1: Loop consisting of bonded steel and aluminum parts.

Global Interactions are Bonded. Therefore, parts are bonded along touching faces. When temperature is increased, the assembly deforms because of different thermal expansion of steel and aluminum.

The same component is inserted into assembly in two configurations, with different materials.

The LOOP assembly consists of two parts created as **Sheet Metal**. They can be analyzed using shell elements or they can be converted to solids and analyzed with solid elements. In this exercise we decide to use solid elements. Deselect option **Mesh all solid bodies with solid mesh** in study Default Options, Mesh. This way we will be able to practice the manual conversion from Shell Body to Solid Body.

Create a **Static** study and follow steps in Figure 10-2 to convert assembly components from Shell Body to Solid Body.

(1)
Select both Sheet Metal bodies and right-click to open the pop-up menu. →

(2) →
From the pop-up menu select **Treat selected bodies as solids**.

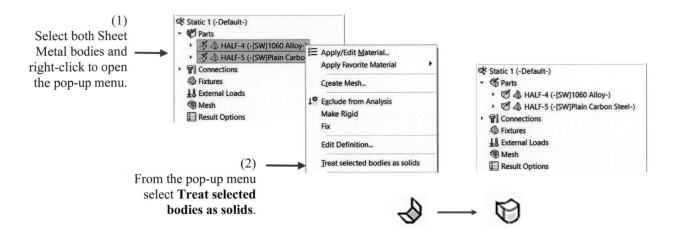

Figure 10-2: Manually changing the body type from Sheet Metal to Solid.

Body symbol changes from Sheet Metal to Solid.

To account for thermal effects, go to **Flow/Thermal Effects** in **Static** study properties and select **Input temperature** option (Figure 10-3).

Input temperature →

Reference temperature →

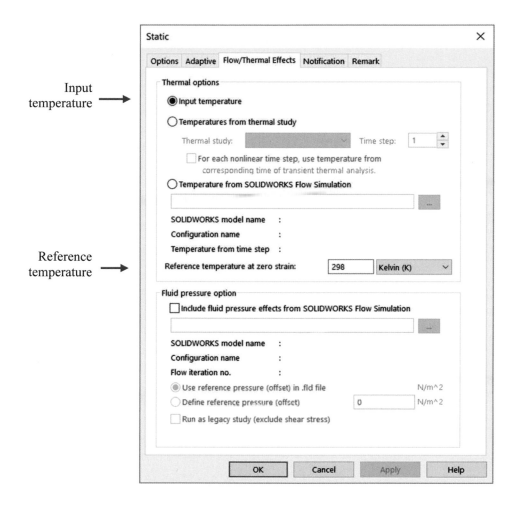

Figure 10-3: The Flow/Thermal Effects tab in the Static study properties window. We instruct the solver to account for the effect of an Input temperature and define the reference temperature at zero strain as 298K.

Input temperature is the default selection.

Before proceeding, let us take this opportunity to review all thermal options available in the study window.

Thermal Option	Definition
Input temperature	Use if prescribed temperatures will be defined in the *Load/Restraint* folder of the study to calculate thermal stresses. This is our case.
Temperature from thermal study	Use if temperature results are available from a previously conducted thermal study.
Temperature from SOLIDWORKS Flow Simulation	Use if temperature results are available from a previously conducted SOLIDWORKS **Flow Simulation** project.

See the book "Thermal Analysis with SOLIDWORKS Simulation and Flow Simulation" for more information on transfer of temperatures from **SOLIDWORKS Flow Simulation** to **SOLIDWORKS Simulation**.

Apply **Fixed** restraints to washer footprints on both sides (Figure 10-1).

To apply a temperature load, right-click the *External Loads* folder and select **Temperature**. This will open the **Temperature** window. From the fly-out menu, select both assembly components and enter a temperature of 500K. This means that assembly temperature will be increased by 202K from the initial 298K (Figure 10-4).

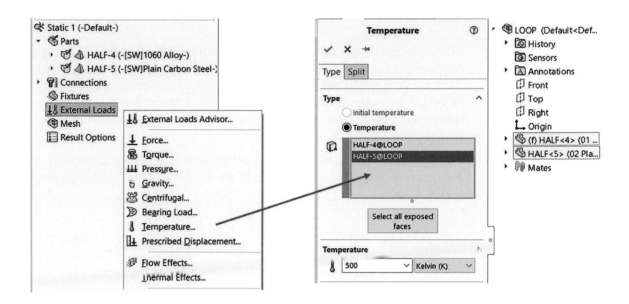

Figure 10-4: A temperature of 500K is applied to both assembly components; components may be selected from the SOLIDWORKS fly-out menu.

Select both components from the fly-out menu.

Mesh the model with a Standard Mesh using 4mm element size and run the study.

Displacement results are shown in Figure 10-5.

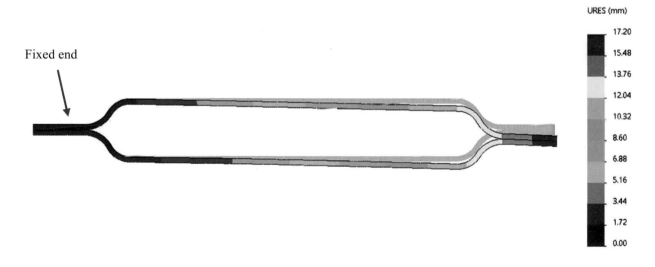

Figure 10-5: Displacement results for the bi-metal beam.

Due to the different thermal expansion ratios of steel and aluminum, thermal strains develop and bend the loop.

The scale of deformation is 1:1. Undeformed shape (gray) is superimposed on the displacement plot.

Review stress results to notice the unavoidable stress singularities along the connections between two assembly components. Stress singularities are present there because of the rapid change in the material properties across the touching faces and because of the sharp re-entrant edges; remember that two assembly components are bonded and act as one part. Stress singularities are also present near the fixed restraint.

The deformation caused by thermal stress can be studied by saving the model in the deformed shape. To save the model in the deformed state, follow the steps in Figure 10-6.

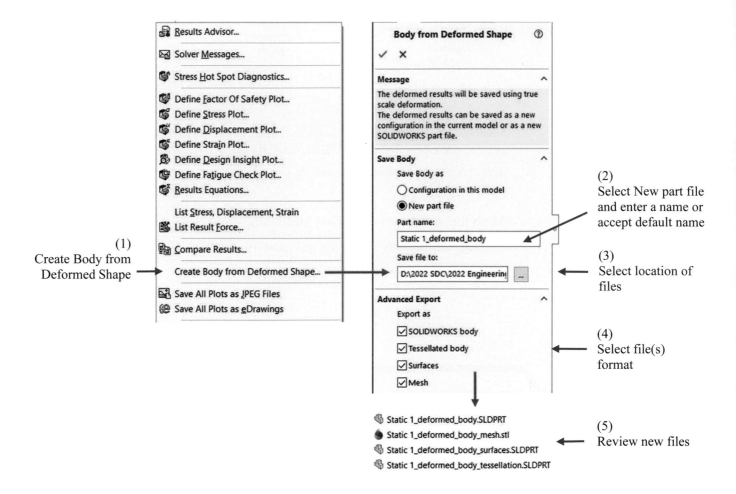

Figure 10-6: Saving a model in the deformed shape.

You do not have to select all formats available in the Advanced Export options.

Selections shown in Figure 10-6 create four files and save the deformed model in four different formats as shown in the table below. Notice that the deformed assembly is saved as a part.

File name	Type of geometry
Static 1_deformed_body.sldprt	Solid bodies
Static 1_deformed_body_tessellation.sldprt	Solid bodies, tessellated
Static 1_deformed_body_surfaces.sldprt	Surface bodies
Static 1_deformed_body_mesh.stl	STL (Stereolithography) mesh

Models in this chapter

Model	Configuration	Study Name	Study Type
LOOP.sldasm	*Default*	*Static 1*	Static

Notes:

11: Buckling analysis of an I-beam

Topics covered

- Buckling analysis
- Buckling load safety factor
- Stress safety factor

Project description

A curved I beam is compressed with a 3500N load, as shown in Figure 11-1. Our goal is to calculate the factor of safety related to the yield stress and the factor of safety related to buckling. This way we will find out what is the deciding mode of failure: yielding or buckling?

Procedure

Open I BEAM model. The material is Alloy Steel with the yield strength of 620MPa. The I-BEAM is loaded by a 5000N force uniformly distributed over the end face. A fixed restraint is applied to the opposite end face.

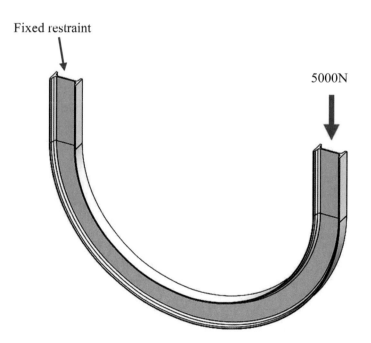

Figure 11-1: The I-beam model.

The I-beam is modeled as a Structural Member.

Before creating any study go to **Simulation**, **Default Options**, **Mesh** and select **Mesh all solid bodies with solid mesh**. This way **Weldment** geometry will be meshed with solid elements with no need of manual conversion of Beam Bodies to Solid Bodies. This approach is opposite to the one used in Chapter 10, where we practiced manual conversions of body type.

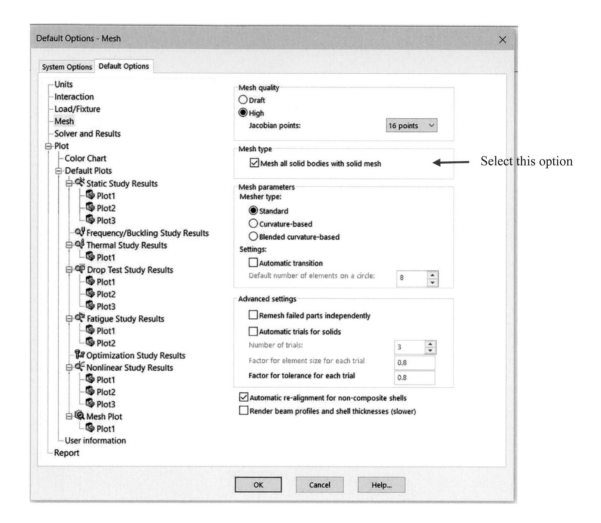

Figure 11-2: Select the option Mesh all solid bodies with solid mesh to use solid mesh without the need to convert Beam bodies manually.

This is a repetition of Figure 4-4.

Create static study *01 static*, apply restraints and load as shown in Figure 11-1, and mesh the model using a **Standard Mesh** with 5mm element. This element size produces element with high turn angle in fillets. We accept this mesh because the poorly shaped elements will be far from the highest stress. You may want to repeat the analysis with a more refined mesh to see that mesh refinement highlights the unavoidable singularities at the supports but does not significantly affect stress levels in the flanges.

Run the solution and review the stress results shown in Figure 11-3.

The results of the static analysis show the maximum von Mises stress is around 230MPa, which is below the yield strength 620MPa of Alloy Steel. Notice that the highest stress 260MPa shown in Figure 11-3 is caused by stress singularity at support.

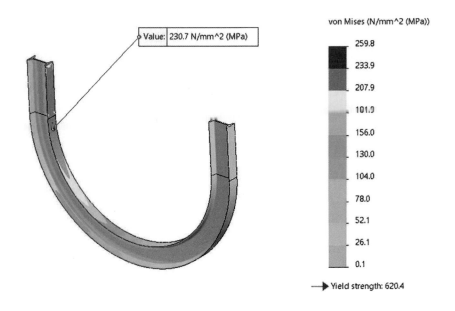

Figure 11-3: Von Mises stress plot shown with superimposed undeformed shape.

Animate this plot and notice that the deformation takes place in the plane of curvature of the model.

Von Mises stress in the probed location (away from the singularity) is approximately 230MPa (Figure 11-3). Therefore, the factor of safety to yield is:

$$FOS_{yield} = \frac{yield\ strength}{\max von\ Mises\ stress} = \frac{620MPa}{230MPa} = 2.7$$

meaning that the beam's material is below yield. Therefore, the results of static analysis indicate that the beam is safe.

As is always the case with slender members, the factor of safety related to material yield strength may not be sufficient to describe the structure's safety. This is because of the possible occurrence of buckling. We need to calculate the factor of safety related to buckling, which requires performing a buckling analysis. Define a buckling study *02 buckling* as shown in Figure 11-4.

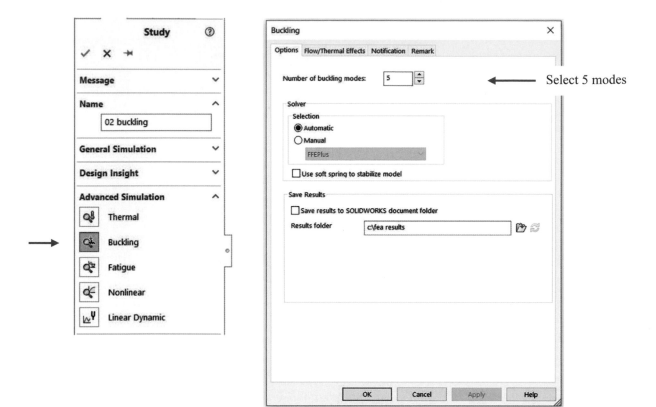

Figure 11-4: Definition of a Buckling study and the Buckling study properties window.

Defining a buckling study includes specifying the number of desired buckling modes. The default number is one but here we ask for five buckling modes.

When defining a buckling study, we need to decide how many buckling modes should be calculated. This is a close analogy to the number of modes in a frequency analysis. In most practical cases, the first buckling mode determines the safety of the analyzed structure. The reason why we specify five modes and not just one will become clear once the results are analyzed.

Copy loads, restraints, and mesh from static study *01 static* to buckling study *02 buckling* and run the analysis.

Once the buckling analysis has been completed, right-click the *Results* folder and select **List Buckling Factor of Safety**; the list corresponds to the five calculated buckling modes.

List
Buckling Factor →
of Safety

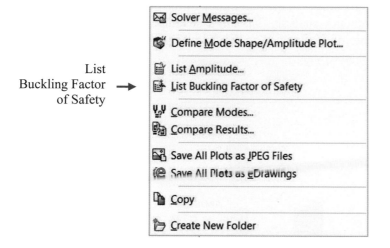
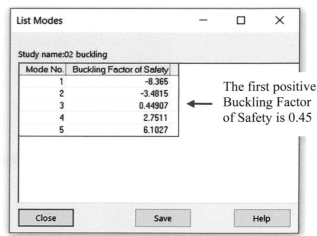

The first positive
← Buckling Factor
of Safety is 0.45

Figure 11-5: Summary of Buckling Load Factors (BLF) corresponding to five calculated buckling modes. The first positive buckling factor of safety is 0.45.

We calculated five buckling modes to introduce the concept of negative Buckling Load Factor of Safety (BLFS). If only one BLFS is requested, Simulation will report the first positive BLFS.

The BLFS is a number by which the applied load has to be multiplied in order for buckling to take place for a given buckling mode. If the BLFS is negative the load direction must be reversed. The list in Figure 11-5 shows two negative BLFS. The third BLFS is therefore the first real BLFS=0.45.

The buckling load can now be calculated as follows:

$$BLF = \frac{Buckling\ Load}{Applied\ Load} = 0.45$$

Buckling Load = Applied Load x *BLF* = 5000N x 0.45 = 2250N

The BLFS is lower than 1, meaning that the I BEAM will buckle under the applied load. Our conclusion is that buckling is the deciding mode of failure. Also notice that high stresses affect the beam only locally, while buckling is global. The onset of yielding does not necessarily mean a structural failure, while buckling most often does mean a structural failure.

We will now review a displacement plot to study the shape of the first positive buckling mode. Displacement plot is called Amplitude plot, the same as in Frequency analysis.

Even though displacement results can be shown in color, they do not provide any useful information. In a buckling analysis, the magnitude of the displacement is meaningless, just like in a frequency analysis. The mode shape plot is shown in Figure 11-6. Remember that numerical values on the color legend are meaningless just like in modal analysis.

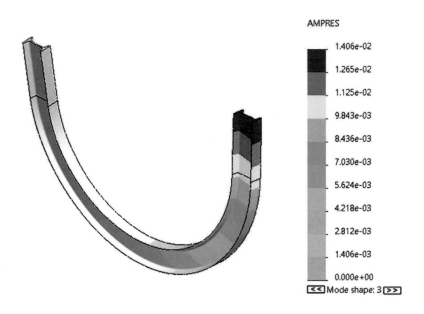

Figure 11-6: The deformation plot provides visual feedback on the shape of the buckled structure. Mode shape 3 is shown.

This plot shows the buckled shape along with the undeformed model. Numbers by the color legend are only relative. Animate the plot to see that the model buckles out of plane.

Figure 11-7 presents an important comparison of the deformed shapes obtained from the static and buckling analyses; displacement plots are used.

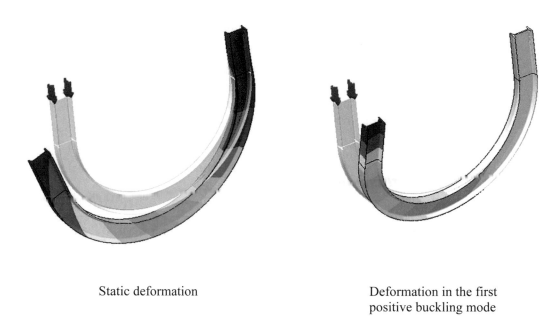

Static deformation   Deformation in the first
                                      positive buckling mode

Figure 11-7: The model deforms in-plane in the static analysis. The same model deforms out of plane in the buckling analysis.

Undeformed shape is identified by load symbol; deformed shape is identified by colors.

A side-by-side comparison of the deformed shapes from the static and buckling analyses shown in Figure 11-7 reveals a very important difference between these shapes. The static analysis plot shows in plane bending for which the I-beam is well designed. The buckling mode plot shows torsion of the I-beam. An I BEAM is poorly designed to resist torque. The low torsional stiffness of the I BEAM makes buckling the deciding mode of failure.

The out of plane deformation may be counterintuitive, making it easy to forget that the I BEAM will fail in buckling rather than in yielding.

We start with a description of the **h-adaptive** solution method. On several previous occasions, we performed mesh refinements to investigate the effect of mesh density on results. We also added mesh controls to analyze stresses more accurately in "difficult" locations. In this chapter we demonstrate that the **h-adaptive** solution method automates the mesh refinement process and to some extent, relieves users from having to make meshing decisions such as selecting element size and applying mesh controls.

Figure 12-2 shows the properties of an **h-adaptive** study. **Target accuracy**, **Accuracy bias** and **Maximum number of loops** control the iterative process of mesh refinement.

h-adaptive →

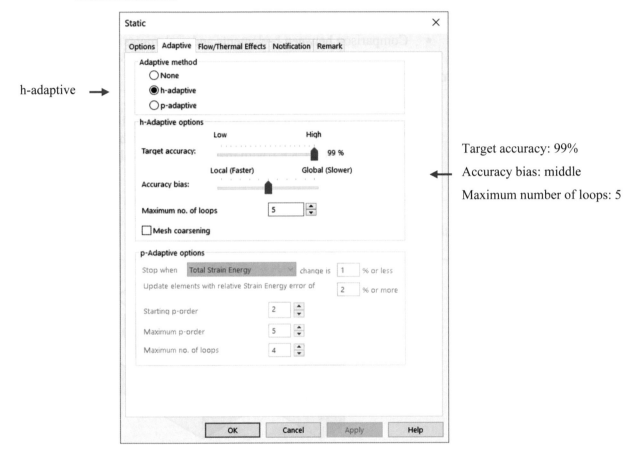

Target accuracy: 99%

Accuracy bias: middle

Maximum number of loops: 5

Figure 12-2: The Adaptive tab in the Static study properties window with the h-adaptive solution method selected.

For the h-adaptive solution presented in this chapter, we use the above settings.

The h-adaptive solution options are explained in the table below.

Setting	Definition
Target accuracy	Sets the accuracy level for the strain energy norm. This is NOT the stress accuracy level. However, a high level of accuracy indicates more accurate stress results.
Accuracy bias	You can move the slider towards Local to instruct the program to concentrate on getting accurate peak stress results using a fewer number of elements. You can also move the slider towards Global to instruct the program to concentrate on getting overall accurate results.
Maximum no. of loops	Sets the maximum number of loops allowed when you run the study. The maximum possible number of loops is five.
Mesh coarsening	Check this option to allow the program to coarsen the mesh in regions with low error during the adaptive loops. The number of elements in consecutive loops may increase or decrease depending on the model and the initial mesh. If this option is not checked, the program does not change the mesh in regions with low errors.

The **h-adaptive** solution method is called "adaptive" because mesh refinement is "adapted" to the stress pattern and the mesh is refined only where it is necessary to produce results satisfying accuracy requirements specified in the **Study** properties. h-elements retain their order; they cannot be upgraded to a higher order.

Now review the **p-adaptive** solution method. If you recall, in chapter 1 we mentioned that **SOLIDWORKS Simulation** can use either first order elements (called draft quality), or second order elements (called high quality). Furthermore, recall that first order elements model a linear (or first order) displacement distribution and constant stress distribution, while second order elements model a parabolic (second order) displacement distribution and linear stress distribution. We now have to amend the above statements.

Aside from first and second order elements, **SOLIDWORKS Simulation** can also work with elements of "floating" order which can change order during the iterative solution process. The highest available order is five. These elements of "floating" order are called p-elements and are available when the **p-adaptive** option is selected as the solution method in the **Study** properties window under the **Adaptive** tab (Figure 12-3). This option is available only for static studies using solid elements.

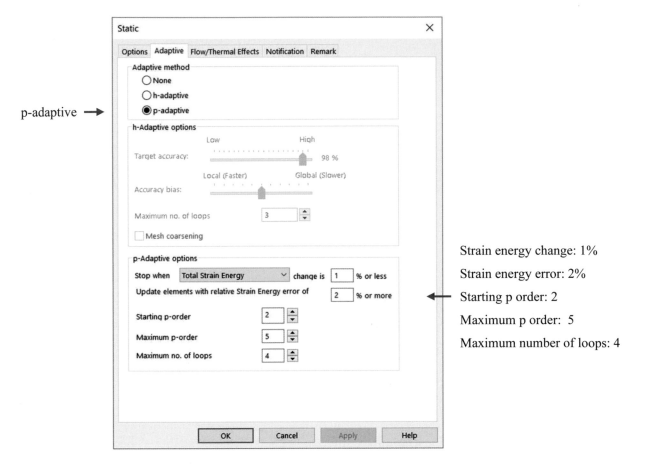

Figure 12-3: The adaptive tab in the Static study properties window with the p-adaptive solution method selected.

The selection "p-adaptive" made in the Adaptive tab in the Static study properties window activates the use of the p-adaptive solution method.

The p-adaptive solution options are shown in Figure 12-3 and explained in the table below:

Setting	Definition
Stop when	Iterations (looping) increase element order until the change in Total Strain Energy (or other measures like RMS resultant displacement or RMS von Mises stress) between the two consecutive iterations is less than the specified value (default is 1%) shown in the p-adaptive options area. If this requirement is not satisfied, then looping will stop when the elements reach the fifth order; this will be the fourth loop.
Update elements with relative Strain Energy error of …	This setting controls which elements are upgraded during the iterative solution. By default, elements with relative Strain Energy error of 2% or more are updated.
Starting p-order	The initial order of elements. Usually the starting p-order is set to 2, which means that all elements are defined to start as second order elements.
Maximum p-order	The actual highest order to be used in p-adaptive solution. The highest order available in **SOLIDWORKS Simulation** is the fifth order.
Maximum no. of loops	Sets the maximum number of iterations (loops) allowed when the p-adaptive study is run. The maximum possible number of loops is four.

Let us pause for a moment and explain some terminology. Refer to Figure 2-16 which explains that h denotes the characteristic element size. While the mesh is refined during the convergence process, the characteristic element size h becomes smaller. Therefore, the mesh refinement process that we conducted in chapters 2 and 3 is called the h-convergence process, and the elements used in this process are called h-elements. An h-adaptive solution is an iterative solution where the h-element mesh is automatically refined in several iterations.

When p-elements are used, the iterative process does not involve mesh refinement. While the mesh remains unchanged, the element order changes from second all the way to fifth order elements. The iterations may also stop sooner if the convergence criterion (here the change in **Total Strain Energy**) is satisfied before the fifth order is reached.

The order of any element is defined by the order of polynomial functions that describe the displacements in the element. Because the polynomial order experiences changes in the p-adaptive solution, the process of consecutive element order upgrade is called a p-convergence process (p stands for polynomial), and the upgradeable elements used in this process are called p-elements.

Not all p-elements are upgraded during the solution process. Which elements are updated depends on the selection made in the field **Update elements with relative Strain Energy error of ___ % or more**. Here we set it to 2%, meaning that only those elements that do not satisfy the above criterion will be upgraded (investigate other criteria as well). Therefore, we say that the element upgrading is "adaptive," or driven by the results of consecutive iterations.

The p-adaptive solution process is analogous to the process of mesh refinement, which also continues until the change in the selected result is no longer significant.

Procedure

We will solve the same problem in five different ways:

1. Using one mesh of h elements (study *standard*)

2. Using one mesh of h elements with mesh controls added (study *standard with mesh control*)

3. Using the h-adaptive solution method (study *h adaptive*)

4. Using the p-adaptive solution method with default **p-adaptive** solution settings (study *p adaptive 01*)

5. Using the p-adaptive solution method with modified **p-adaptive** solution settings (study *p adaptive 02*)

First, solve the model using second order solid tetrahedral h-elements using the default element size. Name the study *standard*. Von Mises stress results with superimposed mesh are shown in Figure 12-4.

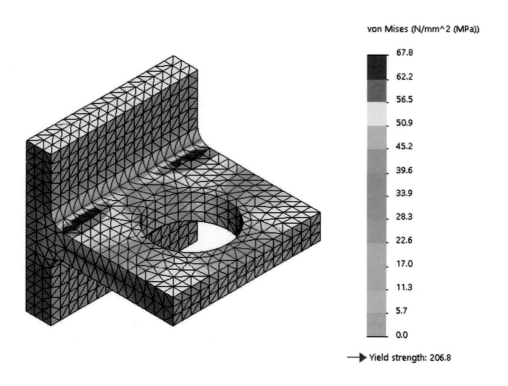

Figure 12-4: Von Mises stress results obtained in a standard study.

The maximum von Mises stress is 67.8MPa.

Now, repeat the analysis using a mesh with default mesh controls applied to both fillets (top and bottom). A default mesh control produces elements the size of one half of the default global element size. Name the study *standard with mesh controls*. Von Mises stress results produced by this study are shown in Figure 12-5.

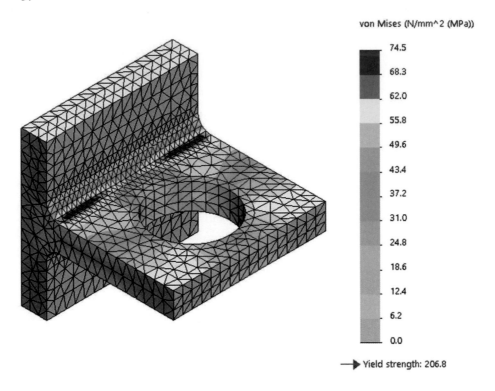

Figure 12-5: Von Mises stress results obtained in study *standard with mesh controls.*

The maximum von Mises stress is 74.5MPa. Notice that the decision of adding mesh controls was made based on results from the previous study.

Now, create a new study *h adaptive* identical to *standard* study (the one without the mesh control) except in the study properties window under the **Adaptive** tab, select **h-adaptive**. Use the settings shown in Figure 12-2: **Target accuracy 99%**, no **Accuracy bias** (slider in the middle), **Maximum no. of loops 5.** Do not use any mesh controls.

As we have previously mentioned, the finite element mesh is refined during the h-adaptive solution. The results are reported for the mesh from the last performed iteration (the most refined one). Run the solution of the *h adaptive* study and display the results as shown in Figure 12-6.

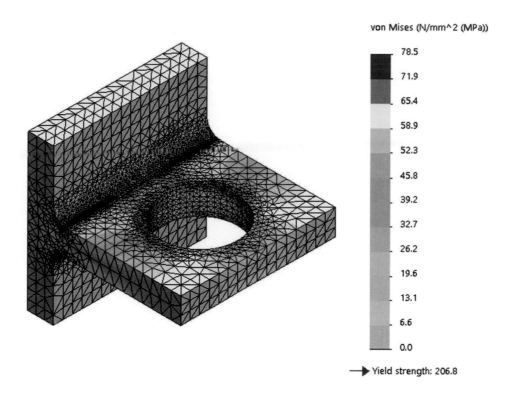

Figure 12-6: Von Mises stress results obtained with an h-adaptive solution. The refined mesh (after solution) is shown.

Notice that the mesh has been automatically refined as compared to the initial mesh visible in Figure 12-4. The maximum von Mises stress is 78.5MPa.

Now, create a study called *p adaptive 01*. In the study window under the **Adaptive** tab, select **p-adaptive**. Use all defaults for a p-adaptive study definition as shown in Figure 12-3. Restraints and Loads can be copied from any of the two previous studies.

Considering that the p-adaptive solution will be used, we can manage with a mesh without bias (no controls need to be applied). Therefore, copy the mesh from the *standard* study. Using higher order elements, which is equivalent to refinement of an h-element mesh, our mesh without bias will still deliver results with acceptable accuracy.

Having solved the study with p-elements, the stress plot produced is shown in Figure 12-7.

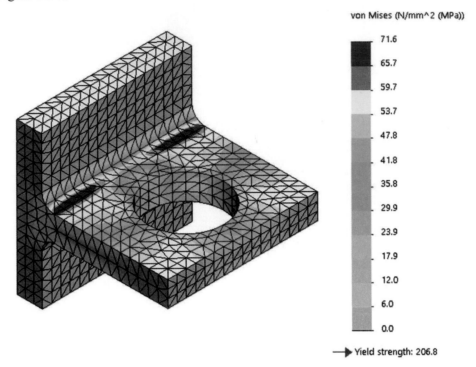

Figure 12-7: Von Mises stress results obtained using the p-adaptive method. The mesh remains unchanged during the p-adaptive solution.

The mesh is identical to the mesh visible in Figure 12-4. The maximum von Mises stress is 71.6 MPa.

To illustrate the iterative nature of **h-adaptive** and **p-adaptive** solutions, create convergence graphs which are available for both types of adaptive solutions. To create a convergence graph, right-click the *Results* folder and select **Define Adaptive Convergence Graph**. For both adaptive studies select **Maximum von Mises Stress** in the **Convergence Graph** window (Figure 12-8). Later try experimenting with other **Convergence Graph** options. **h-adaptive** and **p-adaptive** convergence graphs are shown in Figure 12-9.

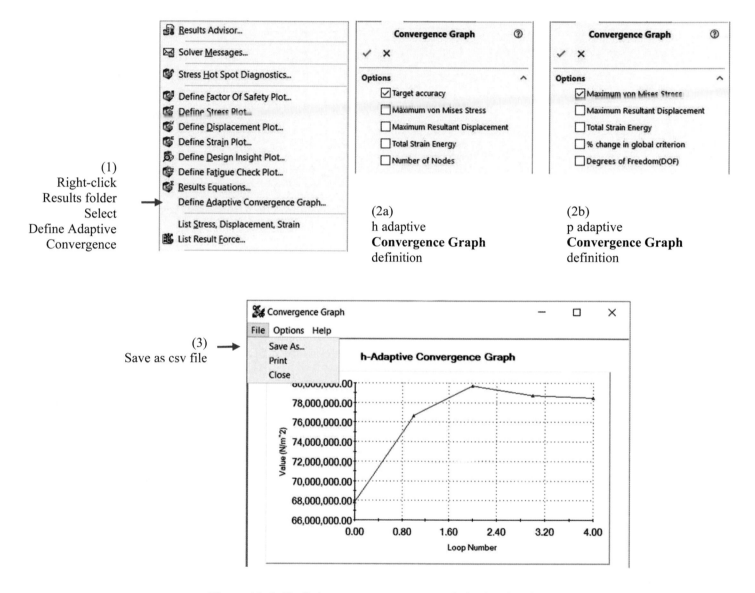

(1)
Right-click
Results folder
Select
Define Adaptive
Convergence

(2a)
h adaptive
Convergence Graph
definition

(2b)
p adaptive
Convergence Graph
definition

(3)
Save as csv file

Figure 12-8: Defining a convergence graph for h-adaptive or p-adaptive solution methods.

Automatically created graphs can be modified using Graph Options or saved as a .csv file and formatted in Excel. In the graphs to follow we will use the second method. The graph window shows h adaptive convergence results; it has been trimmed to fit this page.

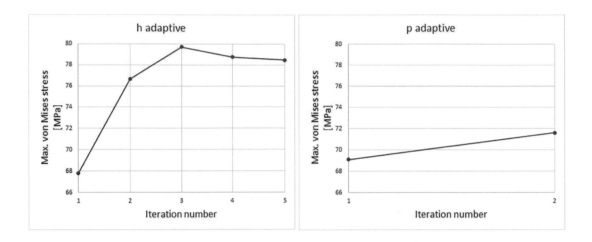

<u>Figure 12-9: Convergence graphs of maximum von Mises stress results obtained in the adaptive solutions.</u>

The h-adaptive convergence graph shows that five iterations were performed during the h-adaptive solutions. The p-adaptive graph shows that only two iterations were performed during p-adaptive solutions.

The maximum number of iterations in the h-adaptive process is five and as Figure 12-9 indicates, all five iterations were performed because we used a very demanding 99% **Target accuracy** specified in the settings of the h-adaptive study. **Target accuracy** indirectly controls the number of iterations performed during the **h-adaptive** solution. Try experimenting with a less demanding **Target accuracy** to observe that the solution will stop without using up all available iterations.

The maximum number of iterations in a p-adaptive study is four. Figure 12-9 shows that only two were performed, so the iterative solution must have stopped due to the convergence requirements being satisfied before all available iterations were used.

Create the last study and call it *p-adaptive 02* (you may copy study *p adaptive 01*). Define much more demanding convergence requirements to force the solver into using all available iterations and reaching the highest available element order (Figure 12-10).

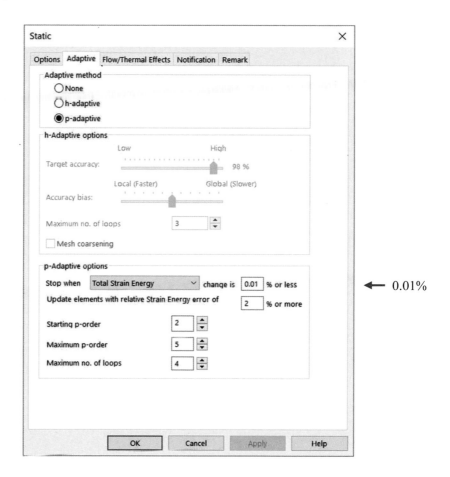

Figure 12-10: p-adaptive solution settings for study *p-adaptive 02.*

A very demanding convergence requirement forces the solver to go up to the highest available p order (5th order) and hence to perform the maximum possible number of iterations (four iterations).

Von Mises stress results of study *p adaptive 02* are shown in Figure 12-11.

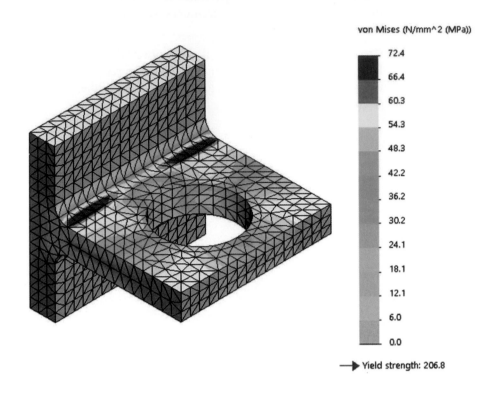

Figure 12-11: Von Mises stress results of study *p-adaptive 02*. Mesh remains unchanged during the p adaptive solution.

The maximum von Mises stress is 72.4MPa.

Convergence of von Mises stress results obtained in the *p adaptive 02* study is shown in Figure 12-12.

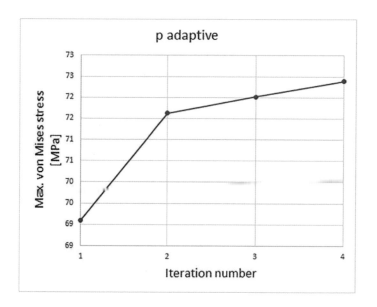

Figure 12-12: Convergence of von Mises stress results obtained in the *p-adaptive 02* study.

The p-adaptive convergence graph shows that four iterations were performed.

When the maximum number of iterations is performed as shown in Figure 12-12, we get no feedback as to whether the solution stopped because the specified accuracy in study properties was achieved, or because the maximum allowed number of iterations was reached. In the h-adaptive solution the maximum number of iterations is 5, whereas in the p-adaptive solution, the maximum number of iterations is 4 and coincides with the highest available element order 5.

A summary of results obtained in all studies is shown in Figure 12-13. These graphs were prepared outside of **SOLIDWORKS Simulation** based on data exported from the **SOLIDWORKS Simulation** graphs.

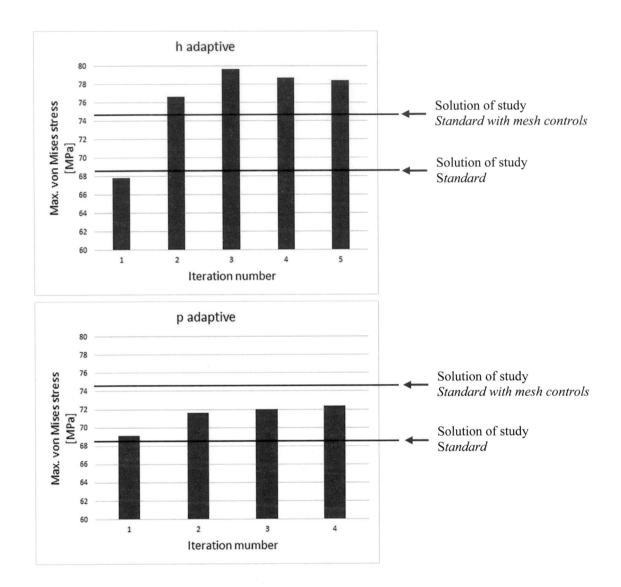

Figure 12-13: Summary of all results.

Top: *The results of two non-adaptive studies shown with horizontal lines, and the results of the h-adaptive study repeated from Figure 12-9 shown with bar graph.*

Bottom: *The results of the two non-adaptive studies shown with horizontal lines, and the results of the p-adaptive study shown with vertical bars repeated from Figure 12-12.*

Examine the bottom graph in Figure 12-13 and notice that the first two iterations are the results of study *p-adaptive 01*. All four iterations 1,2,3,4 are the results of study *p-adaptive 02*. The difference between studies *p-adaptive 01* and *p-adaptive 02* is that in study *p-adaptive 02,* two more iterations were performed because of the more demanding accuracy requirements.

You are encouraged to repeat both h and p convergence studies using an initially more refined mesh and different convergence criteria.

Which one of the three solution methods is preferred? The "regular" method using h-elements (here studies *standard* and *standard with mesh controls*), the h-adaptive solution (*h adaptive* study), or the p-adaptive solution (*p adaptive 01* and *p adaptive 02* studies)?

Experience indicates that second order h-elements offer the best combination of accuracy and computational simplicity. For this reason, the SOLIDWORKS Simulation mesher is tuned to create h-element meshes. This makes h-elements preferable over p-elements. Also, a **p-adaptive** solution is much more computationally intensive than a standard or adaptive h-element solution. For these reasons, the **p-adaptive** solution should be reserved for special cases where solution accuracy must be known according to the settings available in the p-adaptive study.

This leaves us with the choice between standard and **h-adaptive** solution methods. The h-adaptive solution is more computationally demanding and more time-consuming than the standard solution. At the same time, it offers a very important advantage: it relieves users from the need to exercise judgment over meshing choices. In a standard solution, the user must decide if a default mesh is acceptable and how it should be modified by specifying mesh controls and/or global refinement. The **h-adaptive** solution automatically takes care of refining the mesh (both globally and locally).

With increasing computational power and the decreasing cost of computer hardware, the **h-adaptive** solution is becoming the preferred solution method in SOLIDWORKS Simulation.

Both **h-adaptive** and **p-adaptive** methods are great learning tools, leading to a better understanding of element order, the convergence process, and discretization error. For this reason, readers are encouraged to repeat some, if not all, of the previous exercises using both of the adaptive solution methods.

Models in this chapter

Model	Configuration	Study Name	Study Type
BRACKET.sldprt	*Default*	*standard*	Static
		standard with mesh control	Static
		h adaptive	Static
		p adaptive 01	Static
		p adaptive 02	Static

13: Drop test

Topics covered

- ❑ Drop test analysis
- ❑ Stress wave propagation
- ❑ Direct time integration solution

Project description

A ceramic porcelain RING is dropped from a height of 100mm measured from the center of mass, and lands on a flat, horizontal rigid floor (Figure 13-1). We will simulate the impact using a **Drop Test** study.

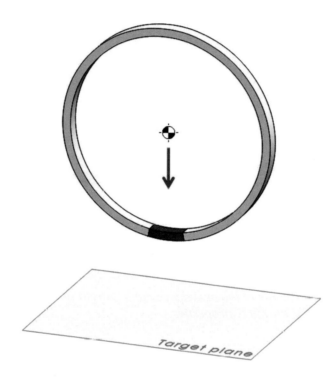

Figure 13-1: The RING dropping on a rigid floor.

The direction of gravitational acceleration (red arrow) is normal to the target plane. The target plane is 100mm below the center of mass.

Examine the CAD model and notice that it consists of two bodies: *Ring* and *Arch20*. Body *Arch20*, shown in red in Figure 13-1, is placed at the bottom where contact will be made. The split into two bodies makes it easier to apply mesh controls.

Procedure

Open the part RING. It has properties of ceramic porcelain material already assigned.

Create a **Drop Test** study *Drop test 1 100mm.* **SOLIDWORKS Simulation** creates two folders in the **Drop Test** study: *Setup* and *Result Options* (Figure 13-2).

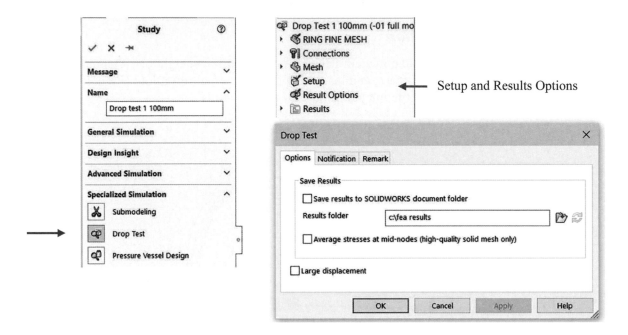

Figure 13-2: Drop Test study definition (left), Drop Test study window with *Setup* and *Result Options* (top right) and Drop Test properties window.

Displacements in the model will not be large; do not select Large displacements option in the study properties.

Right-click the *Setup* folder and select **Define/Edit** to open the **Drop Test Setup** window. Select **Drop height** and **From centroid** and enter 100mm as the drop height. From the fly-out **SOLIDWORKS** menu (not shown in Figure 13-3), select the **Top Plane** to define the line of action of gravitational acceleration as normal to the **Top Plane**. Enter the magnitude of gravitational acceleration as 9.81m/s^2. Finally, select **Normal to gravity** as **Target Orientation**.

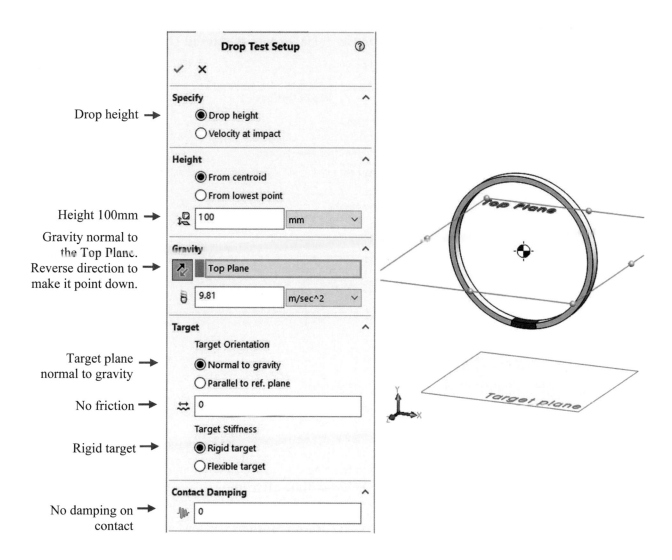

Drop height →

Height 100mm →

Gravity normal to
the Top Plane.
Reverse direction to
make it point down. →

Target plane →
normal to gravity

No friction →

Rigid target →

No damping on →
contact

Figure 13-3: **Drop Test Setup** window.

RING lands perfectly square on a rigid floor.

Right-click the *Result Options* folder to open the **Result Options** window.

Solution Time after
impact 600µs →

250 plots to be created
over the solution time →

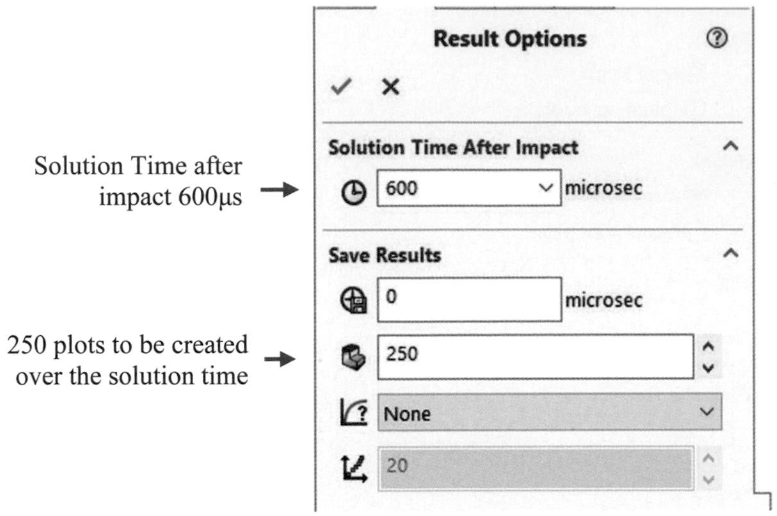

Figure 13-4: Result Options window.

250 is not a time step; this is the number of plots to be created during the event lasting 600µs.

To monitor what occurs to the ring during the first 600 microseconds after impact, enter 600 as the **Solution time after impact**. See **SOLIDWORKS Simulation** help for more information.

In the **Save Results** area of the **Result Options** window, accept the default 0 (microseconds) meaning that results will be saved immediately after impact. Enter 250 for the **No. of plots**.

Since the impact time is very short, it is measured in microseconds. The maximum displacement or stress may occur during this time, or later when the model is rebounding. A sufficiently long solution time needs to be specified to capture the physics of impact.

In preparation for meshing RING model consisting of two bodies, specify **Bonded Component Interaction** and select **Global Interaction** option (Figure 13-5).

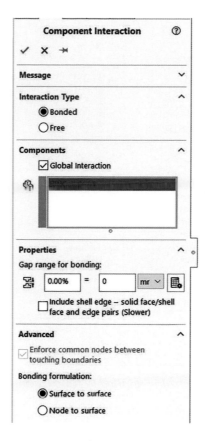

Figure 13-5: Component Interaction.

Bonded contact is specified with the option "Enforce common nodes between touching boundaries" checked by default. This assures smooth transitions between two bodies.

Apply mesh controls 0.5mm and ratio 1.1 to body *Arch20* and mesh the model with 1mm element size as shown in Figure 13-6.

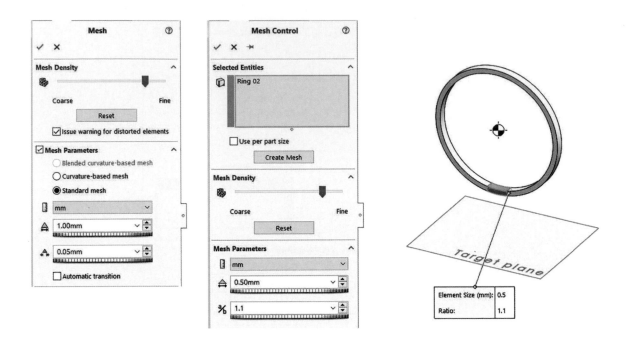

Figure 13-6: Mesh Density and Mesh Control definition.

Mesh Control is applied to Ring 02 body.

Center of mass symbol is shown.

Drop Test uses **Explicit Time Integration** solver. The solution time is calculated based on the number of time steps necessary to cover the event. In our case the solution time is considered long, and a warning message is displayed (Figure 13-7). Select **No** to dismiss this message.

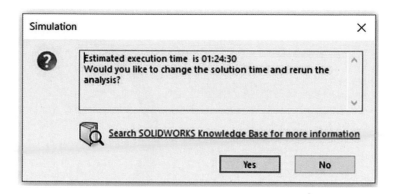

Figure 13-7: Warning message and solution progress windows.

Solution continues after the warning message has been dismissed. Solution time is long because Drop Test solver uses numerically intensive explicit time integration.

Upon completion of the solution, the time history is available in 250 plots. Envelope plots are also available to study the maximum values the drop event.

Create P1 stress envelope plot as shown in Figure 13-8.

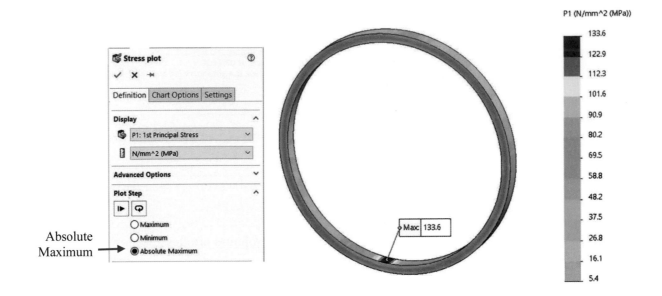

Figure 13-8: The absolute maximum P1 stress is located on the inside face near the contact point.

Compare the highest P1 stress 133.6MPa to the ultimate strength of ceramic porcelain 172MPa. The factor of safety is 1.3.

P1 stress (and not von Mises stress) is used because the RING material is brittle.

Follow steps in Figure 13-9 to construct **Time History** graph of P1 stress.

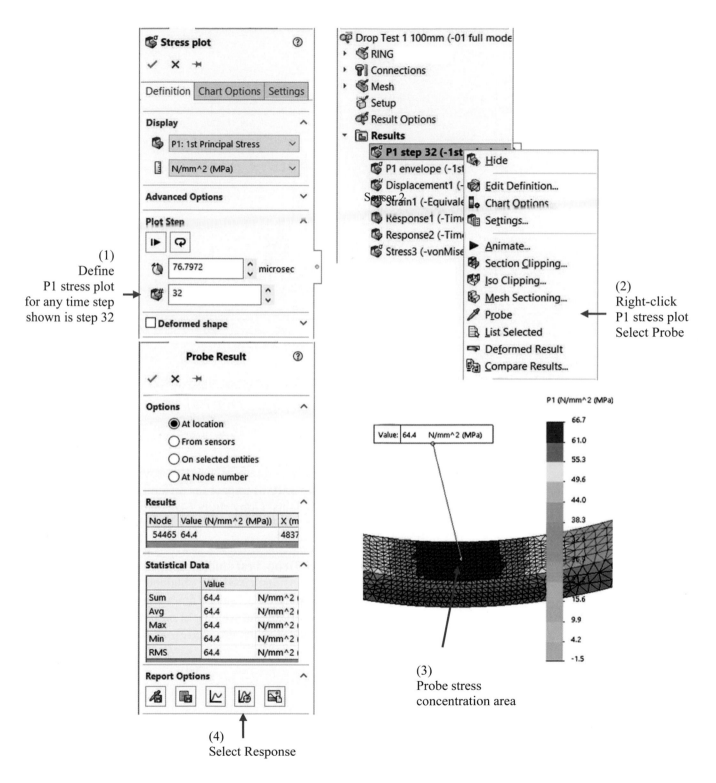

Figure 13-9: Four steps required to construct Time History graph.

Probed location will snap to the nearest node.

P1 **Time History** graph is shown in Figure 13-10.

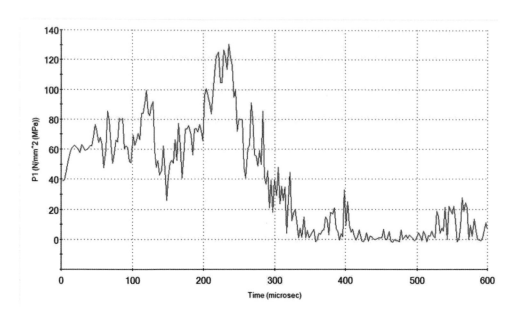

Figure 13-10: Time history of P1 stress during 600µs; study.

Graph refers to probed location shown in Figure 13-9. Results are from study Drop test 1 100mm.

Will the ring break? The **Drop Test** analysis does not directly provide pass/fail results. It is best used to compare the severity of impact for different scenarios. Repeat the analysis for different drop heights. Copy study *Drop Test 1 100mm* into *Drop Test 2 200mm* and *Drop Test 3 500mm* and modify drop height as indicated in the study name.

Figure 13-11 shows a summary of three **Drop Test** studies with drop height 100mm, 200mm and 500mm. All stress results have been obtained under an assumption of a rigid target and no damping on contact. Remembering that the ultimate strength of ceramic porcelain is 172MPa, we may conclude that RING will survive drop from 100mm, it may break when dropped from 200mm and will break when dropped from 500mm.

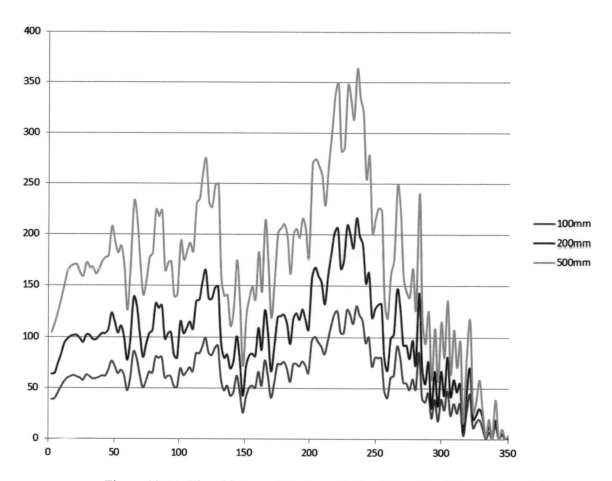

Figure 13-11: Time history of P1 stress during 350μs for different drop heights.

Curve corresponding to drop height 100mm is also shown in Figure 13-10.

Animate displacement or stress plot to see the ring bouncing off the floor. If you run analysis long enough you will see it bouncing more than once. Bouncing would last forever because there is no mechanism of energy dissipation in the study setup.

After completing this exercise, try experimenting with a different **Target Orientation**, **Flexible Target,** and **Damping** (Figure 13-3).

Models in this chapter

Model	Configuration	Study Name	Study Type
RING.sldprt	Default	*Drop Test 1 100mm*	Drop Test
		Drop Test 2 200mm	Drop Test
		Drop Test 3 500mm	Drop Test

14: Selected nonlinear problems

Topics covered

- Large displacement analysis
- Analysis with shell elements
- Membrane effects
- Following and non-following load
- Nonlinear material analysis
- Residual stress

In all previous exercises we assumed that the model stiffness did not change significantly when the model deformed due to the applied load. Consequently, the stiffness only needed to be calculated once, before any load had been applied. Since this stiffness adequately described model behavior during the entire loading process, the model stiffness did not have to be updated and the load could be applied in one single step. The only time we departed from these assumptions was in the analysis of the contact problem.

In this chapter we will discuss problems where the stiffness changes globally rather than locally, as was the case in the contact problem.

If a change of stiffness during the process of load application is caused by a change in model geometry, the problem may be solved with the **Large displacement** option selected in a **Static** study. Other types of nonlinear behavior require a **Nonlinear** study (Figure 14-1).

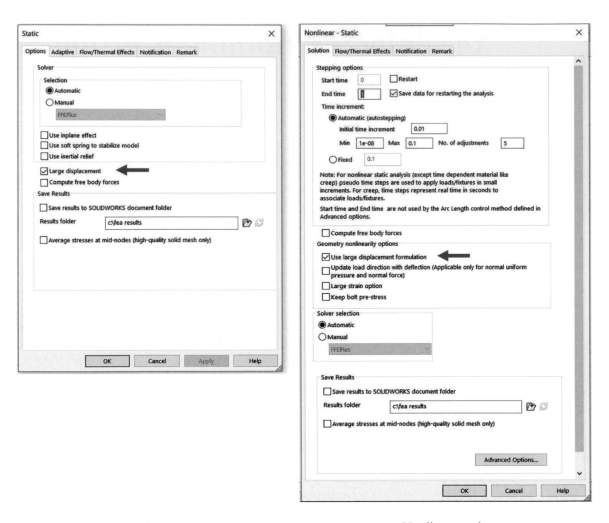

Static study   Nonlinear study

Figure 14-1: Large displacement options in Static study and Nonlinear studies.

Nonlinear study window is cropped to fit this illustration better.

Options shown in Figure 14-1 do not describe all nonlinear analysis capabilities of either study. Contact analysis is available in both Static and Nonlinear if Contact Sets are defined. Nonlinear material is available only in the Nonlinear study.

In NL002 exercise both linear analysis and nonlinear analysis will be run as **Static** studies. The **Nonlinear** study window shown in Figure 14-1 is for information only; it is not used in NL002 exercise.

Nonlinear problems caused by a change of model geometry can be solved either in a **Static** study with the **Large displacement** option selected, or in a **Nonlinear** study. All nonlinear problems may be solved in **Nonlinear** study.

Some limitations apply to a nonlinear geometry analysis conducted in a Static study. You may always select the **Large displacement** option in a **Static** study, whether it is necessary or not. However, this significantly increases solution time.

There is a common misconception that nonlinear analysis is the analysis of nonlinear material only. However, material nonlinearity is just one of many types of non-linearity; some of them will be presented in this chapter.

Open model NL002 and notice that the model contains only surface geometry. **Simulation** will recognize it and will mesh the model with shell elements.

Create two **Static** studies: *01 linear* and *02 nonlinear*. Check the **Large displacement** option in the properties of the *02 nonlinear* study. This is the only difference between these two studies.

When a study is created using a model with surfaces only, we need to define shell thickness. Thickness definition was not required in chapter 4 where a sheet metal model was analyzed, because shell thickness was taken from solid model geometry. Right-click the shell folder and select **Edit Definition** from the pop-up menu. In the **Shell Definition** window, select **Thin** as the shell element type and enter 2mm for the shell thickness, and accept 0 as the shell offset (Figure 14-2). Material properties are transferred automatically from the CAD model.

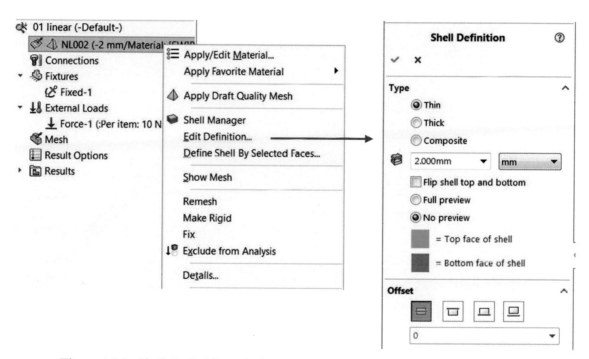

Figure 14-2: Shell Definition window.

Thin shell element has a constant transverse shear stress distribution across thickness. Thick shell has a parabolic transverse shear stress distribution across thickness. Refer to Simulation help for information on Composite shells.

Apply a **Fixed** restraint to the edge of the wide end and 10N normal force to the round split face (Figure 14-3).

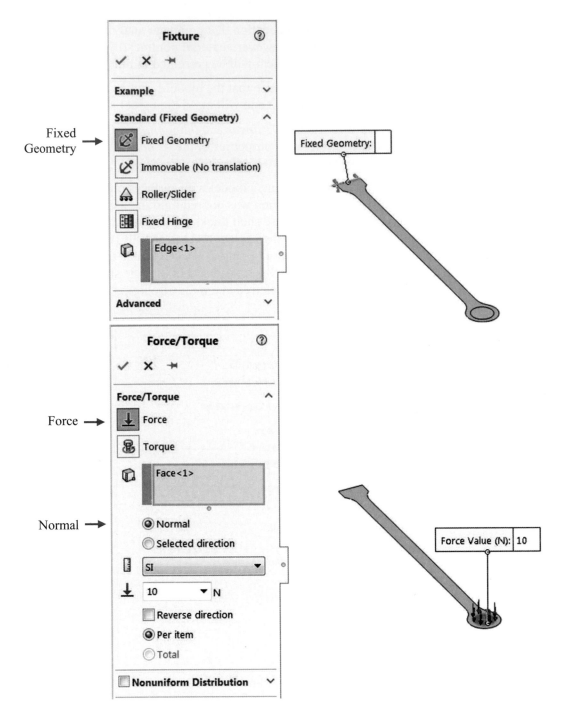

Figure 14-3: Restraints (top) and load (bottom) definition; arrows indicate the required selection.

Fixture window offers the choice "Immovable" which is due to the shell elements being analyzed. In our case, defining the restraint as Immovable would result in a hinge support; therefore, select Fixed Geometry.

Now mesh the model with shell elements using Standard mesh with the default element size and examine the shell element orientation (Figure 14-4).

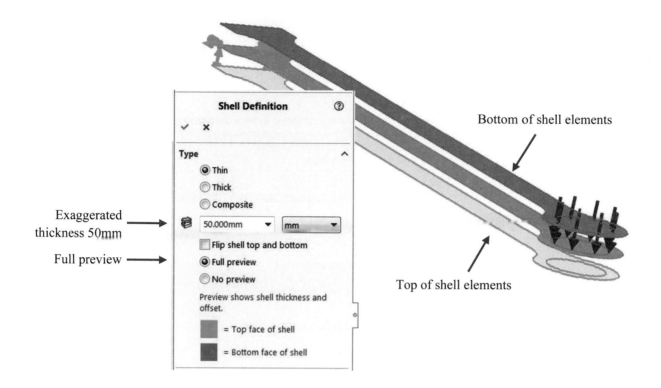

Exaggerated
thickness 50mm

Full preview

Bottom of shell elements

Top of shell elements

<u>Figure 14-4: The shell mesh orientation in NL002 is such that the bottoms of shell
elements are on the load side.</u>

*The mesh is orange on the side where the load is applied. As you remember in
chapter 4, the bottoms of shell elements are by default marked with the color
specified in study Options. The default color is orange.*

*Exaggerated thickness 50mm is used in this illustration for clarity. Experiment
with different thicknesses to see how that changes the plot, and then return to the
correct thickness 2mm.*

Solve the study named *01 linear* and notice that the solution of this study without
the **Large displacement** option displays a warning message (Figure 14-5) that
must be acknowledged to complete the solution process.

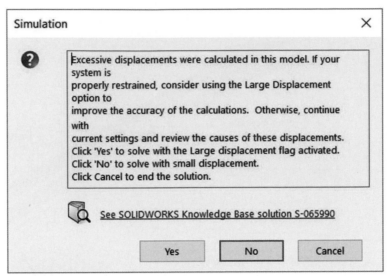

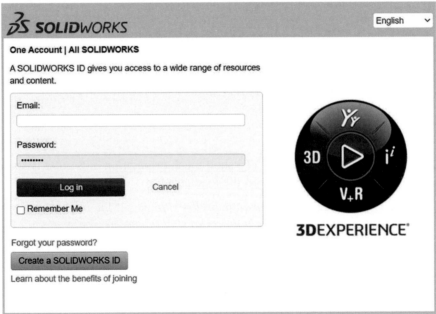

Figure 14-5: A warning message displayed by the solver if large displacements are detected, and the Large Displacement option is not checked (top) and access to SOLIDWORKS Knowledge Base (bottom).

Create a SOLIDWORKS ID to gain access to all SOLIDWORKS related topics.

Once displacements are calculated and before calculating stresses, solver checks the maximum displacement among the total number of the nodes and then compares it to the characteristic length of the model. If the ratio of (maximum displacement)/(characteristic length) > 10%, the solver will prompt the message: "Excessive displacements were calculated…".

Refer to the online SOLIDWORKS Knowledge Base for more information. You will need an account to access the SOLIDWORKS Knowledge Base.

Next solve the study *02 nonlinear* with the **Large displacement** option selected. Solving with the **Large displacement** option is a nonlinear solution, which progresses in iterations. The load is increased in automatically determined steps while the model stiffness is updated according to the progressing displacement (Figure 14-6).

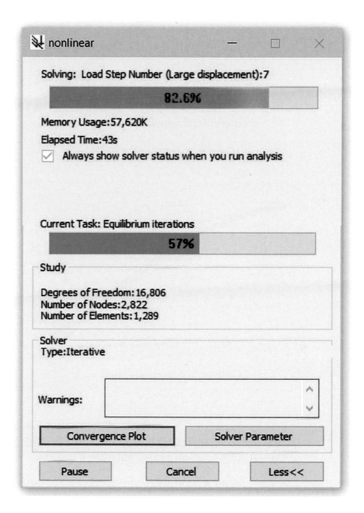

Figure 14-6: The nonlinear solution (with the Large Displacement option) progresses in iterations while the load gradually increases, and the stiffness is updated.

Due to the iterative nature of the Large Displacement solution, the solution time is longer than linear solution.

Create displacement plots obtained from the linear and nonlinear solutions (Figure 14-7). Use a 1:1 scale of deformation for both plots.

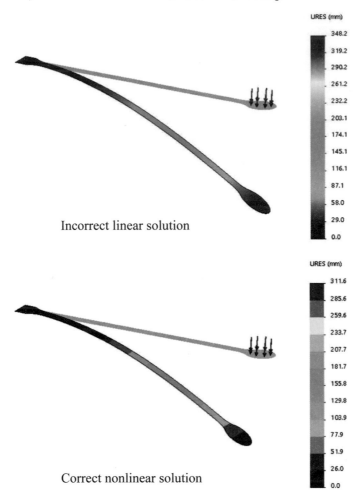

Figure 14-7: Incorrect linear solution (top) and correct nonlinear solution (bottom) obtained with the Large displacement option.

The undeformed model is superimposed on both plots. Notice that the linear solution produces an incorrect shape of deformation. This is seen by following the tip of the beam which travels along a straight line through deformation (the beam appears to be stretching out).

Now create two stress plots based on the nonlinear solution using a 1:1 scale of deformation. Create a von Mises stress plot for the bottoms of shell elements and for the tops of shell elements.

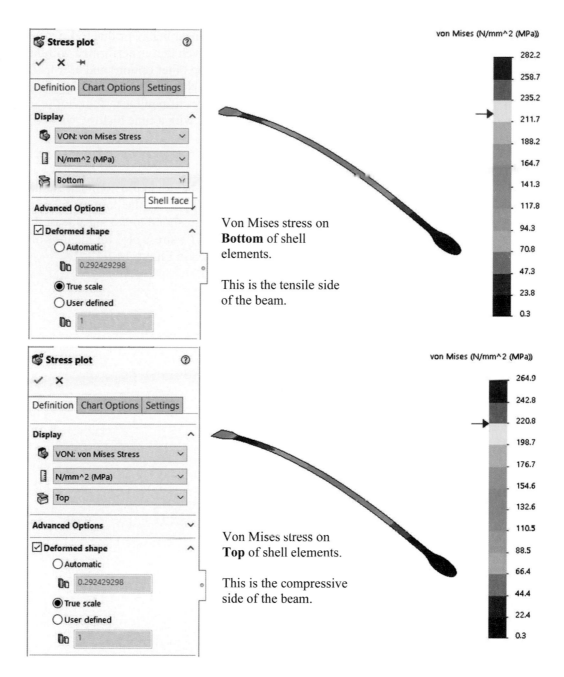

Von Mises stress on **Bottom** of shell elements.

This is the tensile side of the beam.

Von Mises stress on **Top** of shell elements.

This is the compressive side of the beam.

Figure 14-8: Von Mises stress plots for element Bottoms (top illustration) and element Tops (bottom illustration).

Both plots show similar stress magnitudes because von Mises stress is always a positive, scalar value. Tensile and compressive stresses equally contribute to von Mises stress.

Examine the plots in Figure 14-8 to notice that the maximum von Mises stress on the bottom side (tensile side) is 282MPa, and that the maximum von Mises stress on the top side (compressive side) is 265MPa. Both values are above yield which is 221MPa for Plain Carbon Steel. Even though we ran a nonlinear analysis, the only source of nonlinear behavior is the large displacement. Material yield is not modeled. **Static** study can model nonlinear behavior due to **Large displacement** but not due to nonlinear material and even that is performed with limitations; the load is "ramped up" linearly (there is no other option) and user has no control over load increments. Also, the load cannot change direction during the process of load application. To account for other sources of nonlinear behavior (e.g. material yielding) and to be able to apply a load with a time history, we need the **Nonlinear**.

Now, create two more stress plots also based on the nonlinear solution using a 1:1 scale of deformation. Create a P1 stress plot for the bottoms of shell elements. Since the bottoms of shell elements correspond to the top side of the beam, this plot shows tensile stress. Next create a P3 stress plot for the tops of shell elements. These results are summarized in Figure 14-9.

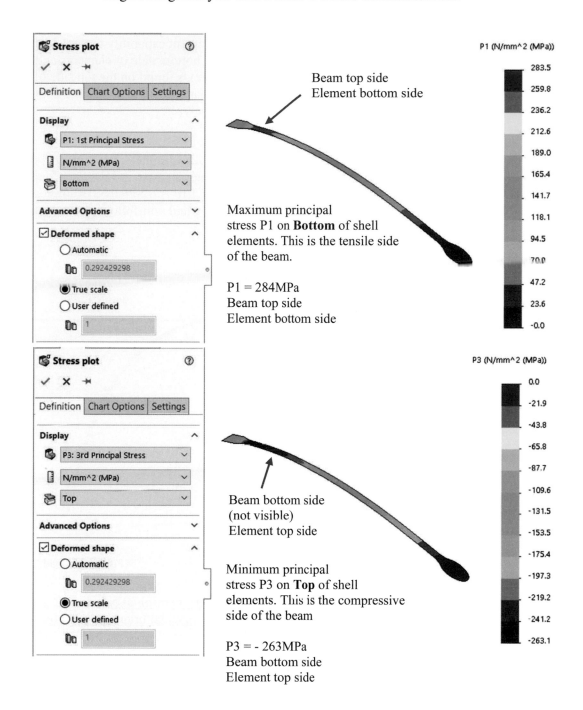

Figure 14-9: P1 stress results for element Bottoms (top illustration) and P3 stress results for element Tops (bottom illustration).

Figure 14-9 clearly illustrates the shell element capability to model bending. The highest P1 stress (tensile) is found on the bottom side of elements and the numerically lowest P3 stress (compressive) is found on the top side of elements; both are in the same location along the beam. The above discussion is of course applicable to both linear and nonlinear analyses.

The load in example NL002 was a non-following load; it retained its original direction and did not follow the deforming structure. A following load requires a **Nonlinear** study and will be used in the CLIP and SPRING examples in this chapter. The difference between following and non-following load is explained in Figure 14-10.

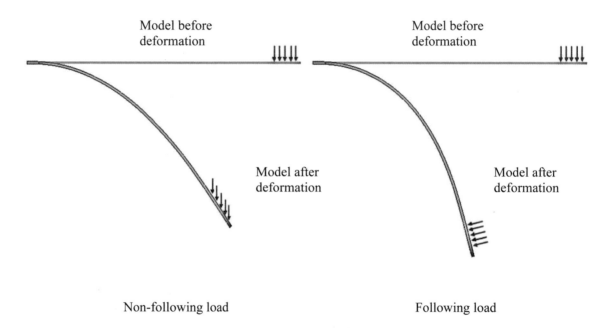

Non-following load   Following load

Figure 14-10: A non-following load retains its original direction. A following load follows the deforming model retaining its original orientation relative to the model.

A following load is not available in Static study; it is available in Nonlinear study.

This illustration has been prepared using model NL002 solid. Solid model makes it easy to show beam thickness.

The next example introduces the analysis of large displacements combined with a contact problem, illustrated by the CLIP model (Figure 14-11).

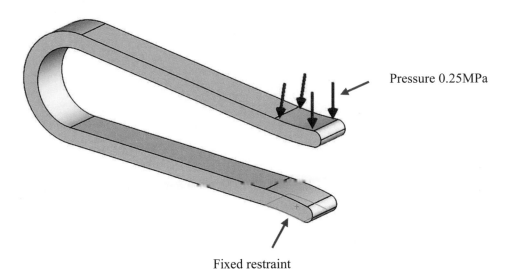

Pressure 0.25MPa

Fixed restraint

Figure 14-11: The CLIP model with load and restraint.

Restraint is applied to the face opposite to the loaded face.

Two surfaces, defined in a contact pair, will experience large displacements before contacting each other. Therefore, there are two sources of nonlinear behavior in this problem: large displacements and contact.

Two studies are used in this exercise: **Static** study called *01 small displacements* and **Nonlinear** study called *02 large displacements*. We use **Static** study without large displacement options only to show an incorrect solution.

In Static study *01small displacement* define a **Local Interaction** between the two surfaces as **Surface to surface** (Figure 14-12).

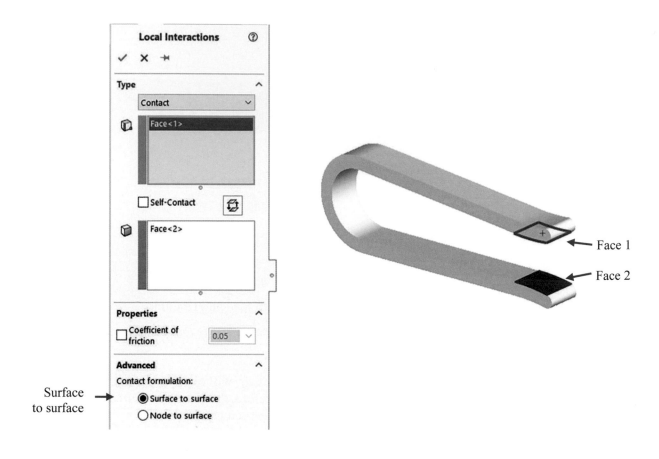

<u>Figure 14-12: Definition of the Contact Set. Surface to surface contact does not require that the faces are initially touching.</u>

Model edges are not shown.

Mesh the assembly with 1.3mm element size, Standard mesh to produce three layers of elements across the thickness. Run solution of **Static** study, ignore solver's warning and do not switch to **Large displacement** solution. Next, create a **Nonlinea**r study *02 large displacement* with the same mesh. Select **Use large displacement formulation** and **Update load direction with deflection**. Run *02 large displacements* study and compare the deformed shapes obtained from two studies (Figure 14-13).

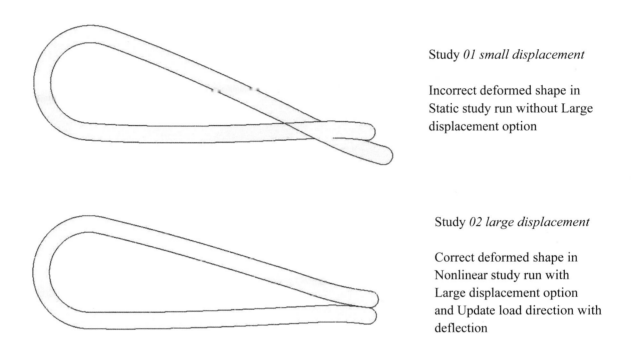

Study *01 small displacement*

Incorrect deformed shape in Static study run without Large displacement option

Study *02 large displacement*

Correct deformed shape in Nonlinear study run with Large displacement option and Update load direction with deflection

Figure 14-13: Deformed shapes from study *01 small displacement* (top) and study *02 large displacement* (bottom).

Both plots use 1:1 scale of deformation.

Linear analysis conducted in study *01 small displacement* produces clearly incorrect results even though **Local Interactions** are defined. Nonlinear analysis conducted in study *02 large displacement* produces correct results. It allows for correct modeling of large displacements and gap closing. Also, it models a following load while Static study cannot model a following load. Therefore, the CLIP problem cannot be solved in **Static** study even with **Large displacement** option.

Review stress results and notice that the element size in the contact area is too large in comparison to the size of the contact area to produce meaningful contact stress results. (Figure 14-14).

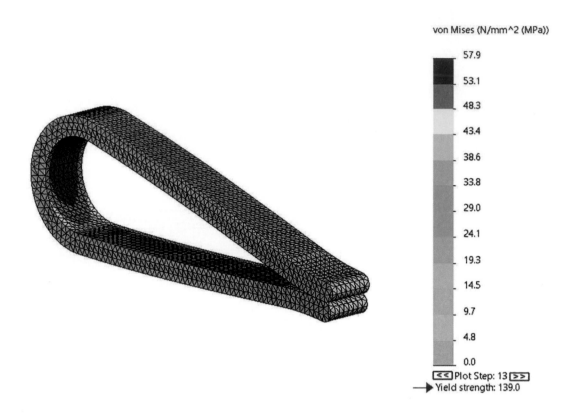

Figure 14-14: Von Mises stress results at gap closure.

Results of the last step (Step 13) are shown.

The deformed shape can be saved as another configuration in the same model or as another part (recall chapter 11). To save the deformed shape as a different model, follow the steps in Figure 14-15.

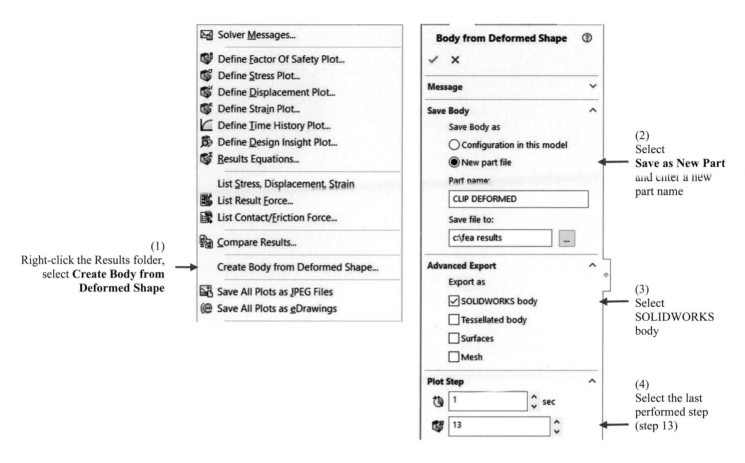

(1)
Right-click the Results folder, select **Create Body from Deformed Shape**

(2)
Select **Save as New Part** and enter a new part name

(3)
Select SOLIDWORKS body

(4)
Select the last performed step (step 13)

Figure 14-15: Deformed model can be saved as a new part; alternatively, a new configuration can be added to the existing model.

Experiment with saving the deformed shape from different Plot Steps.

Open the deformed model CLIP DEFORMED to see that it consists of one imported feature (Figure 14-16).

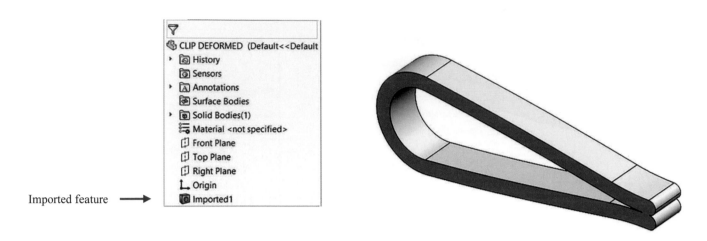

Figure 14-16: Deformed model consists of imported geometry only. Parametric formulation is gone.

The imported feature cannot be modified.

The two previous exercises (NL002 and CLIP) required the **Large displacement** option because the displacements were indeed large. However, problems may exist where displacements are small, but they significantly change the model stiffness. These problems still require solutions with the **Large displacement** option selected. In fact, the term **Large displacement** may be misleading to some users because it implies that this option should only be used if displacements are large.

We will demonstrate that displacements do not have to be large to require a solution with the **Large displacement** option selected. Open the model ROUND PLATE, which is a thin round plate subjected to pressure. Switch to configuration *02 flat section* as shown in Figure 14-17.

Geometry, loads, and restraints are all characterized by axial symmetry. Therefore, a slice with applied boundary conditions can be used to model plate's response to pressure. We'll analyze configurations *02 flat section* and *04 curved section*.

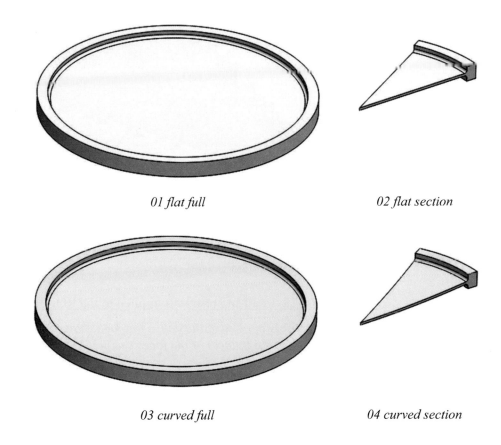

| 01 flat full | 02 flat section |

| 03 curved full | 04 curved section |

Figure 14-17: The ROUND PLATE in four configurations.

Examine model geometry in flat and curved configuration and notice a slight curvature of plate in 03 curved full and 04 curved section configurations.

Radial faces in the section model must remain flat during the process of deformation. **Roller-slider** boundary conditions applied to radial faces in the section model will enforce flatness of these faces.

Create a static study *linear flat* and apply **Roller-Slider** restraints to both flat faces created by the cut. Next, apply a **Fixed Geometry** restraint to the cylindrical face on the plate circumference and a pressure of 0.4MPa to the top face (Figure 14-18).

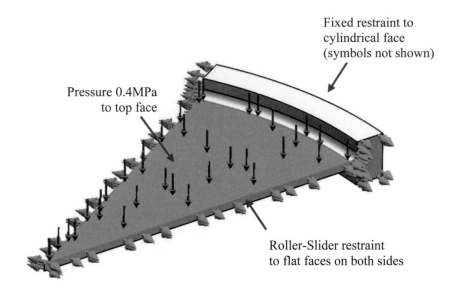

Figure 14-18: Load and restraints applied to the ROUND PLATE model.

The pressure is applied to the entire top face shown in blue. Roller/Slider restraints enforce flatness of the faces created by the cut while the model deforms.

Mesh with element size 2mm to avoid an excessive turn angle in fillets and obtain solution <u>without</u> **Large displacement** option checked. Next, copy study *linear flat* into *nonlinear flat* <u>with</u> **Large displacement** option checked and compare displacement results (Figure 14-19).

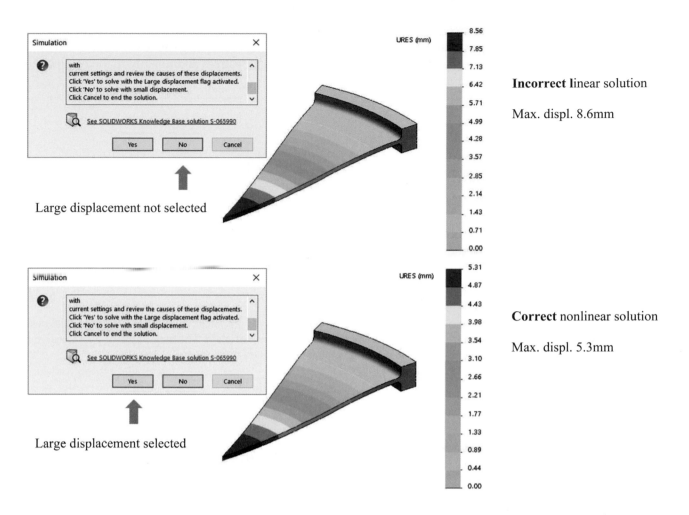

Figure 14-19: Model in *02 flat section* configuration Displacement results obtained without Large displacement option (top) and displacement results obtained with Large displacement option (bottom).

The decision whether to change to the Large displacement solution is left up to the user.

Displacement results shown in Figure 14-19 show that linear and nonlinear results are different by 61%; the nonlinear model is significantly stiffer.

Figure 14-19 shows incorrect linear solution and correct nonlinear solution. However, we should point out that the nonlinear solution obtained in *Static* study with **Large displacement** option activated does not model pressure as a following load. In our case the error is low because of low magnitudes of displacements. Larger magnitudes of displacements would require *Nonlinear* study.

A flat plate under pressure is a classic case where the assumption of small displacements leads to erroneous results. The analysis requires the use of the **Large displacement** option even though displacements may be small in comparison to the size of the model.

The need to use the **Large displacement** option is due to the change of shape (from flat to curved) altering the mechanism resisting the load; a deformed plate can resist pressure with membrane (tensile) stress additionally to the original bending stress (Figure 14-20).

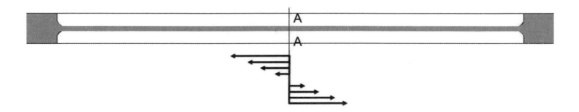

Undeformed plate

Stresses across the thickness are bending stresses with linear variation across plate thickness: compressive stress is on top and tensile on bottom of plate. The magnitude of compressive stress on top is equal to the magnitude of tensile stress on the bottom.

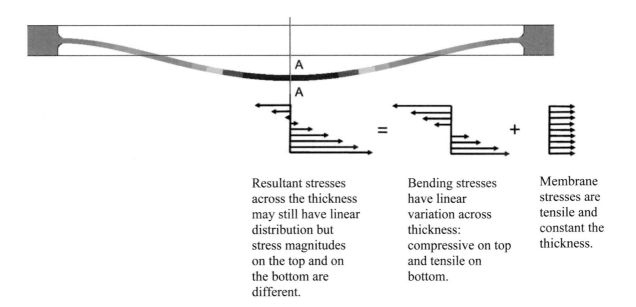

| Resultant stresses across the thickness may still have linear distribution but stress magnitudes on the top and on the bottom are different. | Bending stresses have linear variation across thickness: compressive on top and tensile on bottom. | Membrane stresses are tensile and constant the thickness. |

Figure 14-20: Load resisting mechanism in an undeformed plate (top) and in a deformed plate (bottom).

The stress distribution across the plate thickness does not have to be necessarily linear.

To account for the change of plate stiffness that takes place due to deformation (even though the deformation is small), the stiffness must be updated during the deformation process. This is only possible if a nonlinear analysis is executed. The linear analysis only considers the initial bending stiffness, which is the reason why the linear model is softer than the corresponding nonlinear model.

The demonstrated effect of increasing stiffness during deformation is called membrane stress stiffening. The membrane stress stiffening makes linear and nonlinear solutions very different. That difference is less in a model with initial curvature. Repeat the analysis using ROUND PLATE model *in 04 curved section* configuration. Create study *linear curved* (Large displacement deselected) and *nonlinear curved* (Large displacement selected) and compare displacement results (Figure 14-21).

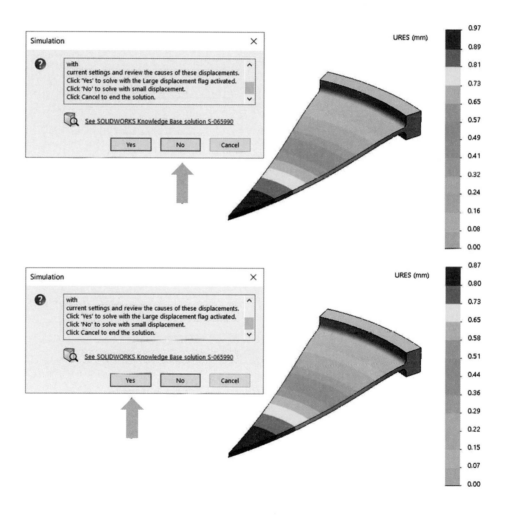

Figure 14-21: Model in *04 curved section* configuration. Displacement results obtained without Large displacement option (top) and displacement results obtained with Large displacement option (bottom).

Linear and nonlinear results are different by 11%.

We will illustrate the same problem of stress stiffening with one more example, similar to LINK described in chapter 5. Open the model LINK02 with the material properties of Nylon 6/10. Due to the material's low modulus of elasticity, the model will experience large displacements which make it easy to show the difference between the correct nonlinear solution using *Static* study with **Large displacement** option and incorrect linear solution with **Large displacement** option selected. Use *01 full model* configuration and apply a 0.2MPa pressure to the top face and hinge restraints to the "eyes" (Figure 14-22).

Create and solve two studies: *01 linear* without the **Large displacements** option and *02 nonlinear* with the **Large displacement** option. Both studies use default solid element mesh. Ignore the solver's message about large displacements and obtain linear solution. Next, obtain nonlinear solution and compare displacements and von Mises stress results from both solutions (Figure 14-22).

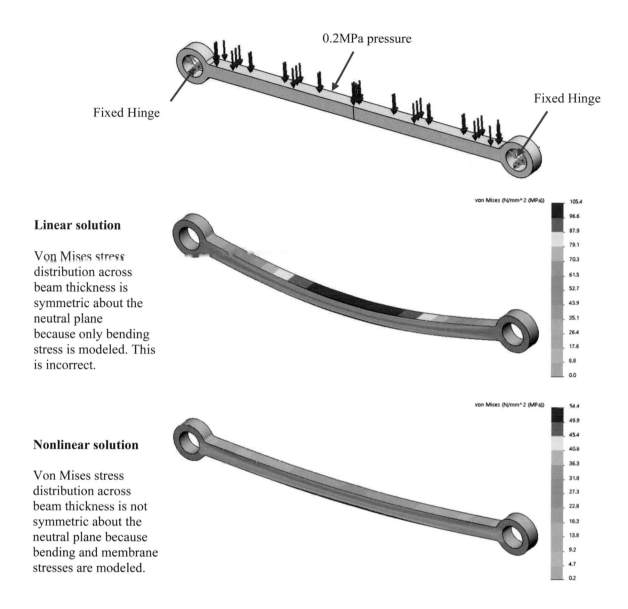

Linear solution

Von Mises stress distribution across beam thickness is symmetric about the neutral plane because only bending stress is modeled. This is incorrect.

Nonlinear solution

Von Mises stress distribution across beam thickness is not symmetric about the neutral plane because bending and membrane stresses are modeled.

Figure 14-22: Loads and restraints on LINK02 model (top); von Mises stress results in linear solution (middle) and nonlinear solution (bottom).

Stress plots use 1:1 scale of deformation. Compare the magnitude of displacements in both solutions.

Linear solution*: the absence of membrane (tensile) stresses in the linear solution is illustrated by the symmetric distribution of von Mises stresses about the neutral bending plane.*

Nonlinear solution*: von Mises stress results in the nonlinear (large displacement) solution shows a non-symmetric distribution of stress proving the presence of membrane (tensile) stresses.*

Another way to demonstrate the presence of membrane stress is to examine a graph of the SX stress distribution across the height of the link. Create a plot of SX stresses and follow steps in Figure 14-23 and Figure 14-24.

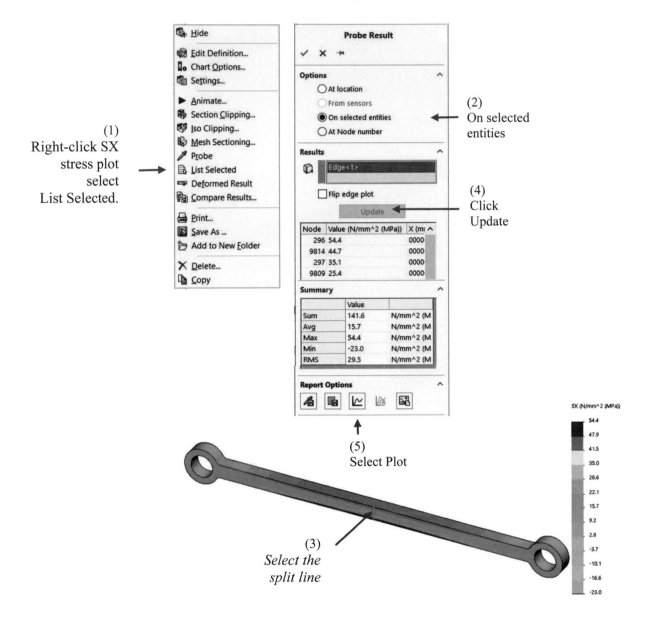

Figure 14-23: Distribution of SX stress component across the thickness can be graphed by following the steps explained in this Figure.

Selecting Plot in step (5) produces graphs shown in Figure 14-24.

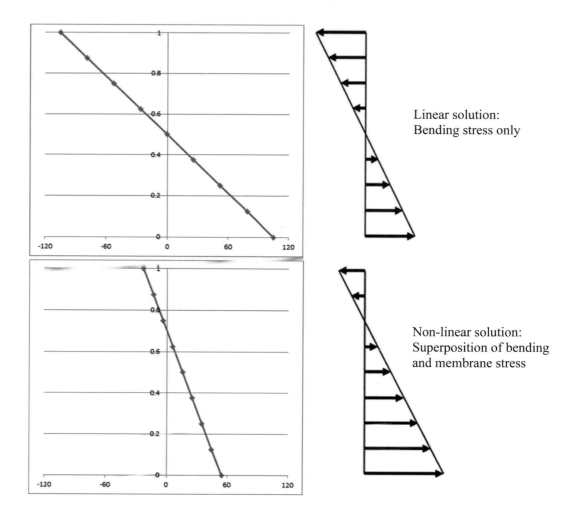

Linear solution:
Bending stress only

Non-linear solution:
Superposition of bending
and membrane stress

Figure 14-24: **Top**: distribution of SX stress across the thickness in the linear model. **Bottom**: distribution of SX stress across the thickness in the nonlinear model.

0 on the ordinate axis denotes the bottom of the beam, 1 denotes the top of the beam.

Refer to Figure 14-25 for an explanation of the differences between linear and nonlinear solutions.

The difference between a nonlinear and linear solution is schematically shown in Figure 14-25.

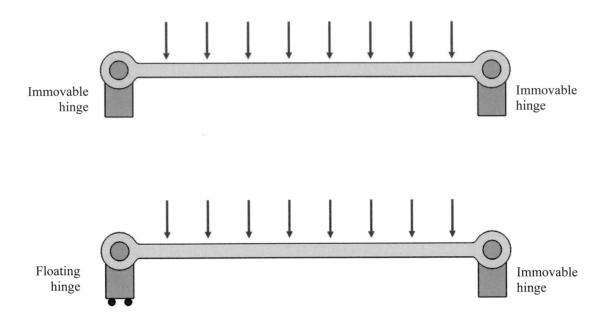

Figure 14-25: The nonlinear solution correctly models hinge supported link where the distance between hinges cannot change (top). The linear solution can only model the configurations where one of the hinges is floating (bottom).

A linear model cannot distinguish between the two configurations shown in this illustration. The floating hinge support is illustrated here by a roller support. You may review this model in LINK02.SLDASM.

The nonlinear solution correctly calculates displacements and stresses in a hinge supported link as shown at the top of Figure 14-25. The linear solution cannot model membrane stresses that develop during the deformation. Consequently, the link is modeled as if one of the hinges had a floating support, symbolically shown in Figure 14-25 as rollers. Notice that horizontal displacements are not modeled in a linear solution. Nonlinear analysis would be always required if horizontal displacements of hinge supported by roller were of interest.

Repeat analysis of LINK02 using configuration *02 half model*. The only difference will be applying symmetry boundary conditions to the face in the plane of symmetry.

All nonlinear problems presented in this chapter owe their nonlinearity to changes in geometry and contact taking place during the loading process. They can be solved using a **Static** study with the **Large displacement** option selected.

Now, we present a problem where nonlinear behavior is caused by nonlinear material. Open part file BRACKET NL, which is similar to the model analyzed in chapter 12. The material is aluminum 1060Alloy, there is a cut along the plane of symmetry to work with one half of the model, and split faces exist in the area where we detected stress concentrations in chapter 12. Use the *02 half model* configuration. Since an h-adaptive solution is not available in a **Nonlinear** study, we draw on our experience with the model and use these split faces to define mesh controls.

Open a **Nonlinear** study with the **Static** option and define its properties as shown in Figure 14-26.

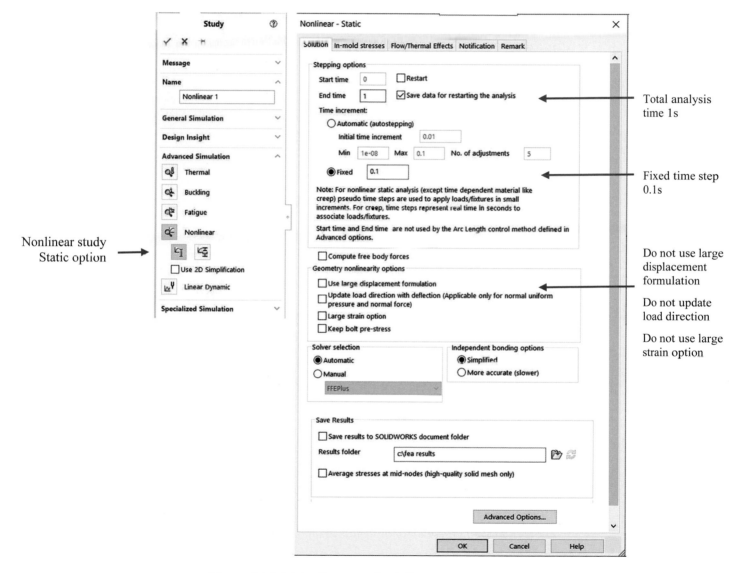

Figure 14-26: Nonlinear study definition and study properties.

Make selections as shown above. We do not expect large displacements so load direction does not need to be updated during the load application process, and a fixed time step can be used. Nonlinear material is the only source of nonlinear behavior in this model.

Following the selections shown in Figure 14-26, the load application will take 1s. Considering the fixed time step of 0.1s, the load will be applied in 10 steps. It is important to remember that in static analysis, time is only used to define the shape of the load history curve. The same results are obtained if the analysis time is 1s and the time steps are 0.1s or if the analysis takes 1000s and a time step of 100s. To differentiate it from "real" time used in a dynamic analysis, the time used to define the load history curve is called pseudo time. The BRACKET NL can also be solved with automatic time stepping.

The ability to control the load time history is an important difference between a **Static** study executed with the **Large displacement** option and a **Nonlinear** study. In a **Static** study executed with **Large displacement**, a load can only be increased linearly in automatically determined steps. Without the **Large displacement** option, a load in a **Static** study is applied just in one step.

Now we define a nonlinear material. We use the simplest type of nonlinear material called the elastic-perfectly plastic material model, von Mises type. Its stress-strain curve is shown in Figure 14-27.

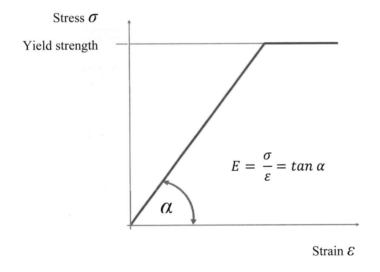

Figure 14-27: Stress-strain curve of an elastic-perfectly plastic material.

According to an elastic-perfectly plastic material model, stress is proportional to strain until von Mises stress reaches yield strength. After that, stress becomes constant, regardless of the magnitude of strain.

The modulus of elasticity of material in the linear elastic range is E = tan

When stress reaches the yield strength, the modulus of elasticity of the material becomes zero; material becomes perfectly plastic.

To define an elastic-perfectly plastic material model, we need to know the material modulus of elasticity describing its behavior in the elastic range and the magnitude of the yield strength. We also need to decide how to determine if yield strength has been reached. In this example we compare von Mises stress to yield strength. Once von Mises stress reaches yield strength, the material modulus of elasticity becomes zero. In **SOLIDWORKS Simulation** this material is called **Plasticity – von Mises**.

No material is defined in CAD model; material definition will be done in Simulation study. Right-click *Solid* folder in *Nonlinear 1* study; select **Apply/Edit Material** to open window shown in Figure 14-28. Change material of BRACKET NL to **1060 Alloy**, **Plasticity – von Mises**.

Figure 14-28: Follow the above steps to define Plasticity-von Mises material model.

Yield strength 27.57MPa is the highest von Mises stress that the model will see.

Define a **Fixed** restraint on the back side and a **Symmetry** restraint to the faces created by the symmetry cut. The total load on the model is 5000N; therefore, apply 2500N to one half of the model geometry, to the flat face as shown in Figure 14-29. The load application requires definition of the load time history. We want to "ramp-up" the load to the maximum value to see the maximum stresses, then we drop it back down to zero to examine the residual stresses. The load definition is shown in Figure 14-29.

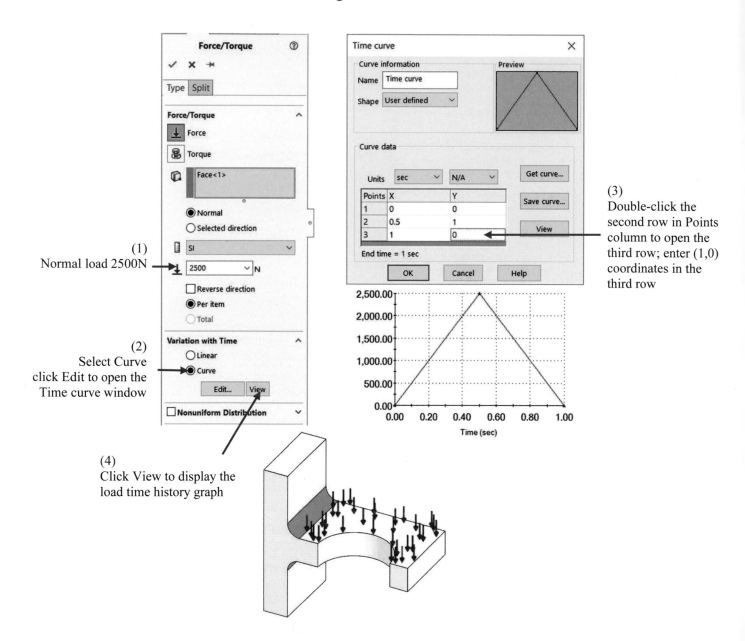

Figure 14-29: Load definition.

Load definition requires load time history. The maximum load 2500N takes place at time t=0.5s. At t=1s, the load returns to zero. Y coordinates in Time Curve definition window are multipliers to load defined in Force/Torque window.

Define **Mesh Control** to two green faces and mesh with **Standard mesh** using settings shown in Figure 14-30.

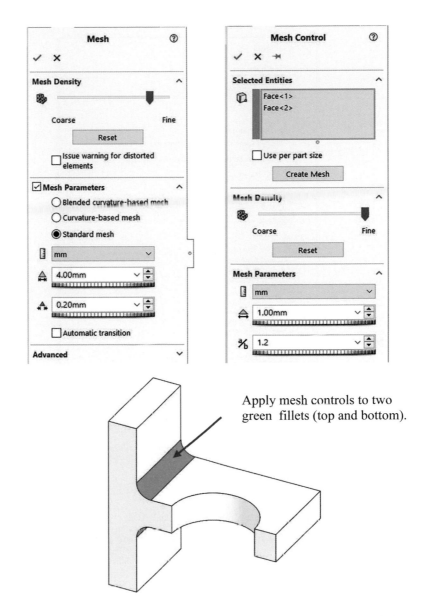

Apply mesh controls to two green fillets (top and bottom).

Figure 14-30: Mesh controls are applied to produce a fine mesh in the locations of the expected stress concentrations.

a/b ratio 1.2 in Mesh Controls specifies a more gradual transition of element size.

If you experience long solution time, use larger elements in **Mesh** and **Mesh Control**.

Execute the nonlinear solution and observe the solution progress in the Solver window. Be prepared for a solution time much longer than a typical linear analysis. The nonlinear solver window is shown in Figure 14-31.

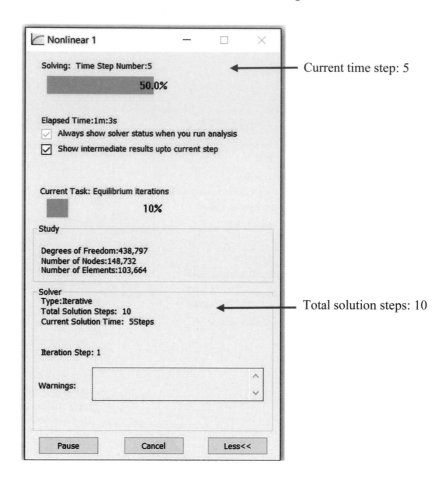

Figure 14-31: Nonlinear solver window.

This window shows the solver performing time step 5 of 10.

Von Mises stress results for the maximum load (time step 5) are shown in Figure 14-32.

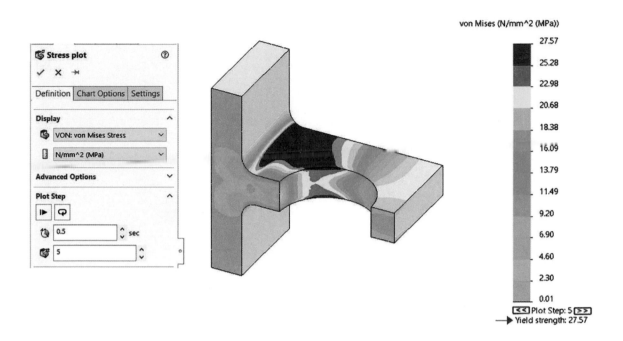

Figure 14-32: Von Mises stresses under the maximum load, time step 5.

Notice that the maximum von Mises stress magnitude is 27.57MPa, which is the yield strength of the elastic-perfectly plastic material used in the analysis.

Yielding occurs in the portions of the model where stress reaches the maximum magnitude. A large size of the yield zone indicates that the mesh controls were not necessary to produce these results.

Von Mises stress results for zero load (time step 10) are shown in Figure 14-33. These are residual stresses left there after completing the load cycle.

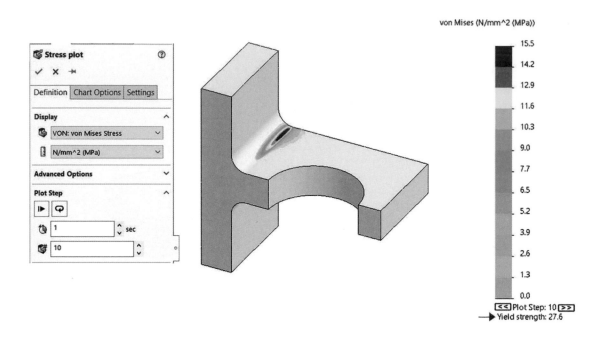

<u>Figure 14-33: Residual von Mises stresses after the load has been removed at time step 10.</u>

Residual stresses reach 12.0MPa in the area that experienced plastic deformation (yield). The small size of the high stress zone indicates that the mesh controls were necessary to produce these results. Retrospectively, we could have applied mesh controls to a smaller area.

Show plot of von Mises stress in step 10 and probe it in the location of the maximum stress. Follow the steps in Figure 14-34 to construct a graph of von Mises stress time history in the location of the maximum residual stresses.

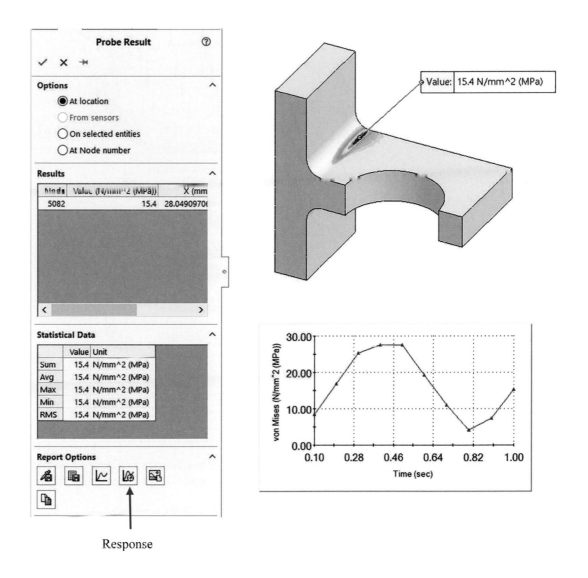

Response

Figure 14-34: Maximum von Mises stress as a function of time.

The maximum von Mises stress is 27.57MPa, which is the yield strength of Plasticity-von Mises material model.

Response graphs may be created by probing any time step plot. We use the last step to probe the location of the highest residual stresses.

Try running the same analysis with 20 time steps to see a more detailed graph, especially a "dip" in stress in step 8 seen in Figure 14-34.

To continue with the theme of elastic-perfectly plastic materials, open the model SPRING, and create a nonlinear study. Apply loads and restraints as shown in Figure 14-35.

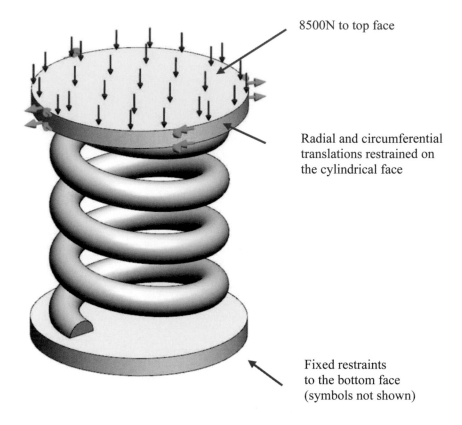

8500N to top face

Radial and circumferential translations restrained on the cylindrical face

Fixed restraints to the bottom face (symbols not shown)

Figure 14-35: Restraint and load of SPRING model.

The top plate is restrained using On Cylindrical Faces restraint; circumferential and radial translations are restrained. The top plate can only move in the direction of load; it cannot tilt.

Use **Linear** variation with Time in the load definition. Linear variation means that the load is a linear function of time.

The CAD model has no material properties assigned. Define Alloy Steel material in Simulation study. Make it the elastic-perfectly plastic material with a yield strength of 620MPa (Figure 14-36).

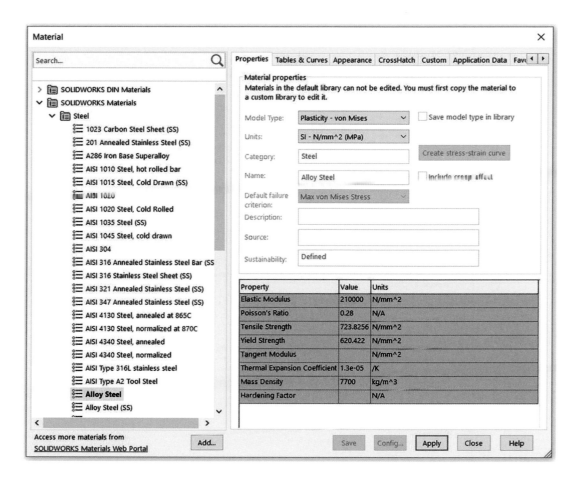

Figure 14-36: Elastic-perfectly plastic material definition in the SPRING model.

The materials yield strength is 620.4MPa.

Recall previous exercises to notice that the material's yield strength shows both in the linear and nonlinear material definitions. In case of a linear material, the yield strength is used only for results interpretation such as making a plot of the factor of safety to yield. In a linear analysis, stress results are not affected in any way by the yield strength value. In a nonlinear material definition, yield strength plays a crucial role. In the case of the Plasticity-von Mises (Elastic-Perfectly Plastic) material model, once von Mises stress reaches the yield strength, the modulus of elasticity changes to zero in that location.

Mesh the model with a **Curvature Based** mesh, with default element size (7.42mm); make the minimum number of elements in a circle 16. In study properties select the option **Use large displacement formulation** (Figure 14-37). There is no need to update load direction with deformation.

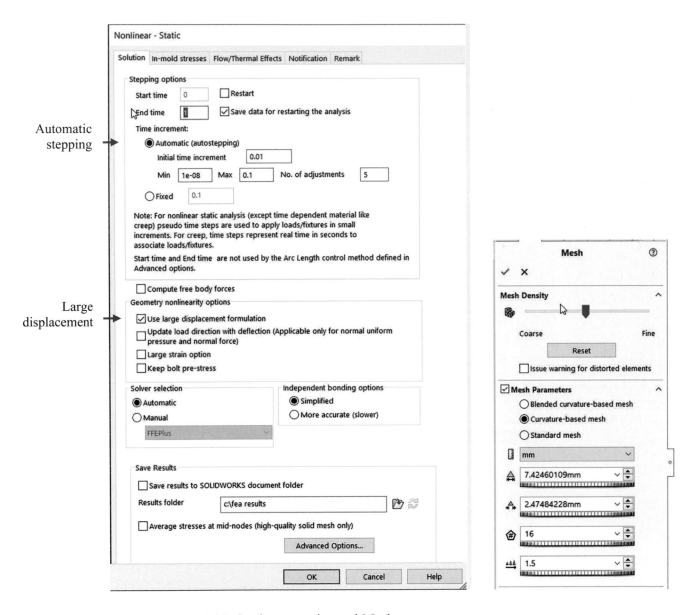

Figure 14-37: Study properties and Mesh parameters.

When Automatic time steps are used, the load increment is selected automatically by the solver.

Use a default maximum element size with a Curvature-based mesh and 16 as the minimum number of elements in a circle.

Run the nonlinear solution and review the displacement results; notice the large displacement that the model has experienced under the load (Figure 14-38). Next, review the von Mises Stress and examine the portions of the model that reached yield strength (Figure 14-39). Display both plots for the last performed time step (this is step 13). Because the load has a Linear Variation with time, the last step has the highest load magnitude.

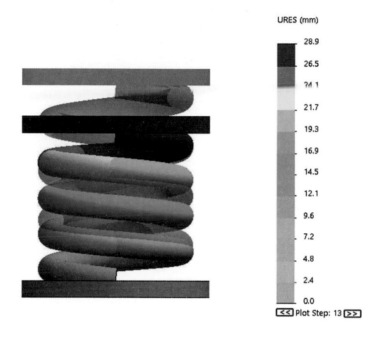

Figure 14-38: The displacement results plot in 1:1 scale of deformation shows that the model experiences large displacement.

The undeformed model is superimposed on the deformed plot.

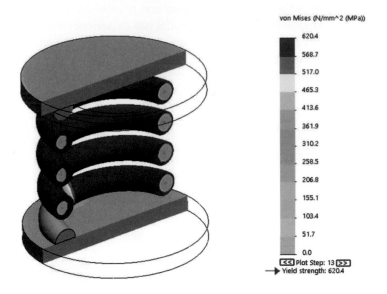

Figure 14-39: Von Mises stress plot shows that large portions of the model have yielded.

Section plot reveals large areas where stress has reached the yield strength of 620.4MPa.

You may want to experiment with the model by increasing the load until the solution crashes, meaning that the entire spring cross section has yielded, and it is no longer capable of resisting the applied load.

Models in this chapter

Model	Configuration	Study Name	Study Type
NL002.sldprt	Default	01 linear	Static
		02 nonlinear	Static
NL002 solid.sldprt	Default	non-following	Nonlinear
		following	Nonlinear
CLIP.sldprt	Default	01 small displacement	Static
		02 large displacement	Nonlinear
CLIP DEFORMED.sldprt	Default		
ROUND PLATE	01 flat full		
	02 flat section	linear flat	Static
		nonlinear flat	Static
	03 curved full		
	04 curved section	linear curved	Static
		nonlinear curved	Static
LINK02.sldprt	01 full model	01 linear	Static
		02 nonlinear	Static
	02 half model		
LINK02.sldasm	01 no rollers		
	02 rollers		
BRACKET NL.sldprt	01 full model		
	02 half model	Nonlinear 1	Nonlinear
SPRING.sldprt	Default	Nonlinear 1	Nonlinear

Notes:

15: Mixed meshing problem

Topics covered

- Using solid and shell elements in the same mesh
- Mesh compatibility
- Manual and automatic finding of contact sets
- Shell Manager

While in previous exercises we used different types of meshes such as solid and shell meshes, they have never been used together in the same model. This exercise introduces the use of different mesh types within the same model and different ways of connecting solid and shell element meshes.

Open the part model FLYWHEEL and examine three configurations shown in Figure 15-1. In configurations *surfaces A* and *surfaces B*, the hub and the rim are connected by spokes modeled as surfaces. In configuration *surfaces C,* they are connected by a solid web of varying thickness.

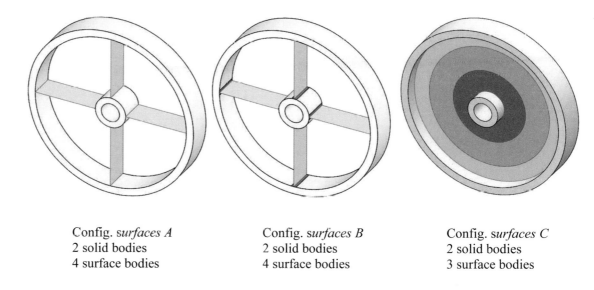

Config. *surfaces A*	Config. *surfaces B*	Config. *surfaces C*
2 solid bodies	2 solid bodies	2 solid bodies
4 surface bodies	4 surface bodies	3 surface bodies

Figure 15-1: FLYWHEEL in three configurations.

In configuration surfaces B, notice eight green split faces where spokes meet the hub.

All configurations use the same material 1060Alloy defined in the Feature Manager. There is one more configuration *solids*, not used in any Simulation study. In *solids* configuration, the web from configuration surfaces C is modeled as solid geometry for visualization purposes.

Move to model configuration *surfaces A* and create a **Frequency** study *surfaces A*. The part folder contains two solid bodies and three surface bodies. The surface bodies do not contain information on thickness; therefore, we must define this parameter. Select all surfaces in the part folder, right-click, and select **Edit Definition** to open the **Shell definition** window. Use the **Thin** shell formulation and enter 8mm as the surface thickness. Shell elements will have thickness 8mm (Figure 15-2).

(1)
Select four
Surface Bodies
and right-click

(2)
Edit Definition

(3)
Thin shell

Enter shell
thickness 8mm

Full preview

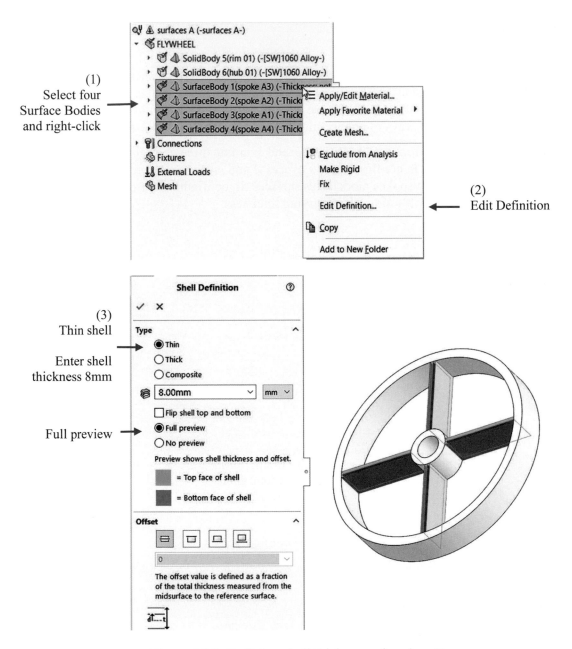

Figure 15-2: Defining shell thickness of spokes: 2mm.

Colors show the orientation of surfaces; the same orientation will apply to shell elements that will mesh these surfaces.

Mesh the model with Standard mesh 4mm and examine the incompatibility between shell and solid element mesh shown in Figure 15-3.

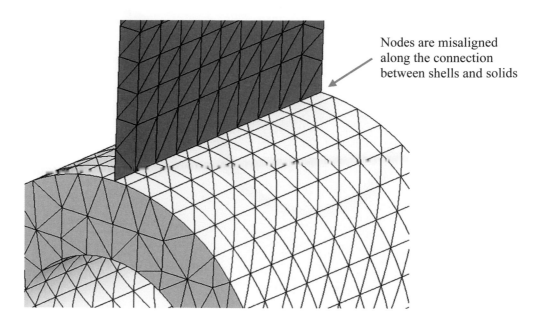

Nodes are misaligned along the connection between shells and solids

Figure 15-3: Shell elements and solid elements are not aligned along the line where spoke meets hub.

Mesh compatibility requires that both corner nodes and mid-side nodes are shared.

Nodes of solid and shell elements, even if they were coincident, could not be shared because the nodes of solid and shell elements have different number of degrees of freedom. Nodes could be coincident but that would create an unintentional hinge. In Simulation, there are different methods of treating the incompatibility between solid and shell elements.

When **Component Interaction** is set as **Bonded**, Simulation rigidly bonds each node of the shell elements to the nearest node of solid element. The stiffness of the connection depends on the element size near the interface. We will use the automatic bonding to complete analysis in *surfaces A* configuration.

Apply a **Fixed** restraint to the hole in the hub and run the solution. The first and the second mode of vibration are shown in Figure 15-4.

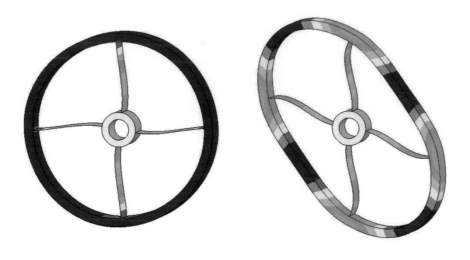

Mode 1: 196Hz   Mode 2: 632Hz

Figure 15-4: The first two modes of vibration in configuration *surfaces A.*

Both modes show in-plane deformation.

An alternative to using automatic bonding is to define **Local Interactions**. This method allows you to control which entities are bonded on both sides of a solid-shell connection.

Move to configuration *surfaces B*. Define Frequency study *surfaces B,* then define the thickness of spokes the same way as in the study *surfaces A* (Figure 15-2). Define eight **Local Interactions** as shown in Figure 15-5.

Bonded

(1)
Select Edge
of the surface
as the source

(2)
Select Face of
the solid
as the target

Face

Edge

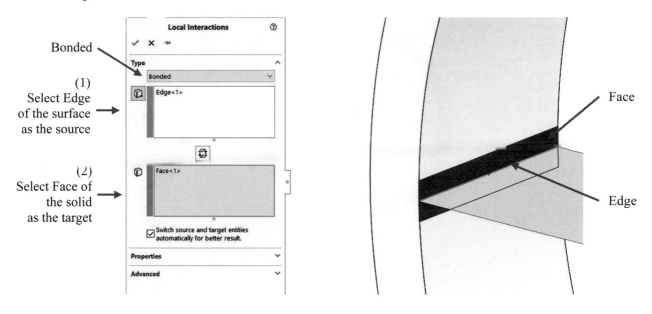

Figure 15-5: Definition of Local Interaction between the Edge of a surface and the Face of a solid.

Repeat this for all eight locations in the model.

Bonded Local Interaction shown in Figure 15-5 bonds nodes of shell elements along the edge to the nodes of solid elements on the face. The face has the width 8mm; this is the same as the thickness of shell elements. The stiffness of the bonded connection closely matches stiffness of spoke to rim connection if the spoke was modeled as 8mm thick solid.

Define the same support as in study *surfaces A,* mesh the model with elements 4mm in size and obtain the solution (Figure 15-6).

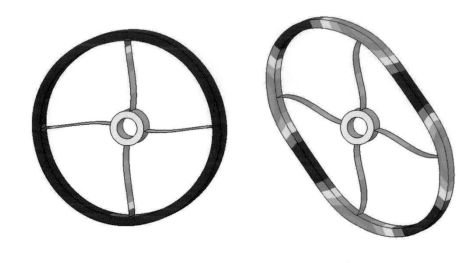

Mode 1: 194Hz   Mode 2: 627Hz

Figure 15-6: The first two modes of vibration in configuration *surfaces B.*

Modal shapes are identical as in study surfaces A. Frequencies are close but not the same as in study surfaces A.

Study *surfaces A* uses automatically created bonds between shell and solid elements with no user control over the extent of bonding. Study *surfaces B* uses **Bonded Local Interaction** where the extent of bonding is controlled by the width of faces on the rim and the hub. Study *surfaces A* produced slightly higher frequencies than study *surfaces B* demonstrating that the extent of automatic bonding was larger, and the resulting connections were stiffer in the study *surfaces A* than those in the study *surfaces B*.

The last exercise in this chapter uses model FLYWHEEL in *surfaces C* configuration shown in Figure 15-7. This design is different from the two previously analyzed configurations; we use it to introduce the **Shell Manager**.

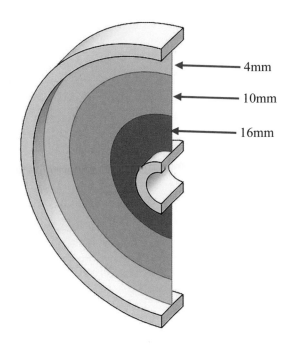

Figure 15-7: FLYWHEEL in *surfaces C* configuration shown here in a section view.

The rim and the hub are connected by a web with three sections of different thickness: 16mm, 10mm,4mm. See configuration solids for a better visualization of web thickness.

The three surfaces forming the web are modeled as three surface bodies. Connections between solids and surfaces will be done automatically, as in configuration *surfaces A*.

Create a **Frequency** study *surfaces C;* define thickness of three surfaces as shown in Figure 15-7.

Practice using **Shell Manager** to review and/or edit surface thickness, material and orientation as shown in Figure 15-8.

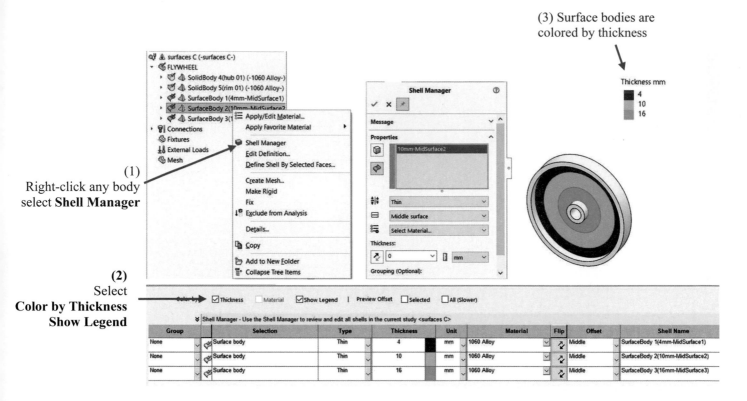

Figure 15-8: Using Shell Manager to review surface thickness.

Shell Manager can be used to define or edit thickness, material, and surface orientation.

Define restraints the same way as in the previous studies, mesh with 4mm mesh and obtain the solution. The first two modes of vibration are shown in Figure 15-9.

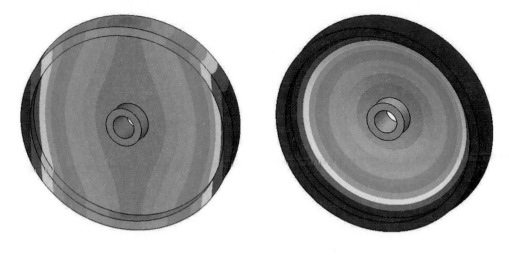

Mode 1: 561Hz   Mode 2: 691Hz

Figure 15-9: Deformation pattern corresponding to the first two modes of vibration in configuration *surfaces C.*

These modal shapes are difficult to visualize in a printed illustration. Review them on screen using animated plots.

Notice that what is shown in Figure 15-9 as mode 2, is listed in the **Results** as mode 3. Modes 1 and 2 are repetitive. The shape is rotated by 90° and the frequency is different only by a discretization error. This is the property of modes of an axisymmetric structure. Refer to "Vibration Analysis with SOLIDWORKS Simulation" for more information.

Models in this chapter

Model	Configuration	Study Name	Study Type
FLYWHEEL.sldprt	*surfaces A*	*surfaces A*	Frequency
	surfaces B	*surfaces B*	Frequency
	surfaces C	*surfaces C*	Frequency
	solids		

16: Analysis of weldments using beam and truss elements

Topics covered

- Different levels of idealization implemented in finite elements
- Preparation of a SOLIDWORKS model for analysis with beam elements
- Beam elements and truss elements
- Analysis of results using beam elements
- Limitations of analysis with beam elements

Project description

Open the ROPS model showing a Rollover Protective Structure (ROPS) used to protect an operator of heavy equipment in case of a roll-over. The model consists of eight hollow rectangular tubes 3" x 2" x 0.25" (Figure 16-1). All tubes are modeled as structural members.

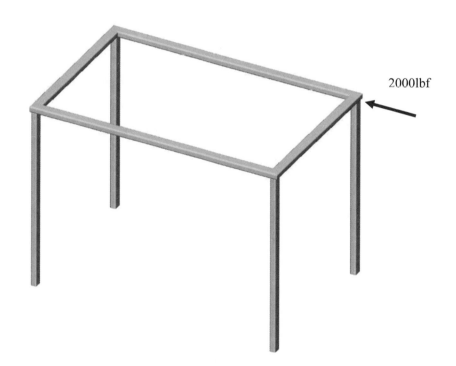

2000lbf

Figure 16-1: A ROPS cage is loaded in one corner with a horizontal load of 2000lbf.

All legs are restrained at the bottom faces; restraints symbols are not shown.

We need to find the displacements and stresses of this structure under a load of 2000lbf as shown in Figure 16-1 with all four legs restrained.

The tube cross section and details of corner treatment and trims are shown in Figure 16-2.

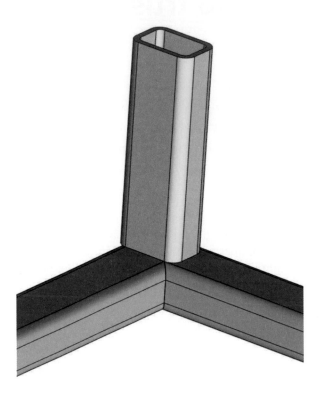

Figure 16-2: A detail of the corner; all tubes are 3″x2″x0.25″ with a 0.5″ radius.

Corner treatments and trims are applied in the SOLIDWORKS model using Weldment tools. The weld bead is not modeled. The yellow leg is shown in a section view to better fit this page.

Due to thin walls and complicated geometry in the corners, this model is not suitable for meshing with solid or shell elements. Even if we were ready to accept long meshing and solution times, the stress results in the corners would be useless because of numerous sharp re-entrant edges causing stress singularities.

To avoid these problems, the model can be meshed and analyzed with beam elements. Before we proceed with analysis, we need to explain what beam elements are and how they compare to solid and shell elements.

The differences between solid, shell and beam elements are summarized in the following table.

Element type	Idealization of geometry intended to be meshed with this element	Assumptions on stress distribution in the element
Solid	No idealization; solid elements are created by meshing 3D solid geometry.	No assumptions on stress distribution need to be made in any direction.
Shell	Surfaces must be created; shell elements are created by meshing surfaces. Thickness is not present in the geometry and must be entered as a numerical value in the shell element definition.	Assumptions on stress distribution across thickness are made. In-plane stresses are assumed to be distributed linearly across the thickness. Transverse shear stresses are either assumed to have uniform distribution across element thickness (thin shell formulation), or to have parabolic distribution (thick shell formulation).
Beam	Curves must be created; beam elements are created by meshing curves. Curves represent beam geometry mathematically and do not physically model the cross section. In CAD terminology this is often called wire frame geometry.	Assumptions about the stress distribution must be made in two directions perpendicular to the curve. These assumptions are the same as in beam theory: bending stresses are distributed linearly in both directions, and both axial and shear stresses are constant.

In summary, solid elements are a natural choice for meshing models with approximately the same size in all three dimensions, shells are a natural choice to mesh models such as sheet metal, and beams are a natural choice to mesh structural members.

Beam cross section geometry is only used to define beam element properties such as the area and second moments of inertia of the beam cross-section. It does not become part of the finite element model.

The information about beam cross sections is retrieved from the **SOLIDWORKS** model which must be created as a **Weldment**. It is important to understand that the solid geometry of a **Weldment** is not meshed when beam elements are used. What is meshed is the underlying wire frame geometry (Figure 16-3).

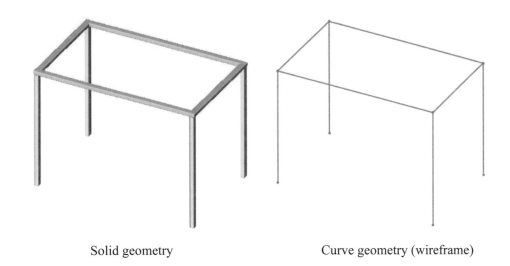

Solid geometry   Curve geometry (wireframe)

Figure 16-3: Solid model and the underlying wire frame geometry.

Solid geometry is used only to define beam cross sections. Beam elements are created by meshing curves. You can think of beam elements as lines with assigned beam cross section properties.

Corner treatments and trims have no relevance in beam element models. Before creating the study, change to configuration *02 no end treatment* where trims are suppressed (Figure 16-4).

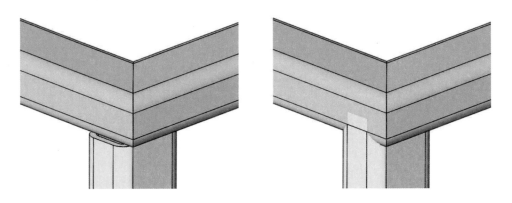

Trim applied
Configuration *01 complete weldment*

No trim
Configuration *02 no end treatment*

Figure 16-4: Corner treatment and trims have no relevance in beam element models. Both geometries will produce the same finite element model when meshed with beam elements.

End treatments will be suppressed for analysis. Miter corner treatments are retained in 02 no end treatment configuration because they do not interfere with creation of beam elements.

Procedure

Having examined the ROPS part, move to configuration *02 no end treatment* and create a **Static** study. **Simulation** recognizes the weldment geometry and anticipates that we intend to use beam elements. It creates a **Solid Body** for each structural member present in the geometry and places them in folders corresponding to **Cut-Lists** present in the **SOLIDWORKS** model (Figure 16-5).

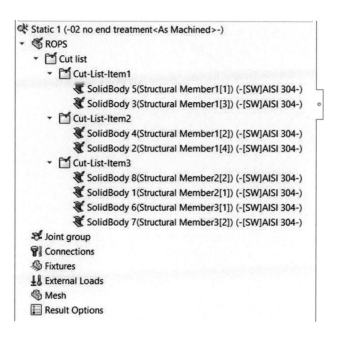

Figure 16-5: Study definition with beam elements; numbering of Solid Bodies may differ.

By default, structural members are meshed with beam elements. You may change it by right-clicking and selecting Treat as Solid. The material is imported from the SOLIDWORKS model.

In Figure 16-5, notice the lack of symbol indicating element order. The order of beam element cannot be changed between 1st and 2nd. Transverse translations in beam elements are described always by 3rd order polynomials.

Recall Figure 4-2 which shows icons denoting geometry intended for **Solid element** and **Shell element** meshing. We may now append two more: geometry intended for **Beam element** meshing and for **Truss element** meshing (Figure 16-6).

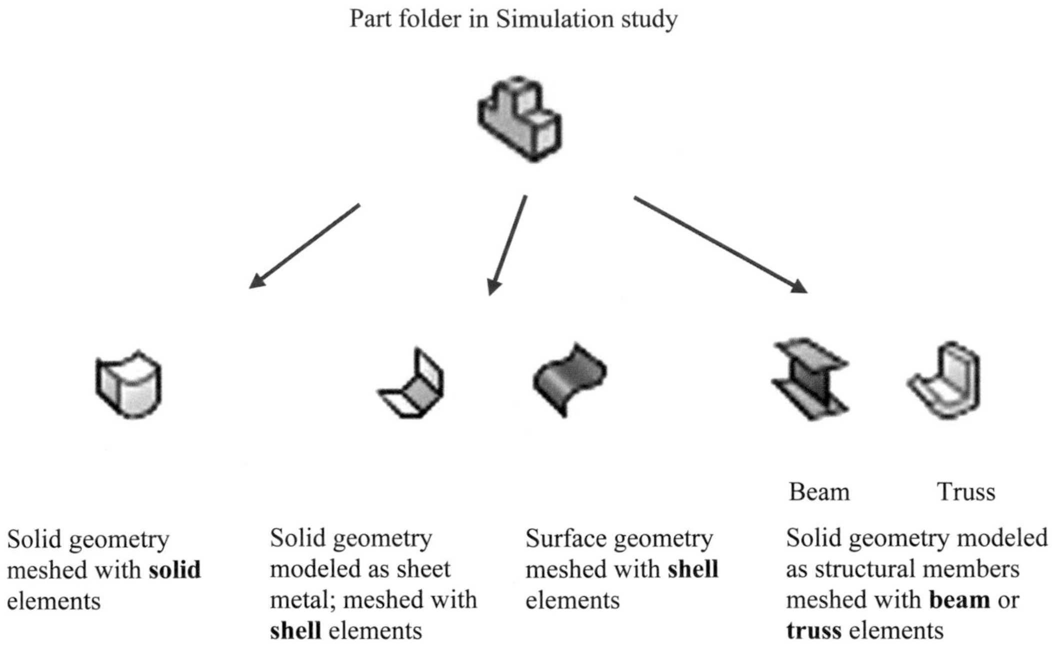

Part folder in Simulation study

			Beam	Truss
Solid geometry meshed with **solid** elements	Solid geometry modeled as sheet metal; meshed with **shell** elements	Surface geometry meshed with **shell** elements	Solid geometry modeled as structural members meshed with **beam** or **truss** elements	

Figure 16-6: A part folder in Simulation study may contain all three types of geometries.

Differences between beam and truss will be explained later in this chapter.

In many cases the automatic designation of geometry to one of three geometry types can be changed. **Beams** may be replaced by **Solids** and **Surfaces,** and sheet metal parts may be replaced by **Solids**. This is accomplished by right-clicking the geometry folder component and selecting the appropriate choice from the pop-up menu.

In this exercise we accept the default assignment of all parts to **Beam** elements. As Figure 16-5 indicates, the geometry folder holds eight bodies and they are all intended for meshing with beam elements.

The next step is the definition of connectivity between the soon to be created beam elements.

Right-click the *Joint Group* folder and select **Edit** from the pop-up menu to open the **Edit Joints** window which shows automatically created joints. If necessary, these automatically created joints may be edited in this window. Accept the default selection **All** beams and click **Calculate** to create joints automatically (Figure 16-7).

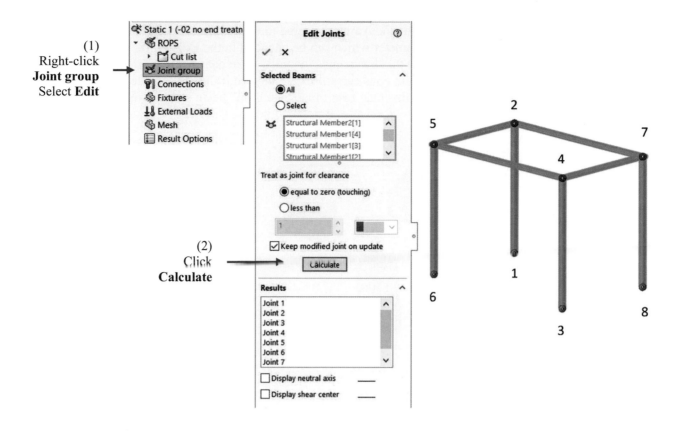

Figure 16-7: Eight joints are automatically created between beams. Numbers indicate joint position.

No further action is required in this case. Working with a simplified geometry in configuration 02 no end treatments, facilitated automatic joint creation. Joint numbering in your model may be different from the one shown in this illustration.

Joints (or beam ends) are connected to each other only if they are contained within the pinball diameter which can be changed in the **Edit joints** window. It is recommended that beam ends intended to be connected are made coincident. Non coincident beam ends can still be connected if they fall inside the pinball diameter, but this may result in a "patch-up" beam element mesh (Figure 16-8).

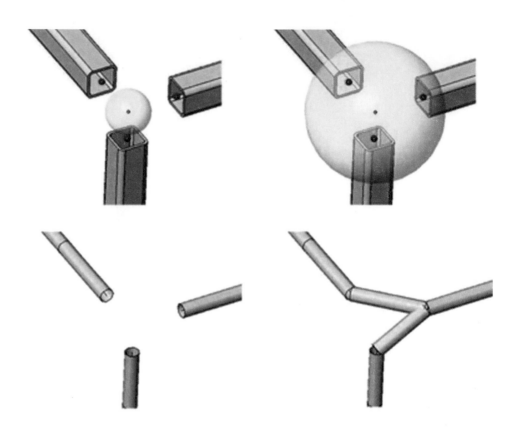

Imaginary pin ball does not contain joints.
Beam element mesh is disconnected.

Imaginary pin ball contains joints.
Beam element mesh is connected.

Figure 16-8: Joints (beam ends) are connected only if they are contained within the imaginary pinball not visible in the model display. Beam elements are shown as thin tubes.

Beam elements are graphically depicted as round tubes even though the beam elements are in fact just lines. The tube diameter is always the same, regardless of the actual cross-section size, shape, and orientation.

This is for illustration only and is not related to ROPS exercise.

Disjoined structural members can be connected if their ends fall within the volume of the pinball (Figure 16-8 top right). However, the beam element mesh (Figure 16-8 bottom right) will then contain automatically created connecting elements. This may create unpredictable results.

Apply fixed restraints to all four joints at the free ends (at the bottom) of the vertical members (Figure 16-9).

Fixed Geometry →

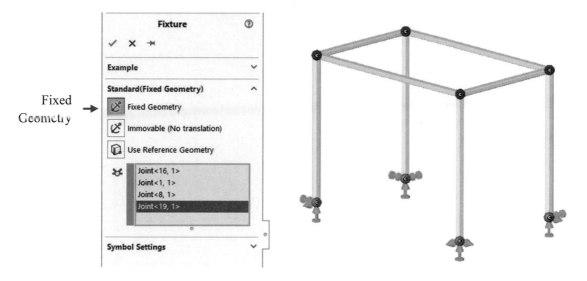

Figure 16-9: Fixed restraints applied to the free ends of vertical members.

These restraints eliminate six DOFs: three translations and three rotations. Joint numbering in your model may be different from the one shown in this illustration.

The size of Restraint symbol has been modified in this illustration.

Notice that beam elements have six degrees of freedom per node and therefore can distinguish between **Fixed** and **Immovable** restraints. Here, we need to use a **Fixed** restraint.

Apply a 2000lbf load to the corner as shown in Figure 16-10.

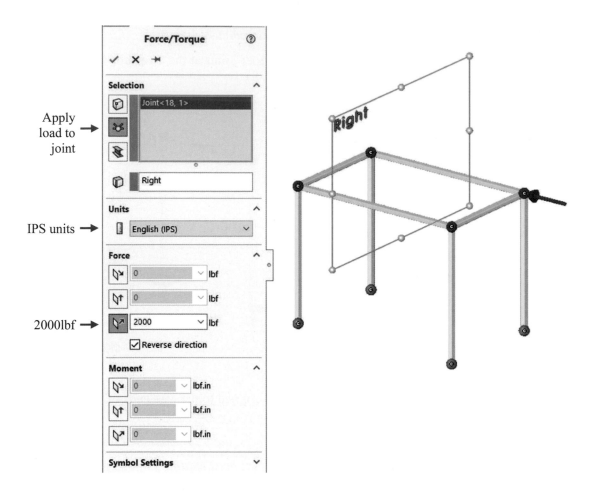

Figure 16-10: Force load applied to the corner joint.

The size and color of the Force symbol has been modified in this illustration.

Notice that beam elements have six degrees of freedom per node and therefore can be loaded with a force load as well as with a moment load. Here we apply force load only.

Now create the beam element mesh using default settings. The beam element mesh is shown in Figure 16-11.

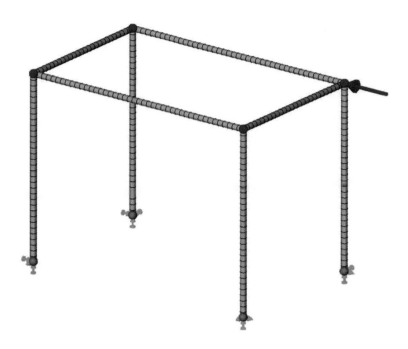

Figure 16-11: A beam element mesh is created from curves (here straight lines) used in the SOLIDWORKS model to define Structural Members.

A beam element is a line with cross-section properties taken from the Structural Member cross section geometry. This is schematically illustrated in Figure 1-8. A beam element mesh is schematically shown as round tubes. The tube diameter shown does not depend on the actual size of the beam cross section.

Load and restraints symbols are shown.

Run the solution and create a displacement plot (Figure 16-12) as well as a stress plot (Figure 16-13).

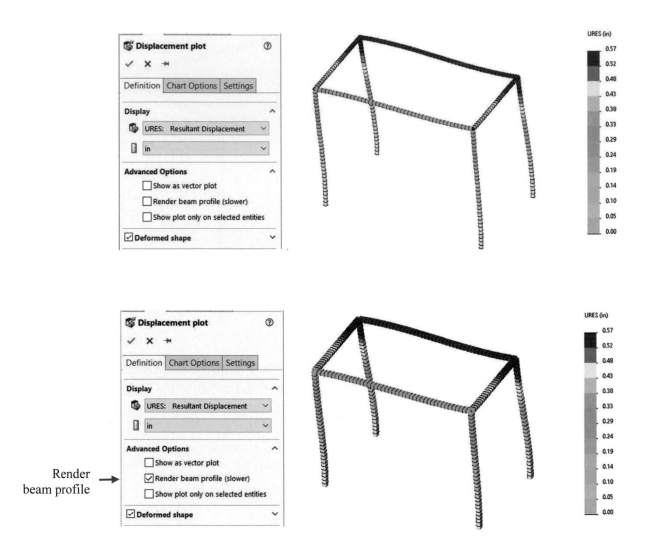

Figure 16-12: Resultant displacement results of ROPS model.

The maximum displacement is 0.57". The beam profile may be rendered by selecting the Render beam profile in plot Advanced Options. This applies also to stress results.

Figure 16-13: A stress plot of the "Upper bound Axial and bending" stresses.

The maximum stress is 24ksi; this is below the material yield strength of 30.2ksi.

Notice that the **Axial and bending** stress is NOT the von Mises stress. To understand what it is, we need to review stress result options available for beam elements.

The software provides the following options for viewing beam stresses and beam diagrams (refer to Figure 16-14):

- Axial: Uniform axial stress = P/A

- Bending about local direction 1: Bending stresses due to moment M1 about axis DIR1

- Bending about local direction 2: Bending stresses due to moment M2 about axis DIR2

- Upper bound axial and bending

- Torsional

- Shear stress in direction 1 (DIR1)

- Shear stress in direction 2 (DIR2)

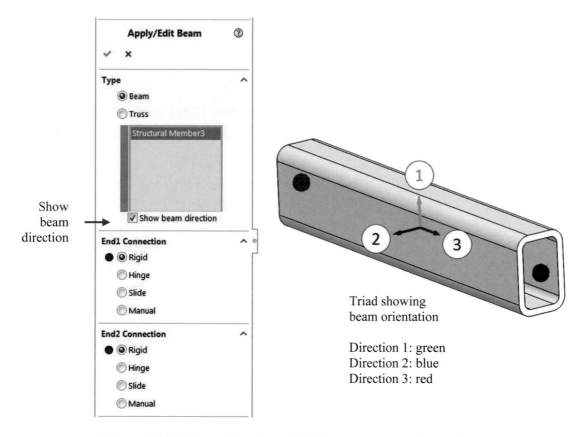

Figure 16-14: Beam directions DIR 1, DIR 2, DIR 3 for the beam cross section used in this exercise.

The origin of triad is in the centroid of the cross-section. DIR 3 is along the length of beam from End 1 to End 2; DIR 1 is parallel to the longer edge of cross section; DIR 2 completes the right-handed triad. Oversized numbers are used to label the triad to improve readability of this illustration.

Use TUBE model to re-create Figure 16-14.

The beam orientation, presented in Figure 16-14, can be shown by selecting **Show Beam direction** in the **Apply/Edit Beam** window (Figure 16-14, 16-15).

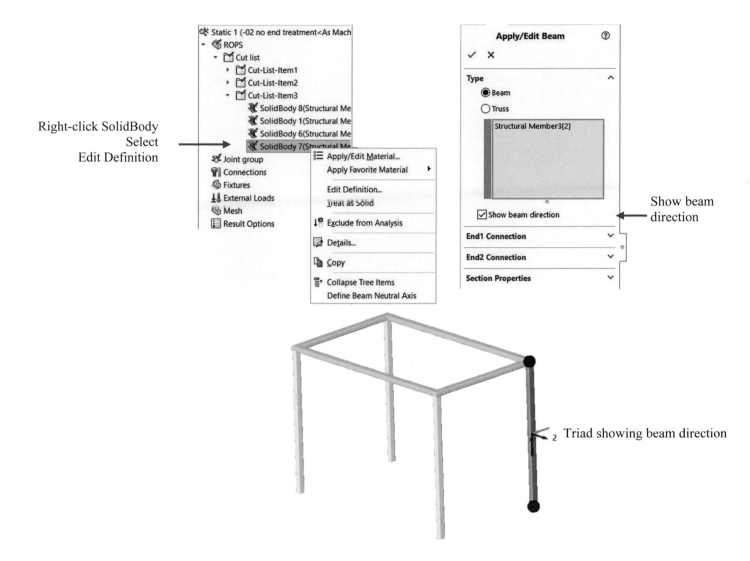

Right-click SolidBody
Select
Edit Definition

Show beam
direction

Triad showing beam direction

Figure 16-15: Apply/Edit Beam window offers ample tools to review beam orientation and Section Properties.

Right-click the selected structural member and select Edit Definition from the pop-up menu to open the Apply/Edit Beam window. Select Show beam direction.

For easier selection, the Cut List may be deleted from the geometry folder in a Simulation study. Beam ends are color coded to show End1 and End2. A triad shows the beam orientation. Connectivity and section properties can be modified.

Stress plots are different depending on the selection made in display settings. If **Render beam profile** is selected, then the stress distribution on the beam profile is shown. If **Render beam profile** is not selected, then only the upper bounds of bending stress components can be shown (Figure 16-16).

Render beam profile is selected.
Plot of Axial and bending stress is available

Render beam profile is not selected.
Plot of Upper bound axial and bending stress is available

Figure 16-16: Stress results are different depending on how the beam profile is shown.

If a beam profile is not shown (right,) then only the upper bounds of axial and bending stress components can be presented.

The **Axial and bending** stress is calculated by combining axial stress and bending stresses due to moments M1 and M2. This is the default selection in the **Stress Plot** window.

Simulation offers ample ways of analyzing beam element results such as a **Beam Diagrams** (Figure 16-17).

To open **Beam Diagrams** window right-click *Results* folder and select **Define Beam Diagrams** from the pop-up menu.

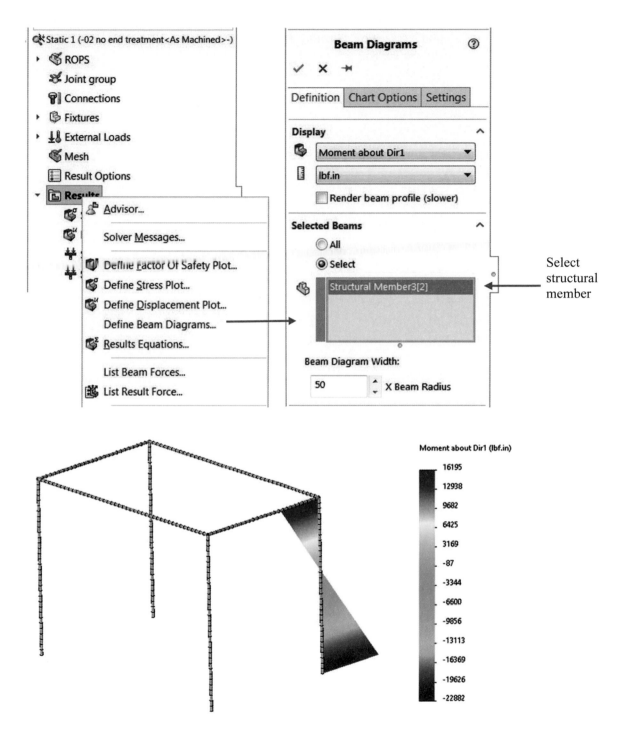

Figure 16-17: Bending moments in DIR1; bending takes place about the DIR1 axis shown in Figure 16-14.

Notice that the beam directions are defined relative to the beam, not to the global coordinate system.

Repeat this exercise with Immovable restraints and observe a very different Beam Diagram plot; there will be no moment at the supports. Moment plot is placed in the plane of bending moment vector.

Using the pop-up menu shown in Figure 16-17 you may also review beam forces and stresses by selecting **List Beam Forces** to display the window shown in Figure 16-18.

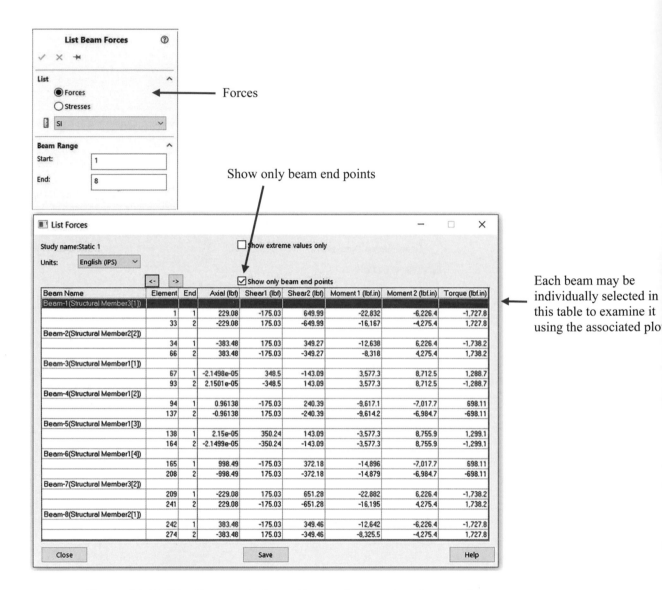

Figure 16-18: List Forces window offers tools to review force and stress components in the beams.

When Show only beam end points is selected, numbers are color coded (red and blue) and match the colors of the ends of beam elements that are displayed along with this window.

We will now analyze the OUTRIGGER assembly shown in Figure 16-19.

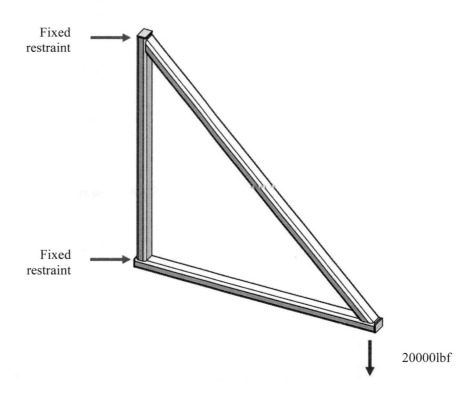

Figure 16-19: The truss is made out of rectangular hollow tubes size 4 x3x0.25"
with 0.5" radius. Locations of restraints and load are also shown.

Model is shown in configuration 01 mfg that includes endcaps shown in yellow.
Hardware required to apply load and restraints is not modeled.

Endcaps would interfere with meshing using beam elements mesh because they
would be meshed with solid elements. We will use configuration *02 analysis*
where endcaps are suppressed.

Switch to *02 analysis* configuration, create Static study *01 beams* and delete *Cut
List* folders from the parts folder. This is done only for convenience of clear
illustrations.

Next, right-click the *Joint Group* folder, select **Edit**, and verify that three joints
have been correctly calculated (Figure 16-20).

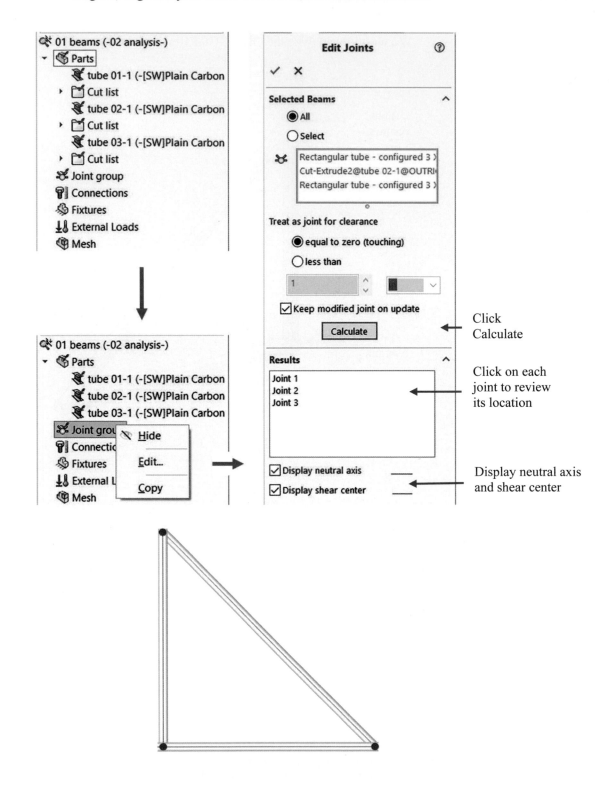

Figure 16-20: Three joints were calculated; this has been done for practice only because joints, in this model, are correctly calculated automatically. Neutral axes and shear centers are indicated by thin black and red lines.

The model view has edges removed to show the neutral axis and shear center lines. The neutral axis and shear center line are coincident in this model.

Apply restraints and load to the joints as shown in Figure 16-21.

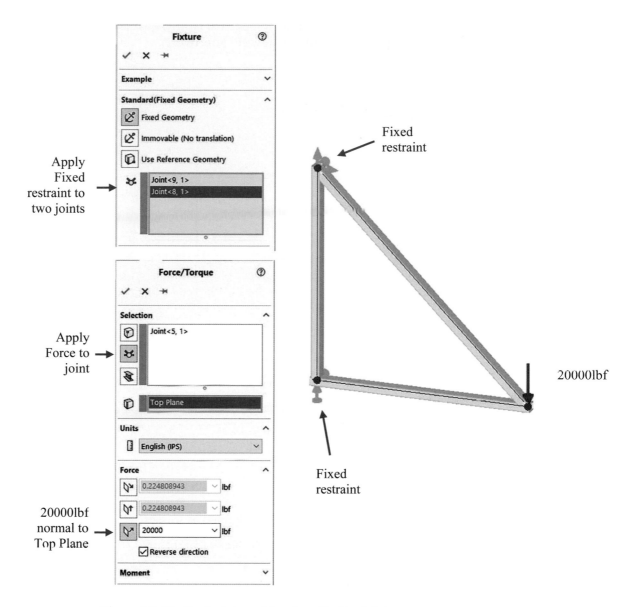

Figure 16-21: Restraints and load applied to joints.

Reverse direction to make the load point down.

Mesh and solve the model then copy the completed study *01 beams* into *02 trusses* and delete again all cut lists.

Select all structural members in the parts folder and right-click to open a pop-up menu. Select **Edit definition** to open the **Apply/Edit beam** window. Select **Truss** in the **Apply/Edit beam** window (Figure 16-22).

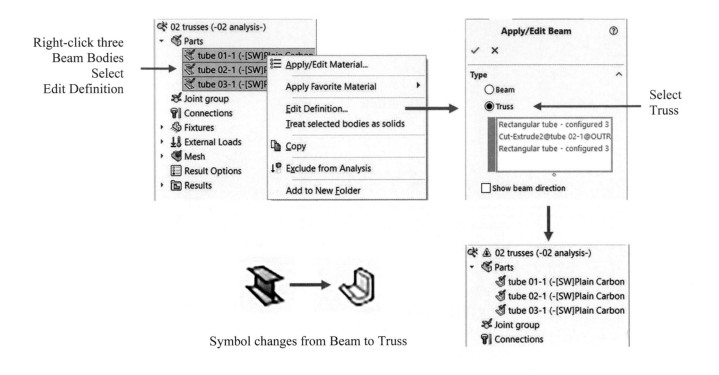

Right-click three
Beam Bodies
Select
Edit Definition

Select
Truss

Symbol changes from Beam to Truss

Figure 16-22: All beams are now defined as trusses.

Notice the change of symbol from Beam to Truss.

This redefines connectivity between beams from rigid to pin joints. While beams can be loaded with any combination of forces and moments, trusses can be only loaded with axial force. Trusses behave as tension/compression springs and are meshed with only one element. See Figure 16-23 for a comparison between beam and truss element meshes.

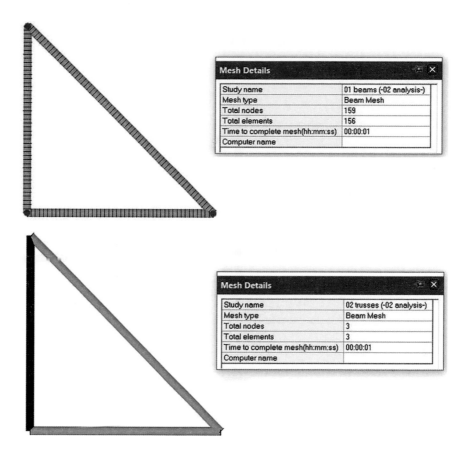

Mesh Details	
Study name	01 beams (-02 analysis-)
Mesh type	Beam Mesh
Total nodes	159
Total elements	156
Time to complete mesh(hh:mm:ss)	00:00:01
Computer name	

Mesh Details	
Study name	02 trusses (-02 analysis-)
Mesh type	Beam Mesh
Total nodes	3
Total elements	3
Time to complete mesh(hh:mm:ss)	00:00:01
Computer name	

Figure 16-23: Finite element mesh in model with beam elements (top) and with truss elements (bottom).

Verify the number of nodes and elements with Mesh Details.

Truss elements bchave as if they were pin jointed and can be loaded only with axial loads.

Obtain the solution of study with truss elements and compare the displacement results between the two studies (Figure 16-24).

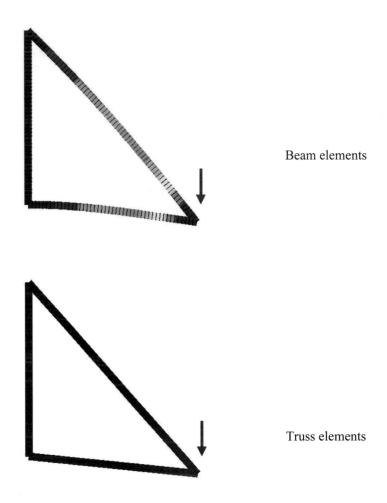

Beam elements

Truss elements

Figure 16-24: The displacement results in the model with beam elements (top) and truss elements (bottom).

Beam profiles are not rendered in these illustrations.

Continue OUTRIGGER exercise by analysis of stress results and observe that truss elements may only produce tensile or compressive stresses in the axial direction (Figure 16-25).

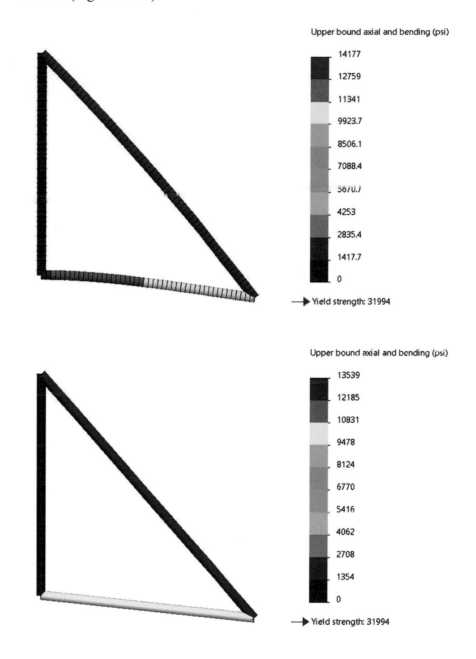

Figure 16-25: Stresses in beam element model (top) and truss element model (bottom).

Beam profiles are not rendered in these illustrations.

Bending of structural members observed in the beam element model proves that beam elements are rigidly connected to each other and transmit bending moments. Conversely, truss elements are connected by pin joints; they cannot transmit bending. Deformation of truss elements can only take the form of stretching and compressing, and therefore deformed truss elements remain straight. Remember that buckling is not modeled in either of these static studies.

An attempt to run the truss element model with an out-of-plane load (load with component in z direction) produces an error. Copy study *02 trusses* into *03 trusses* and add an out-of-plane load of any magnitude, then try solving the study. A warning message will be displayed as shown in Figure 16-26.

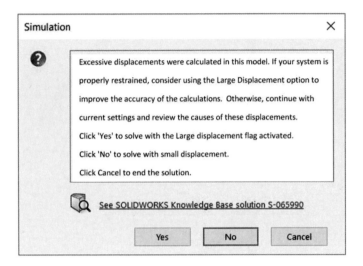

Figure 16-26: An attempt to run the truss element model with out of plane load brings up the above message.

"Excessive displacement" message is produced even if the out-of-plane load component is very small. The window has been modified to show the complete message in one window without scrolling.

An out-of-plane load causes rotation of the OUTRIGGER as a rigid body. The entire model can spin about the line passing through the supports. This happens because nodes of truss elements do not have rotational degrees of freedom and cannot generate moment reactions.

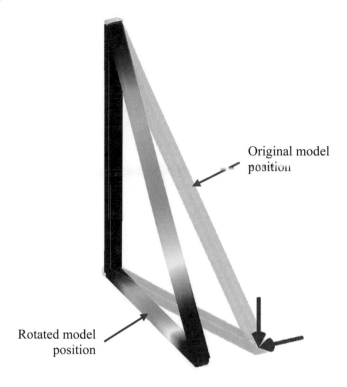

Figure 16-27: An out of plane load rotates model about the axis connecting the support points. Restraints cannot prevent this motion if truss elements are used.

A very small scale of deformation must be used to make this illustration legible.

Notice that in both studies (beam elements and truss elements) the vertical member takes no load at all. Confirm this by excluding the vertical member from analysis.

Models in this chapter

Model	Configuration	Study Name	Study Type
ROPS.sldprt	*01 complete weldment*		
	02 no end treatment	*Static 1*	Static
TUBE.sldprt	*Default*		
OUTRIGGER.sldasm	*01 mfg*		
	02 analysis	*01 beams*	Static
		02 trusses	Static
		03 trusses	Static

17: Review of 2D problems

Topics covered

- Classification of finite element types
- 2D axisymmetric element
- 2D plane stress element
- 2D plane strain element

Recall Figure 1-10 that lists basic elements used in SOLIDWORKS Simulation: solids, shells and beams which represent different levels of idealization possible in 3D space. Solids were introduced in chapter 2, shells in chapter 4 and beams in chapter 16. In this chapter we introduce two dimensional (2D) elements. Before proceeding we will summarize once again properties of the elements we are already familiar with.

Solid elements

Solid elements mesh volumes and therefore all three dimensions are fully represented. Notice that the volume is called "solid" in CAD terminology. This "solid" should not be confused with the term "solid" in FEA which corresponds to solid elements.

The displacement field in a solid element is three dimensional (3D). No assumptions on the displacement field or stress distribution in any direction are made.

Nodal displacements have three components: translation in the x, y, and z directions. Loads and restraints can be applied in these three directions. Nodes of solid elements have three degrees of freedom.

Shell elements

Shell elements mesh surfaces that have one dimension (thickness) collapsed. Thickness is assumed to be small in comparison to other dimensions. Shell elements model displacement and stress fields in two in-plane directions. The stress distribution across the missing dimension (thickness) is assumed to have a linear distribution.

Nodal displacements have six components: translation in the x, y, and z directions, and rotation about the x, y, and z directions. Loads and restraints can be applied in each of these six directions. Nodes of shell elements have six degrees of freedom.

Beam elements

Beam elements mesh curves (wireframe) as they have two dimensions collapsed. It is assumed that the beam cross section is small in comparison with the length. Beam elements model displacements and stresses in one (in-line) direction. Stress in the two missing dimensions must be assumed. These assumptions are based on beam theory.

Nodal displacements have six components: translation in the x, y, and z directions, and rotation about the x, y, and z directions. Nodes of beam elements have six degrees of freedom. Loads and restraints can be applied in these six directions.

Solid, shell and beam elements all belong to the class of 3D elements. In this chapter we introduce **two dimensional elements**. Consider an axisymmetric model VASE with axisymmetric loads (pressure) and axisymmetric restraints (fixed to the bottom face) as shown in Figure 17-1. Notice that due to the axial symmetry of the geometry, loads and restraints, all points located on any radial cross section perform displacements only in two directions: radial and axial translation. There is no translation in the circumferential direction and there are no rotations. Displacements in the model are fully described by radial and axial translations of nodes in radial cross section.

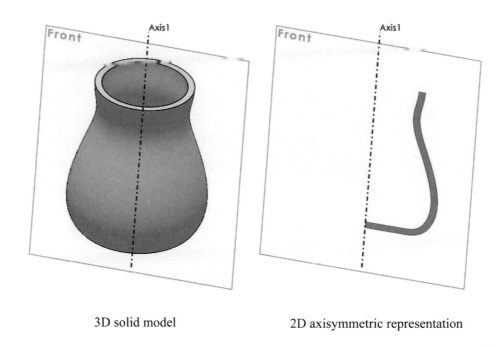

3D solid model   2D axisymmetric representation

Figure 17-1: Axisymmetric 3D solid model and 2D axisymmetric representation.

In 2D axisymmetric representation the model is reduced to a flat radial slice.

An axisymmetric model with axisymmetric loads and restraints may be simplified to a radial cross-section meshed with elements that have only two degrees of freedom per node. These elements are called 2D axisymmetric elements.

We will use 2D axisymmetric simplification to analyze model COVER. Three model configurations will be used to study the importance of nonlinear effects depending on the shape of the model. Review all configurations before proceeding (Figure 17-2). The material is 6/10 Nylon, the same in all configurations.

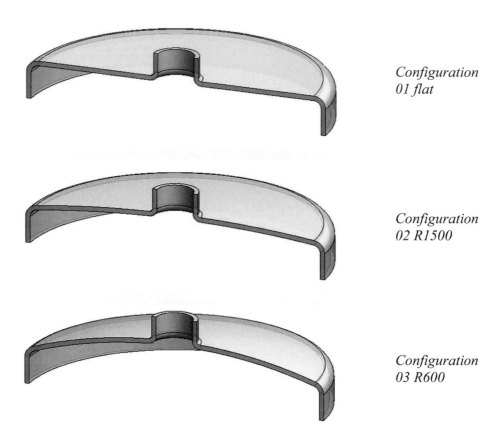

*Configuration
01 flat*

*Configuration
02 R1500*

*Configuration
03 R600*

Figure 17-2: COVER in three configurations.

All configurations are shown in section views.

In configuration *01 flat* the COVER is flat.

In configuration *02 R1500* the COVER is spherical with radius 1500mm.

In configuration *03 R600* the COVER is spherical with radius 600mm.

Move to configuration *01 flat* and create a **Nonlinear** study *01 flat* using 2D Axisymmetric simplification (Figure 17-3).

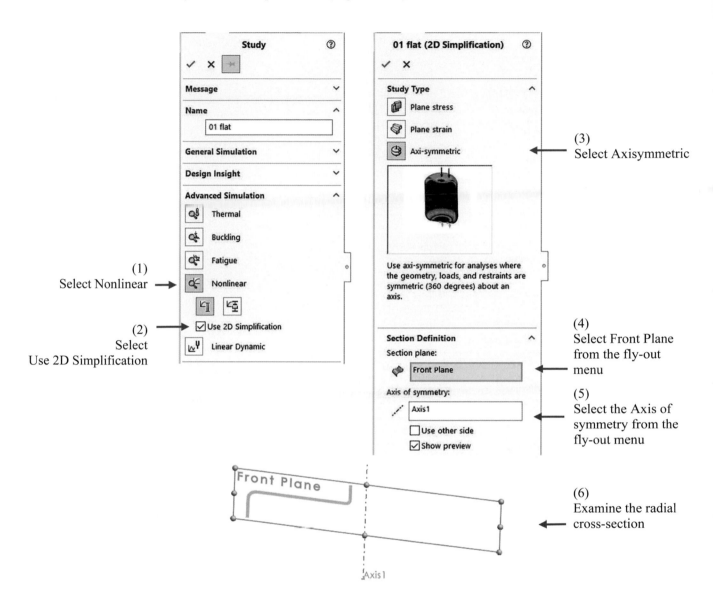

Figure 17-3: A nonlinear study definition with 2D axisymmetric elements.

The model is represented by a radial cross-section. The fly-out menu is used for the selection of the plane and the axis. The fly-out menu is not shown in this illustration.

Apply the restraint and pressure load as shown in Figure 17-4.

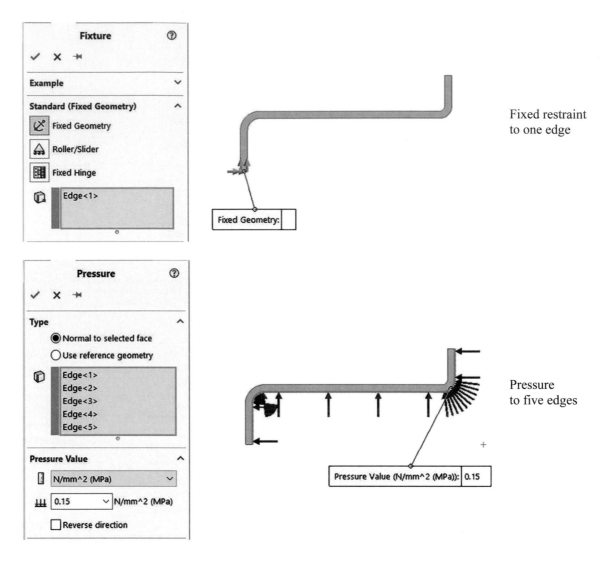

Figure 17-4: The restraint (top) and the pressure load (bottom) defined in the 2D axisymmetric model.

Pressure and restraint are applied to edges but represent pressure and restraints applied to the full model.

Use **Nonlinear** study setting as shown in Figure 17-5 and mesh the model using Standard mesh with default element size.

Use large displacement formulation

Update load direction with deflection

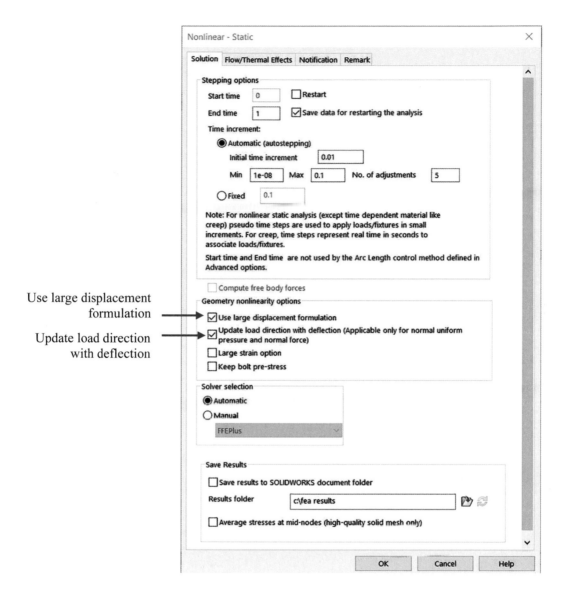

Figure 17-5: Nonlinear study settings.

Select Use large displacement formulation and Update load direction with deflection.

Selecting option Update load direction with deflection means that pressure load follows the deforming model; it remains normal to the deforming face. The only source of nonlinear behavior in this model is the changing shape. This is geometric nonlinearity.

Start the solution and acknowledge the message shown in Figure 17-6.

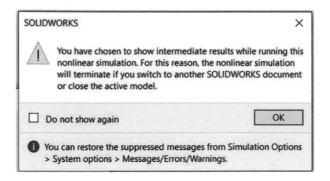

Figure 17-6: Nonlinear solver message.

Show intermediate results option is selected in Simulation Default Options, Solver and Results.

Review displacement and stress results as shown in Figure 17-7.

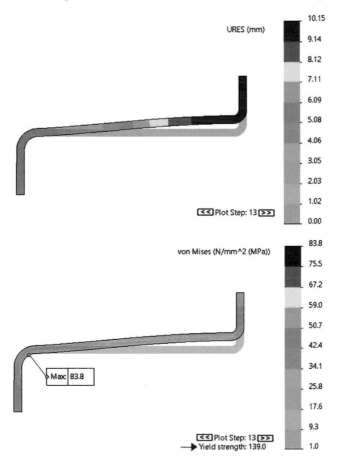

Figure 17-7: Displacement and von Mises stress results in configuration *01 flat for the last step (step 13).*

Both plots are shown in 1:1 scale of deformation. The undeformed shape (magenta color) is superimposed on deformed plots.

Probe displacement plot anywhere along the top edge and display **Response Graph** (Figure 17-8). Save graph in csv format and export it to Excel.

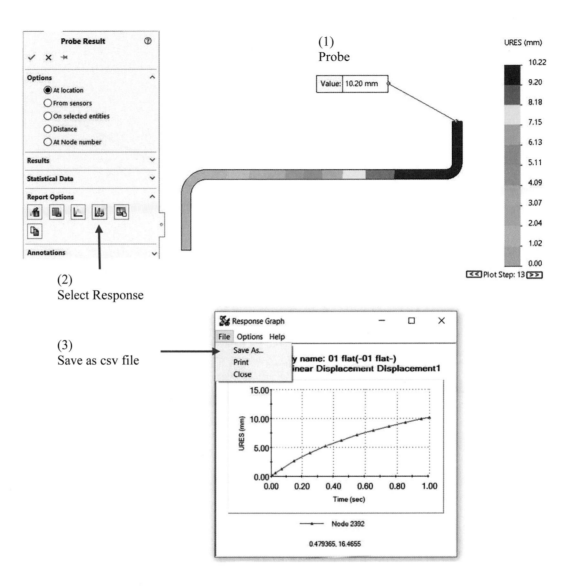

Figure 17-8: Construction of Response Graph and saving the graph for export to Excel.

You may probe displacement plot at any step; it does not have to be the last performed step.

Repeat the analysis in configurations *02 R1500* and *03 R600*. To summarize results, collect results of all three studies in Excel and construct graph like the one shown in Figure 17-9.

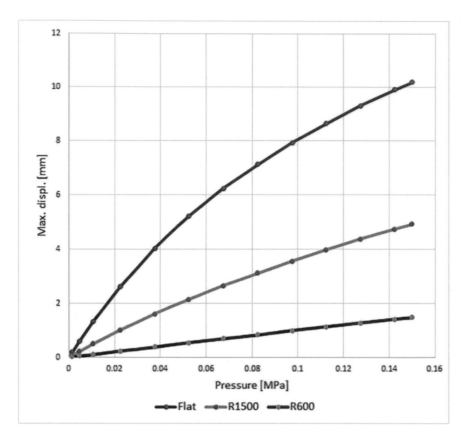

Figure 17-9: The maximum resultant displacement as a function of pressure in three configurations.

The blue curve is a repetition of the graph shown in Figure 17-8.

As can be seen in Figure 17-9, the nonlinear effects are strong in *01 flat* configuration. Configuration *02 R1500* displays a mild nonlinearity and configuration *03 R600* displays negligible nonlinearity. This is because the wall curvature allows the membrane stresses to develop from the beginning of the loading process. Therefore, the change in stiffness is less dependent on the progressing deformation.

As shown in the menu in Figure 17-3, there are other types of 2D elements. Here is the summary:

2D plane stress elements are intended for thin models restrained and loaded in plane. A constant stress across the thickness is assumed.

2D plane strain elements are intended for thick models restrained and loaded in plane. Constant strain across the thickness is assumed.

2D axisymmetric elements are intended for axially symmetric models with axisymmetric restraints and loads.

We will now introduce 2D plane stress elements using PERFORATED PLATE model. Follow the steps in Figure 17-10 to create a **Static** study *01 2D* with a 2D Plane Stress simplification, thickness 1mm.

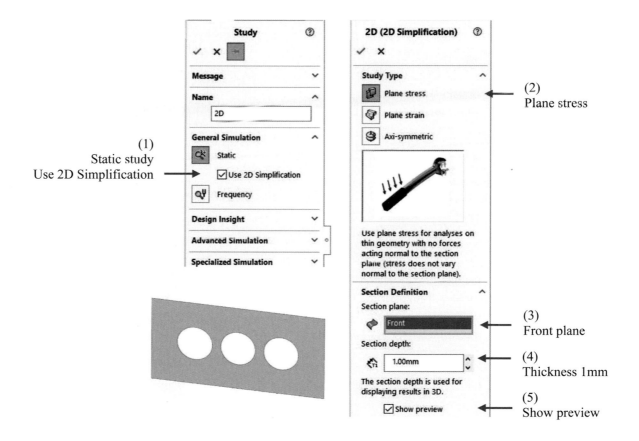

Figure 17-10: Static study definition with 2D plane stress elements.

The model is represented by a planar surface.

Apply restraints and loads as shown in Figure 17-11.

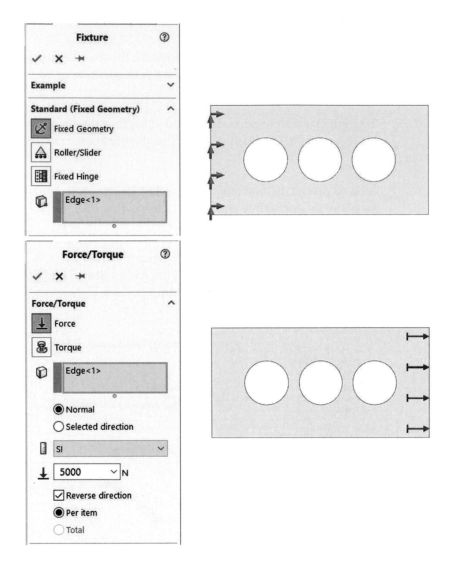

Figure 17-11: Restraint (top) and load (bottom) defined in the 2D plane stress model.

The restraint and load are applied to the edges of the 2D model that correspond to faces of the 3D model.

Create 2D mesh with 1mm element size and run *2D* study.

Now, repeat the analysis without 2D simplification. Create **Static** study *02 3D*; use element size 1mm to produce solid elements with a low aspect ratio. Compare results of *01 2D* study and *02 3D* study (Figure 17-12).

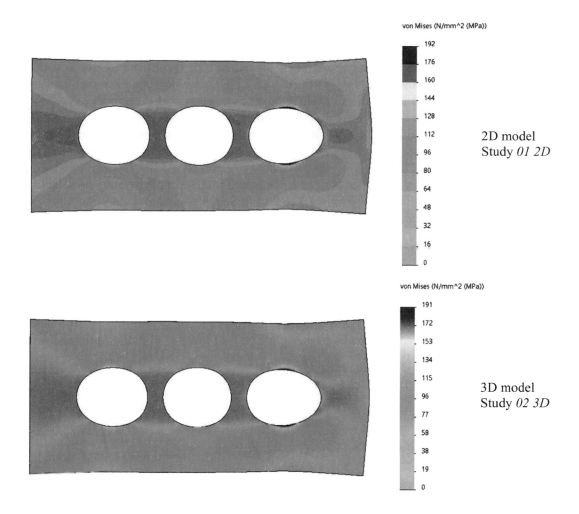

Figure 17-12: Von Mises stress plots in the 2D plane stress model (top) and the 3D solid model (bottom) show almost the same stress results.

Deformed plots are shown. Both plots are shown in Front view. Rotate the model in study 02 3D to see the model thickness.

Von Mises stress results from four studies are summarized in Figure 17-15. Results indicate that convergence does take place.

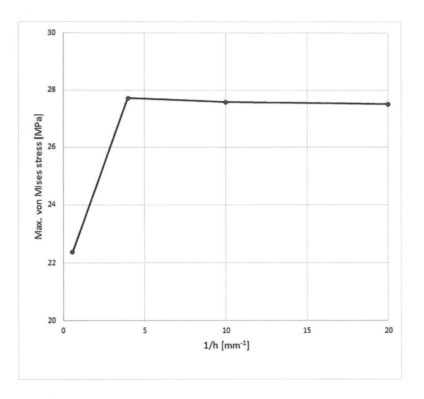

Figure 17-15: Maximum von Mises stress as a function of mesh refinement 1/h in four studies.

h is the element size defined by the mesh control except for the first study 01 2D where it is the default element size.

To fully appreciate the advantages of the 2D Simplification you may want to repeat this analysis using solid elements and observe long solution times.

It is important to remember that the 2D model would not be suitable for analysis of buckling which may be the dominant mode of failure in such a thin plate. Analyze buckling using a 3D model.

To introduce 2D plane strain elements, we will perform an analysis of contact stress between two plates in model CONTACT. The model consists of two identical plates ready to touch each other on the curved faces and connected by a U-shape clamp. The material for all parts is Alloy Steel. Our objective is to study contact stress when the model is loaded with 15000N compressive load. There is a small gap between top and bottom cylindrical faces in the unloaded model. These faces will come in contact under the load (Figure 17-16).

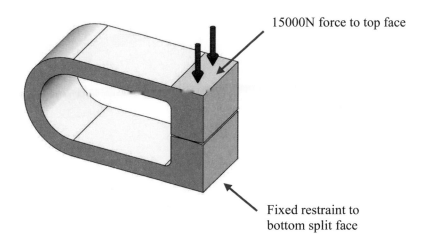

15000N force to top face

Fixed restraint to
bottom split face

Figure 17-16: Load and restraint in CONTACT model.

Load and restraint are applied to split faces. The loaded face is shown in blue; the bottom face is not visible.

Treating this as a 2D plane strain problem we assume constant strain along the thickness; therefore, all models shown in Figure 17-17 will give the same results when treated as 2D plane strain problem. CONTACT model has three configurations with different thickness as shown in Figure 17-17.

7500N   15000N   30000N

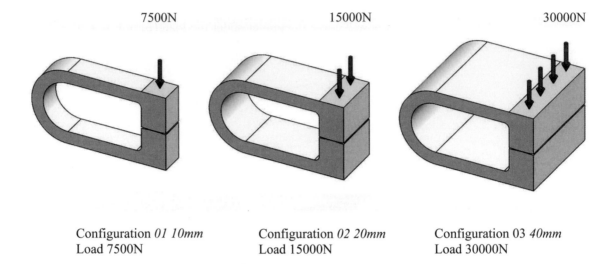

Configuration *01 10mm*   Configuration *02 20mm*   Configuration 03 *40mm*
Load 7500N   Load 15000N   Load 30000N

<u>Figure 17-17: All the above three models will give the same results when treated as 2D plane strain problems.</u>

Load per unit of width is the same in the above three models: 750N/mm.

Any model configuration could be used if load is defined as 15000N and thickness is defined as 20mm in study properties (Figure 17-18).

Follow steps in Figure 17-18 to create a study with 2D plane strain model. Follow the steps in Figure 17-19 to apply restraint and load.

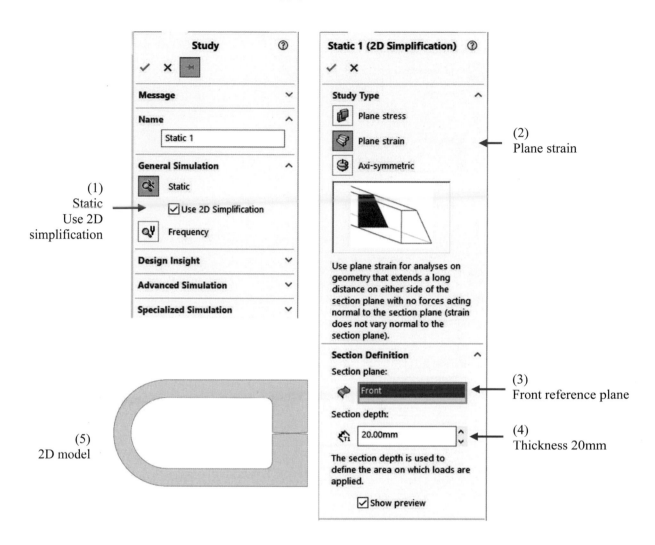

Figure 17-18: Static study definition with 2D plane strain elements.

The model is represented by a planar surface.

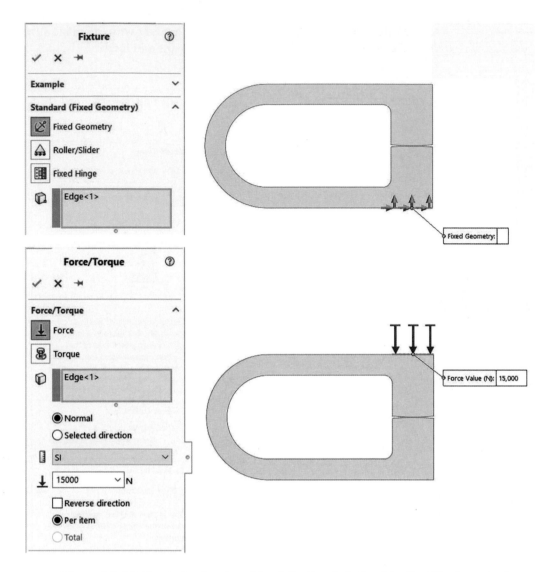

Figure 17-19: Restraint (top) and load (bottom) defined in the 2D plane strain model.

The restraint and load are applied to the edges of the 2D model that correspond to the faces of the 3D model. 15000N means here 15000N per each 20mm of thickness.

Define a **Contact Set** between the two cylindrical faces as shown in Figure 17-20.

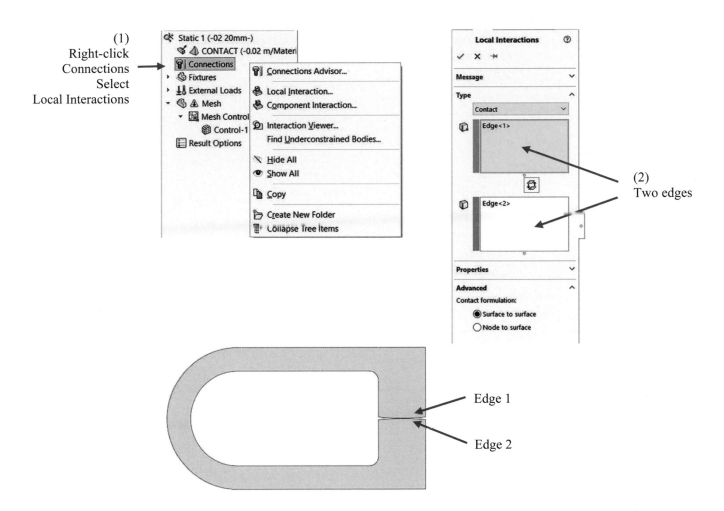

Figure 17-20: Contact Set definition.

A surface to surface type contact must be selected always when faces are not touching initially but may come into contact after the load is applied.

We now define mesh. Adequate mesh density in the contact area is of paramount importance in any contact stress analysis. It is the responsibility of the user to make sure that there are enough elements in the contact area to properly model the distribution of contact stresses. In this exercise we use a default global element size and apply mesh controls 0.05mm and ratio a/b = 1.1 to edges as shown in Figure 17-21.

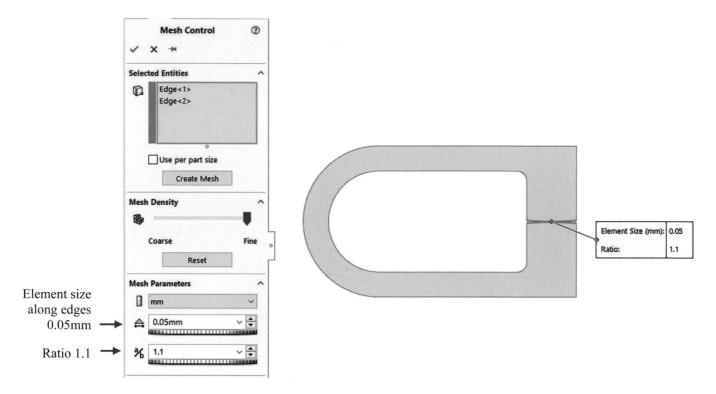

Element size along edges 0.05mm →

Ratio 1.1 →

Figure 17-21: Element size along the edges is 0.05mm; a/b ratio 1.1.

Mesh Control is defined on the round edges where contact will take place.

Low a/b ratio produces smooth transition between small elements along the controlled entities and large elements (default size) everywhere else in the model.

Small size of the contact area that will develop under the load necessitates small element size. Solving this as a 3D problem would result in a model with a very large number of elements. This, combined with iterative solution always required for any contact problem which by its nature is non-linear, would result in a long solution time. 2D representation significantly reduces the numerical complexity of the problem.

Von Mises stress results are shown in Figure 17-22.

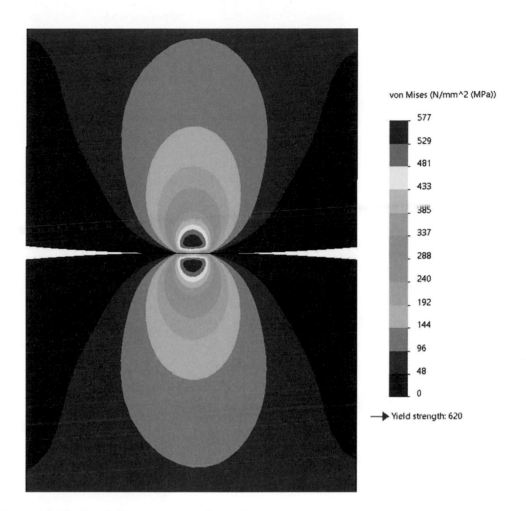

Figure 17-22: Von Mises stress results in the contact area.

The maximum stress 577MPa is located below the surface.

The **Factor of Safety** (FOS) plot based on the **Maximum Shear Stress** is shown in Figure 17-23. The **Factor of Safety** plot based on the **Maximum von Mises Stress** is shown in Figure 17-24. Notice that the **Maximum Shear Stress** criterion gives the FOS = 1 while the **Maximum von Mises Stress**, being less conservative, gives the FOS = 1.1.

Step 1   Step 2   Step 3

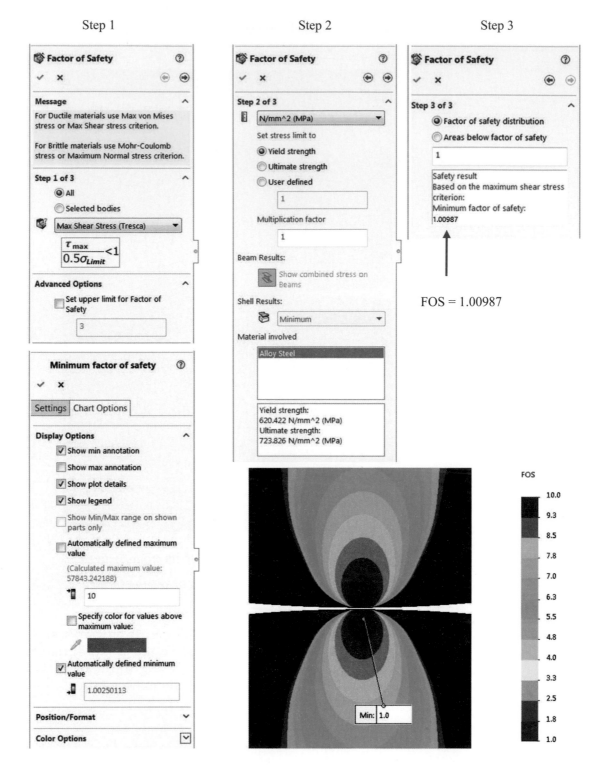

Figure 17-23: Factor of safety based on the Maximum Shear Stress.

A factor of safety plot is created in three steps. The range of the factor of safety 0-10 is defined in Chart Options. Notice that stress distribution closely matches the location of maximum shear stress in the Hertz contact problem.

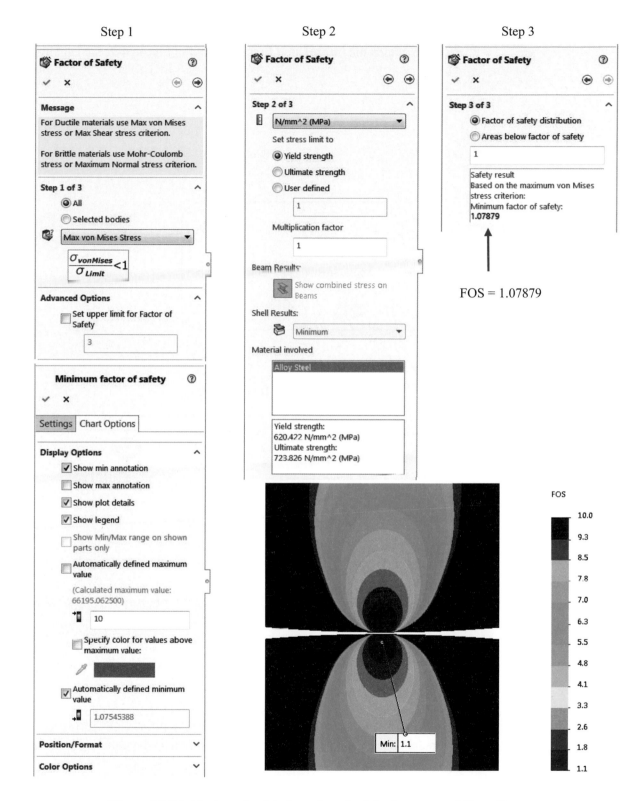

Figure 17-24: Factor of safety based on the Maximum von Mises Stress.

A factor of safety plot is created in three steps. The range of the factor of safety 0-10 is defined in Chart Options.

Figure 17-25 presents a summary of the elements available in **Simulation**. This figure expands the information first presented in Figure 4-2, then again in Figure 16-6.

3D elements

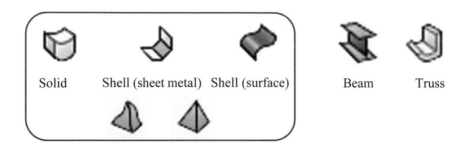

Solid   Shell (sheet metal) Shell (surface)   Beam   Truss

2D elements

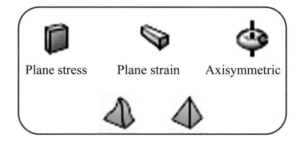

Plane stress   Plane strain   Axisymmetric

Figure 17-25: Summary of elements and their symbols in SOLIDWORKS Simulation.

See chapter 16 for an explanation of differences between beam and truss which both belong to the class of Beam elements.

Solid elements, shell elements and all types of 2D elements can be 1st or 2nd order; the element order is indicated by different tetrahedron symbols (Figure 17-25). Element order is not selectable for Beam elements where the order of transverse translation is always 3rd order.

There are other types of elements in SOLIDWORKS Simulation such as gap elements used in contact problems, mass elements and rigid link elements. These elements do not have a pictorial representation.

Models in this chapter

Model	Configuration	Study Name	Study Type
VASE.sldprt	*01 solid*		
	02 shell		
COVER.sldprt	*01 flat*	*01 flat*	Nonlinear
	02 R1500	*02 R1500*	Nonlinear
	03 R600	*03 R600*	Nonlinear
PERFORATED PLATE.sldprt	*Default*	*01 2D*	Static
		02 3D	Static
ANGLE BRACKET.sldprt	*Default*	*01 2D*	Static
		02 2D	Static
		03 2D	Static
		04 2D	Static
CONTACT.sldprt	*01 10mm*		
	02 20mm	*Static 1*	Static
	03 30mm		

Notes:

18: Vibration analysis - modal time history and harmonic

Topics covered

- ❑ Modal Time History analysis (Time Response)
- ❑ Harmonic analysis (Frequency Response)
- ❑ Modal Superposition Method
- ❑ Damping

What is dynamic analysis?

In preparation for the dynamic analysis exercises, we need to clarify an important terminology issue. The term "Dynamic Analysis" applies to an analysis of unrestrained or partially restrained bodies (mechanisms) as well as to restrained bodies such as structures. "Dynamic Analysis" within the scope of FEA deals only with the vibration of deformable bodies about the position of equilibrium. A more appropriate term to use would be "Vibration Analysis," but the term "Dynamic Analysis" is well entrenched in the FEA literature. We will use the terms Vibration Analysis and Dynamic Analysis interchangeably.

Dynamic analysis with SOLIDWORKS Simulation

All types of analyses that we have discussed so far have assumed that the load is not a function of time. We will now lift this restriction to introduce two common types of dynamic analyses: **Modal Time History** and **Harmonic**. **Modal Time History** is also known as a **Time Response** analysis and **Harmonic** analysis is known as a **Frequency Response** analysis. **Modal Time History** and **Harmonic** analyses both belong to the category of linear analyses.

Other types of dynamic (or vibration) analysis available in SOLIDWORKS Simulation are **Random Vibration** and **Response Spectrum**. **Random Vibration** will be introduced in chapter 19.

Readers interested in vibration analysis are referred to "Vibration Analysis with SOLIDWORKS Simulation" published by SDC Publications.

Modal superposition method

The review of dynamic analyses needs to be preceded by a description of the modal superposition method on which both Time Response and Frequency Response analyses are most often based. The modal superposition method represents a dynamic response of a vibrating structure by using the superposition of responses that characterize a single Degree Of Freedom (1DOF) system. The natural frequencies of these 1DOF systems correspond to the natural frequencies of the analyzed structure. The number of DOF contributing to a dynamic response is equal to the number of modes calculated by a pre-requisite modal (frequency) analysis. How many modes should then be calculated to represent dynamic responses using the modal superposition method? The first few modes are the most important, but the exact number of required modes is not known prior to analysis. One should use a convergence process to demonstrate that increasing the number of modes past a certain number no longer significantly affects results.

The modal superposition method is not always a prerequisite for dynamic analysis. Other methods like the Direct Integration method do not require modal analysis. **SOLIDWORKS Simulation** uses the Direct Integration method in the **Drop test** study and **Nonlinear Dynamic** study.

Modal Time History (Time Response) analysis

In a **Modal Time History** analysis, the applied load is an explicit function of time, mass and damping properties, all of which are taken into consideration, and the vibration equation appears in its full form:

$$[M]\ddot{d} + [C]\dot{d} + [K]d = F(t)$$

Where:

[M]	mass matrix
[C]	damping matrix
[K]	stiffness matrix
F(t)	vector of nodal loads; this vector is a function of time
d	unknown vector of nodal displacements

A **Time Response** analysis requires the definition of a damping coefficient which is most often expressed as a percentage of critical damping. Readers are referred to (1) as listed in Chapter 24 for selected numerical values of damping coefficients.

A **Time Response** analysis is used to model events of a short duration. A typical example would be an analysis of a structure's vibrations due to an impact load or acceleration applied to the base (called base excitation). Results of the **Time Response** analysis will capture both the response during the time when the load is applied, as well as the free vibration after the load has been removed.

Harmonic (Frequency Response) analysis

Harmonic analysis assumes that the load is a function of frequency, here shown as sin(ωt), rather than being directly dependent on time as is the case of a Time Response analysis.

$$[M]\ddot{d} + [C]\dot{d} + [K]d = F(\omega)sin(\omega t)$$

Where:

[M]	mass matrix
[C]	damping matrix
[K]	stiffness matrix
F(ω)	vector of nodal loads, this vector is a function of frequency
d	unknown vector of nodal displacements

A **Frequency Response** analysis models a structure's response to forced excitation or base excitation (excitation applied to the support) that is a harmonic function of time, here shown as a sinusoidal function. It is assumed that the excitation frequency changes very slowly, hence the alternative name **Steady State Harmonic Response** is often used for this type of analysis. A Frequency Response analysis also uses the modal superposition method and requires that damping be defined, usually as a percentage of critical damping.

A typical application of a **Frequency Response** analysis is a simulation of a shaker table, which will be demonstrated later in this chapter.

Both examples presented in this chapter feature discrete systems where mass and stiffness are separated. These examples are intuitive and have simple analytical solutions that can be found in any introductory textbook on vibration analysis. Vibrational analysis of a distributed system is presented in Chapter 19.

Single Degree of Freedom Oscillator (SDOF)

We use SDOF assembly to run five studies as shown in Figure 18-1.

Study name	Study type	Load
01 Modal	Frequency	None
02 Time Response	Modal Time History	Time dependent force, maximum amplitude 20N Linear damping 20Ns/m
03 Frequency Response	Harmonic	Harmonic base excitation amplitude 1mm Modal damping 5%
04 Frequency Response	Harmonic	Harmonic base excitation amplitude 1mm Modal damping 2%
05 Frequency Response	Harmonic	Harmonic base excitation amplitude 1mm Modal damping 10%

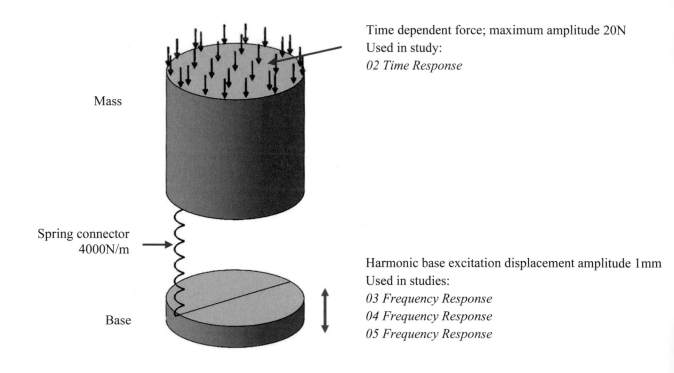

Time dependent force; maximum amplitude 20N
Used in study:
02 Time Response

Mass

Spring connector
4000N/m

Harmonic base excitation displacement amplitude 1mm
Used in studies:
03 Frequency Response
04 Frequency Response
05 Frequency Response

Base

Figure 18-1: Single degree of freedom oscillator SDOF.

Verify that the mass of the large cylinder is 10kg. The spring connector connects vertices on Base and Mass.

The spring is very soft in comparison to stiffness of Base and Mass parts. Therefore, the first mode of vibration will be associated with deformation in spring only, while Mass part oscillates as a rigid body. Therefore, the off-center location of the spring connector does not have any effect on the first mode of vibration of SDOF model.

Create a **Frequency** study called *01 Modal*. Define a **Spring Connector** as shown in Figure 18-2.

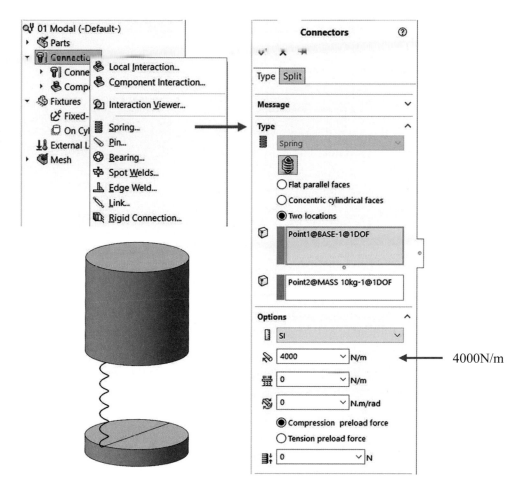

Figure 18-2: Spring connector definition.

Right-click the Connections folder; select Spring to open the Connectors window. Define a Spring Connector between vertices on the two components. Spring connector stiffness is 4000N/m.

Define restraints as shown in Figure 18-3.

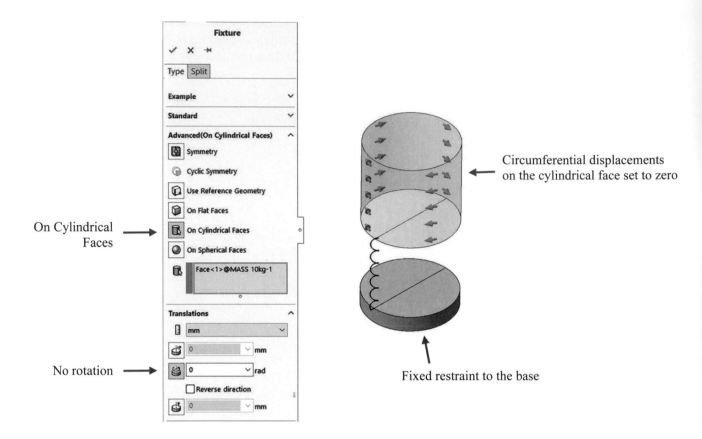

Figure 18-3: Restraints definition: eliminate circumferential displacements on the cylindrical face to prevent the mass from rotating.

Mass part is shown in transparent view. Fixed restraints applied to the bottom face of Base are not shown. The Fixed restraints may be applied to all faces of the base.

The restraints shown in Figure 18-3 allow the Mass only to move up and down. The *Base* is fully restrained and serves only to provide an anchor point to the spring.

Mesh the model with the default mesh size. Run *01 Modal* study and verify that the first natural frequency is 3.25Hz (Figure 18-4).

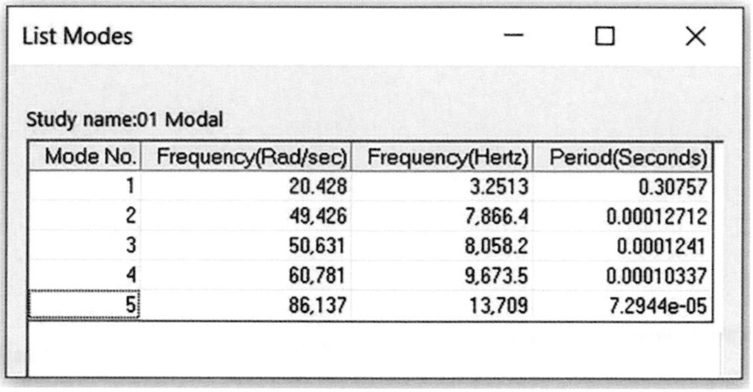

Mode No.	Frequency(Rad/sec)	Frequency(Hertz)	Period(Seconds)
1	20.428	3.2513	0.30757
2	49,426	7,866.4	0.00012712
3	50,631	8,058.2	0.0001241
4	60,781	9,673.5	0.00010337
5	86,137	13,709	7.2944e-05

Study name:01 Modal

Figure 18-4: Modal frequency results for the *Modal* study.

Modes 2-5 correspond to deformation of the cylinder, not the spring. Higher modes are not related to oscillations of a Single Degree of Freedom system.

Recall from the theory of vibration that the natural frequency ω, of a Single Degree of Freedom Oscillator, is:

$$\omega = \sqrt{\frac{k}{m}} = \sqrt{\frac{4000}{10}} = 20 \; rad/s$$

To express the same in Hz:

$$f = \frac{\omega}{2\pi} = \frac{20}{2\pi} = 3.18 \; cycles/s$$

Vibration period *T* is:

$$T = 1/f = 0.31s$$

Results of *01 Modal* study closely match these analytical results.

Now, create a **Modal Time History** study called *02 Time Response* as shown in Figure 18-5.

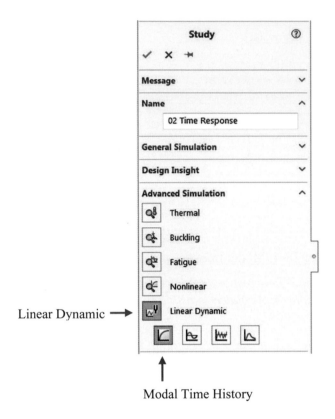

Figure 18-5: Defining a Modal Time History study.

A Modal Time History study is created by selecting a Linear Dynamic study with the Modal Time History option.

You can copy restraints (one at a time) from the modal study, but the **Spring Connector** must be defined in **Modal Time History** since its definition has an option to include damping. Damping can also be defined as modal damping and not explicitly in the spring connector. Modal damping specifies damping as a fraction of critical damping.

Oscillations no longer occur when the damping value is critical or above critical. The critical damping in SDOF model is:

$$c_{cr} = 2\sqrt{km} = 400\,\frac{Ns}{m}$$

To define damping as 5% of critical damping we can either enter 20 Ns/m as linear damping in the **Spring-Damper Connector** window, or as 0.05 in the **Global Damping** window (Figure 18-6).

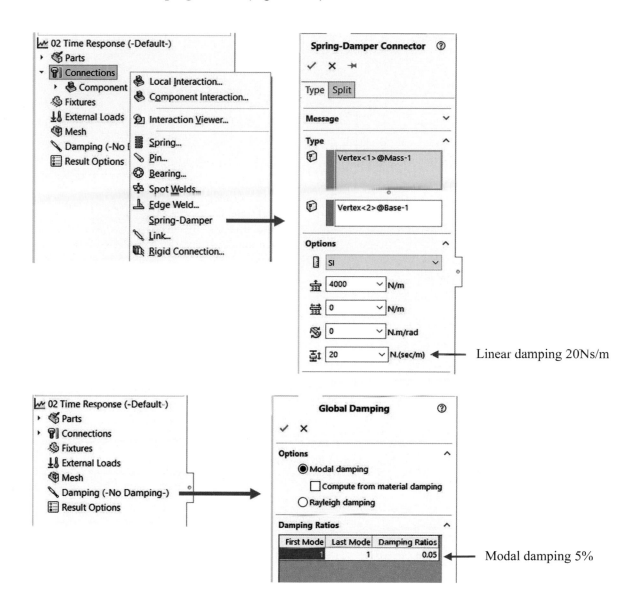

Linear damping 20Ns/m

Modal damping 5%

<u>Figure 18-6: Damping can be defined as a linear damping (top) or as a modal damping (bottom). The entries in both windows define the same damping. In this example we use an explicit damping definition.</u>

Modal damping makes it possible to define damping individually for each mode. Since we base this analysis on one mode only, we define damping for this single mode.

Dynamic analysis requires a load defined as a function of time. To apply a 20N force as shown in Figure 18-1 and define its time history, follow the steps explained in Figure 18-7.

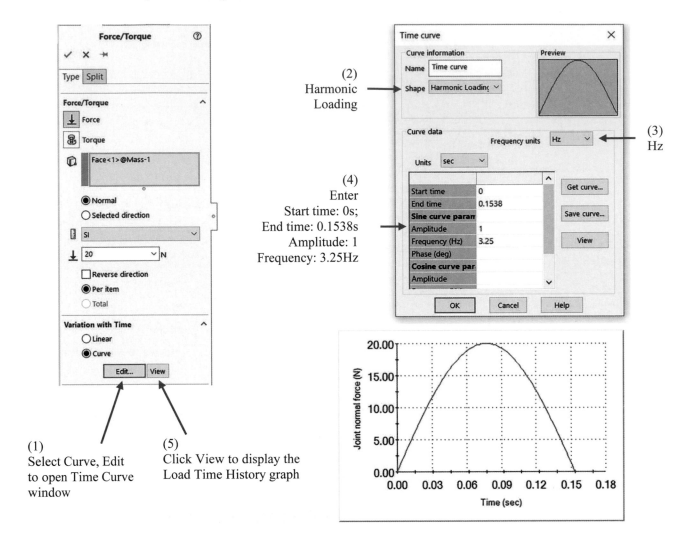

Figure 18-7: Defining load as a function of time. The force takes 0.1538s to reach the maximum of 20N and then drop back to zero. The load remains zero for the rest of analysis.

The duration 0.1538s equals to one half of vibration period of the SDOF.

Notice that neither the entry in the **Force** window or the values defining the **Time curve** defines the load time history on their own. The corresponding values are multiplied to calculate force magnitude as a function of time. Also, it is important to note that **Time curve** window defines the load time history but not the duration of analysis. The duration of analysis is defined in study properties Figure 18-8.

In the **Frequency Options**, define the **Number of frequencies** as 1. This is because our objective is to analyze the Single Degree of Freedom oscillator which only has one natural frequency. In the **Dynamic Options** define **End time** as 5s and **Time increment** as 0.0125s. This way, the dynamic response will be analyzed during the first 5s counting from the beginning of force application. The dynamic response will be evaluated every 0.0125s in 400 steps.

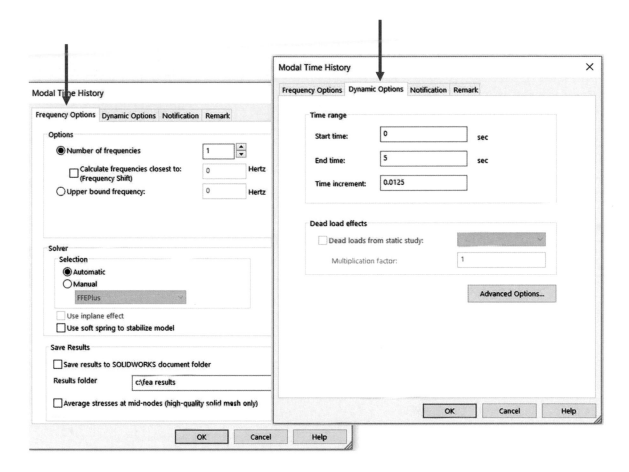

Frequency Options   Dynamic Options

Figure 18-8: Frequency Options definition window and Dynamic Options definition window.

The duration of the load is 0.1538s (Figure 18-7), while the duration of analysis is 5s. Once the load disappears the system enters free vibration.

Define a **Workflow Sensitive Sensor** as shown in Figure 18-9.

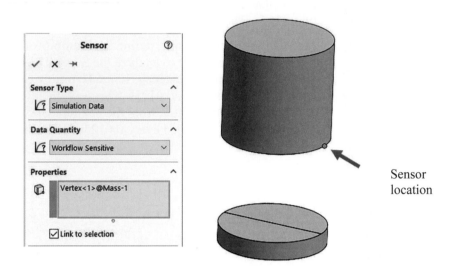

Figure 18-9: Sensor definition in the vertex where the spring is attached.

Sensor definition is done in Feature Manager, not in Simulation study. This figure is for your information only. The model comes with sensor defined.

Right-click the *Results Options* folder to define the **Results Options,** as shown in Figure 18-10.

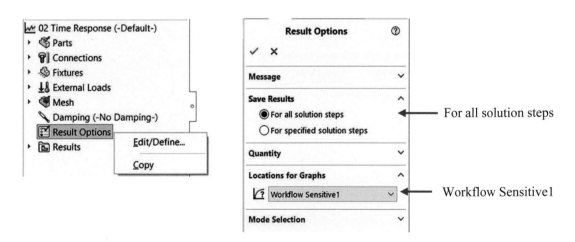

Figure 18-10: Results Options definition.

Make the indicated selections in preparation for graphing results.

Mesh with a default mesh and run the *02 Time Response* study observing that each solution is completed in 400 steps as specified in the study properties (Figure 18-8). Right-click the *Results* folder and follow the steps illustrated in Figure 18-11 to create a graph showing displacement in the sensor location as a function of time.

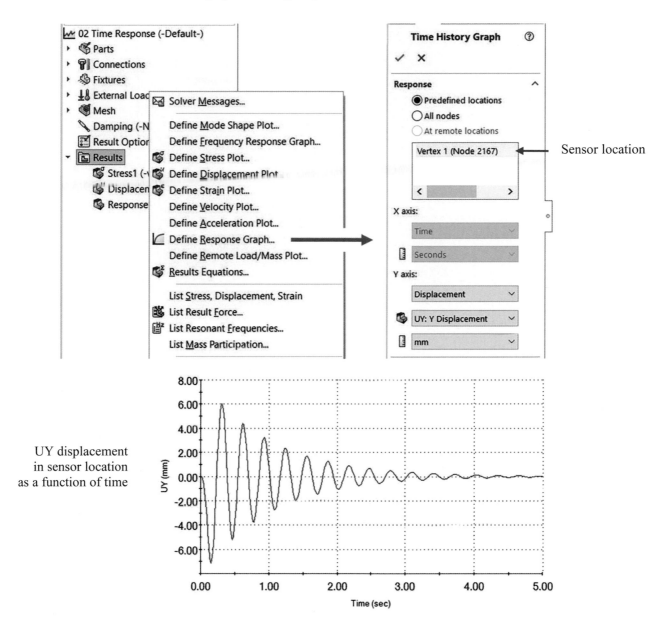

UY displacement in sensor location as a function of time

Figure 18-11: Displacement of the Mass during the first 5s after load application.

Notice that after 0.1538s, the load becomes zero and the SDOF performs free damped oscillations.

Modal Time History requires results of a **Frequency** analysis, a **Frequency** analysis is always run prior to a **Modal Time History**. Within **Modal Time History**, you may select to run only the **Frequency** analysis (Figure 18-12).

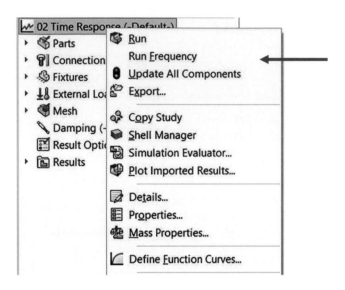

Figure 18-12: A Modal Time History study gives an option to run Frequency study without subsequent dynamic analysis.

This pop-up menu is invoked by right-clicking the Time Response study folder.

Since a **Frequency** analysis is always run prior to a **Modal Time History** analysis, **Modal Time History** results include the same results that are available in a **Frequency** study. To verify this, define a mode shape plot or review the list of modal frequencies. Notice that you will see only one frequency as specified in the **Modal Time History** properties (Figure 18-8).

In continuation of the SDOF analysis, create a **Harmonic** study called *03 Frequency response* (Figure 18-13).

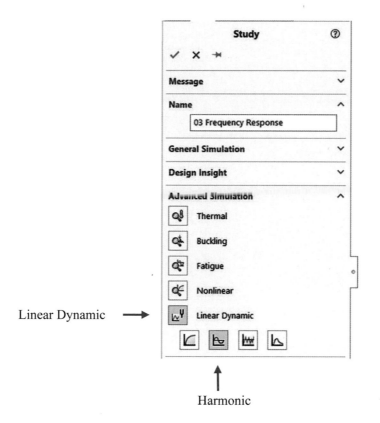

Figure 18-13: Defining a Harmonic study.

A harmonic study is created by selecting a Linear Dynamic study with the Harmonic option.

We will investigate the dynamic response of the SDOF under a harmonic base excitation with displacement amplitude 1mm. The range of frequency of base oscillation is 0Hz to 10Hz. This range includes the system natural frequency 3.25Hz. The displacement amplitude remains constant while the frequency of oscillations increases from 0Hz to 10Hz.

Define the *03 Frequency Response* with study properties as shown in Figure 18-14.

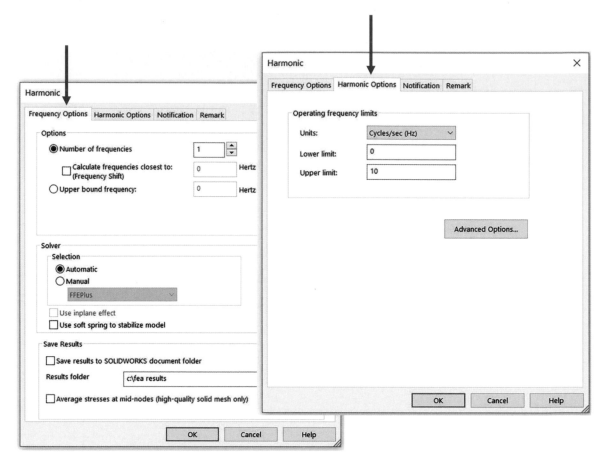

Frequency Options   Harmonic Options

Figure 18-14: Properties of study *Frequency Response.*

The left window defines the number of frequencies as 1 because the analyzed system only has one degree of freedom. The right window defines the range of oscillation frequency from 0 to 100Hz.

Define the same restraints as in the *02 Time Response* study. Linear damping is not available in Harmonic study; **Spring-Damper Connector** is not available either. Therefore, we will use **Spring Connector** and **Modal damping**. Definition of Spring Connector is identical to definition of Spring-Damper Connector except that linear damping is not defined. **Modal Damping** is defined in the window called **Global Damping** (Figure 18-6); these two terms are exchangeable.

Define the same **Results Options** as in study *02 Time Response.*

Define base excitation is shown in Figure 18-15.

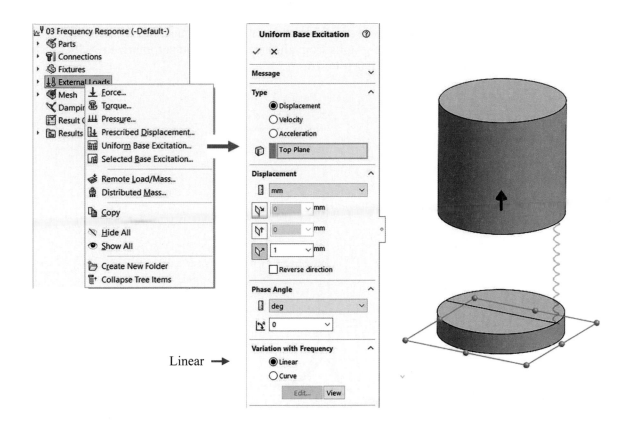

Figure 18-15: Definition of base excitation.

The base oscillates between -1mm and +1mm with a frequency changing slowly from 0Hz to 10Hz as defined in study properties.

Variation with frequency **Linear** means that amplitude of the oscillation does not change with frequency.

Run *03 Frequency Response* study with the default mesh and define a **Response Graph** of the UY displacement component. Copy study *03 Frequency Response* to *04 Frequency Response* and change damping to 2%. Finally, copy *04 Frequency Response* to *05 Frequency Response* and change damping to 10%. Collect response graphs from all three studies and summarize results in a graph like the one in Figure 18-16.

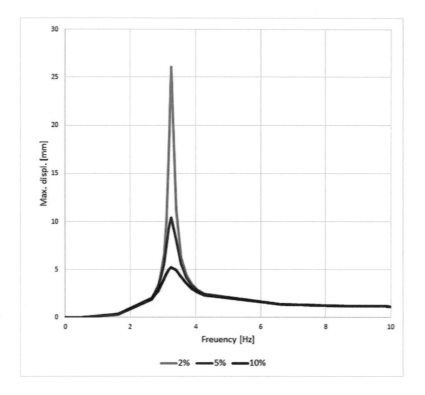

Figure 18-16: Amplitude of vibration as a function of the base excitation frequency for different modal damping ratios.

The amplitude of vibration is measured from the neutral position, not between negative and positive peaks.

As demonstrated in Figure 18-16, damping strongly affects the amplitude for excitation frequencies close to the resonant frequency. It has no effect for excitation frequencies much lower or much higher than the resonant frequency.

The sharp peak in amplitude magnitude visible in Figure 18-16 is the *SDOF* response under an excitation frequency equal to the natural frequency of the system (this is known as resonance). The amplitude of vibration in resonance is controlled only by damping.

Now, open the assembly model BUMPER (Figure 18-17). The assembly features massive steel rings interlocked with rubber spacers. As opposed to SDOF that is a discrete system, BUMPER is a distributed system; all components contribute both to stiffness and inertia.

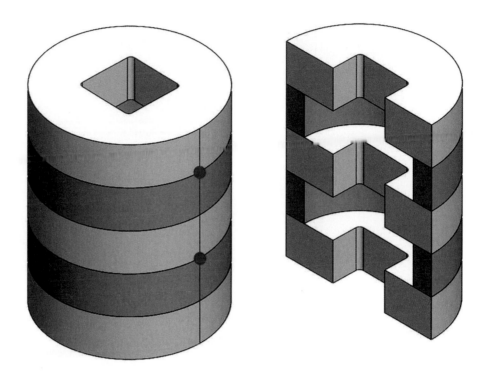

Figure 18-17: BUMPER assembly in a full view and in a section view.

Steel rings have rectangular holes; mass of each ring is 75kg. Steel rings are connected by rubber rings (brown) to form a stack. Red dots mark sensor locations.

Define **Workflow Sensitive** sensors in location indicated by red arrows in Figure 18-17. These sensors will be used to construct displacement response graphs.

Define **Fixed** support to the bottom face of the lowest steel ring. Define **Roller-Slider** restraints to the flat faces of the square holes in all rings. Roller-Slider restraints allow the rings to move only in the vertical direction.

We will analyze vibration response of bumper to **Base Excitation** with displacement amplitude 1mm and frequency changing between 0-100Hz. The analysis will be conducted in two steps. First, we will run modal analysis (study *01 modal*) to find out how many natural frequencies are within the frequency sweep 0-100Hz. Next, we will copy the **Frequency** study *01 modal* into **Harmonic** (study *02 Frequency Response*) to conduct a frequency response analysis.

Obtain solution of *01 Modal* study with default mesh and five modes specified in study properties and review results (Figure 18-18).

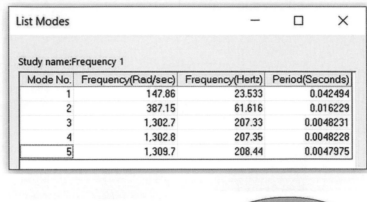

Mode No.	Frequency(Rad/sec)	Frequency(Hertz)	Period(Seconds)
1	147.86	23.533	0.042494
2	387.15	61.616	0.016229
3	1,302.7	207.33	0.0048231
4	1,302.8	207.35	0.0048228
5	1,309.7	208.44	0.0047975

Study name:Frequency 1

List Modes

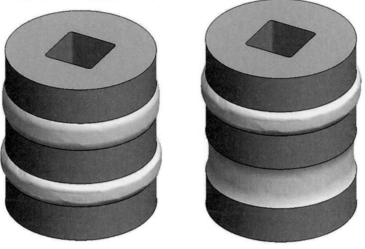

Mode 1: 25.3Hz   Mode 2: 61.6Hz

Figure 18-18: Frequencies of the first five modes (top) and shapes of the two modes with frequencies within the range 0-100Hz.

Modal shape plots have colors deselected.

Deformations in first two modes are aligned with the direction of the base excitation and modal frequencies are within the range of frequencies of the base excitation. Therefore, these two modes must be considered for the subsequent **Harmonic** study. Higher modes do not have to be considered. They will not be excited because their frequencies are above 100Hz; that is the highest frequency of the base excitation.

Copy **Frequency** study *01 Modal* into **Harmonic** study *02 Frequency Response* (Figure 18-19).

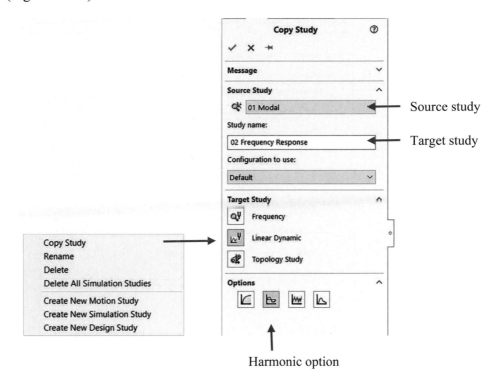

Figure 18-19: Copying Frequency study into Harmonic study.

Right-click 01Modal study tab, select Copy Study, enter target study name, and select Linear Dynamic with Harmonic option.

Define **Frequency Options** and **Harmonic Options** as shown in Figure 18-20.

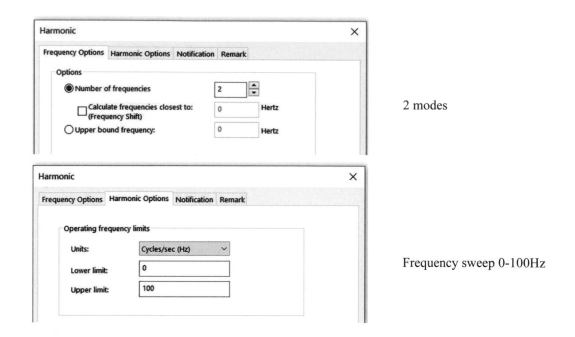

2 modes

Frequency sweep 0-100Hz

Figure 18-20: Settings of *02 Frequency Response* study.

Two modes will be considered in the 02 Frequency Response study.

Fixtures do not have to be defined in *02 Frequency Response* study because they were copied from *01 Modal* study. To define a **Base Excitation**, follow steps shown in Figure 18-21.

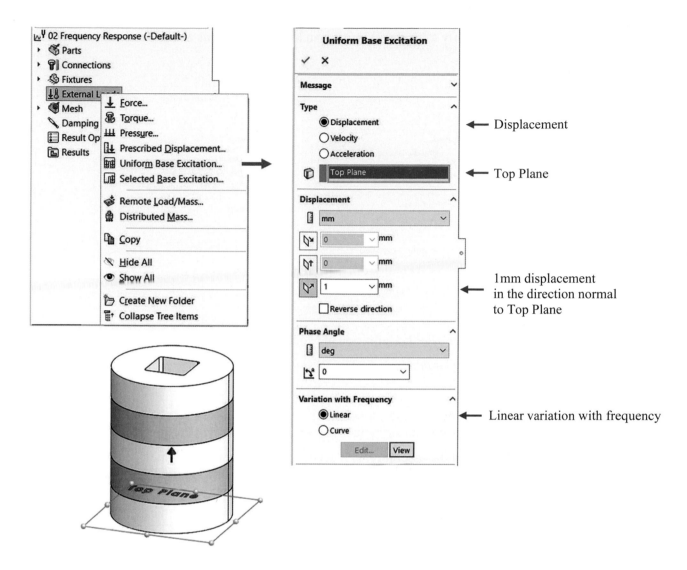

Figure 18-21: Defining Uniform Base Excitation.

1mm is displacement measured from the neutral position.

Uniform Base Excitation is applied to all restraints active in the specified direction. In this case, it is applied only to fixed restrains at the bottom. Roller-Slider restraints present in the model act in the direction normal to the excitation. Therefore, they are not affected by the base excitation.

Linear Variation with Frequency means that the amplitude of displacement remains constant while the frequency changes.

We use displacement excitation to simulate a shaker table test.

Specify a **Modal Damping** and **Results Options** as shown in Figure 18-22.

Modal Damping 5%   Definition of Workflow Sensitive sensor

Figure 18-22: Definition of Modal Damping and Result Options.

A modal damping of 0.05 (5% of critical damping) is defined for modes 1 and 2.

Modal Damping 5% is assumed to be the same for both modes. **Workflow Sensitive** sensor includes two locations shown in Figure 18-17.

Mesh the assembly with default mesh and run the solution of *02 Frequency Response* study. Create a response graph following the steps shown in Figure 18-23. Export data to Excel and construct a graph similar to the one shown in Figure 18-24.

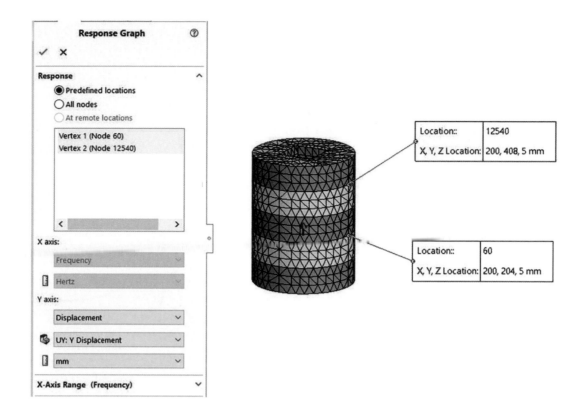

Figure 18-23: Definition of displacement amplitude Response graph.

Select both sensors and UY displacement component. Node 12450 is the location on the top steel ring, node 60 is the location on the middle steel ring.

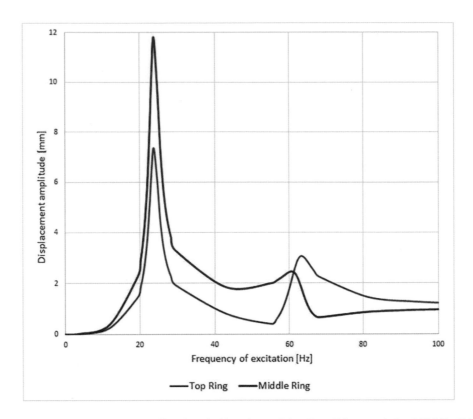

Figure 18-24: The amplitude of vibration of the Top Ring and the Middle Ring as a function of the excitation frequency.

The response graph in Figure 18-24 clearly shows that there are two peaks of amplitude response which correspond to the natural (resonant) frequencies of the model.

Repeat this exercise for different values of **Modal Damping** to study the effect of damping on the vibration amplitudes of the two masses.

Models in this chapter

Model	Configuration	Study Name	Study Type
SDOF.sldasm	Default	01 Modal	Frequency
		02 Time Response	Modal Time History
		03 Frequency Response	Harmonic
		04 Frequency Response	Harmonic
		05 Frequency Response	Harmonic
BUMPER.sldasm	Default	01 Modal	Frequency
		02 Frequency Response	Harmonic
		03 Frequency Response	Harmonic
		04 Frequency Response	Harmonic

Notes:

19: Analysis of random vibration

Topics covered

- Random vibration
- Power Spectral Density
- RMS results
- PSD results
- Modal excitation

Random vibration

Random vibrations are non-periodic. Knowing the history of random vibration, we can predict the probability of occurrence of acceleration, velocity, and displacement magnitudes but cannot predict the precise magnitude at a specific time instant.

Random vibration is composed of a continuous spectrum of frequencies. The huge amount of time history data makes it impractical to solve random vibration problems using tools of time response analysis.

For most structural vibrations, the excitation such as force or base acceleration alternates about zero. Consequently, mean values characterizing the excitation as well as responses to that excitation such as displacement or stress are equal to zero. For this reason, results of a random vibration analysis are given in the form of Root Mean Square (RMS) values.

To explain the concept of an RMS value, refer to the graph in Figure 19-1 which shows the acceleration time history (acceleration as a function of time) of random vibration expressed in units of gravitational acceleration [G].

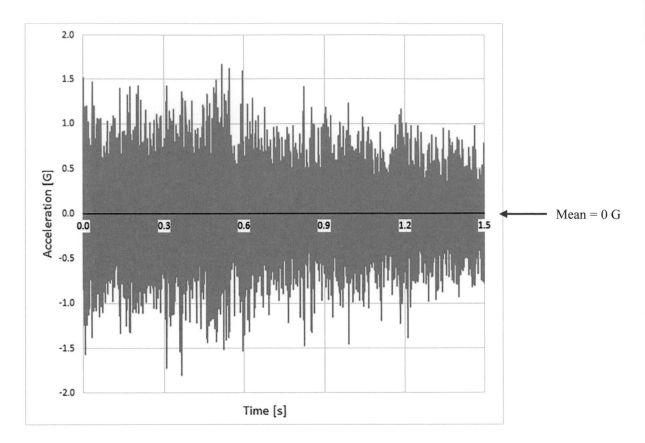

Figure 19-1: An example of acceleration time history data collected during 1.5s.

Considering the sampling rate of 5000 samples per second, this time history contains 7500 data samples.

The mean value of the acceleration time history shown in Figure 19-1 is zero.

However, if we multiply the function by itself, we obtain a function with a positive mean value. Its mean will no longer be zero and this squared function will be well suited to characterize the acceleration time history. This mean value of square acceleration time history is the mean square value and has units of $[G^2]$.

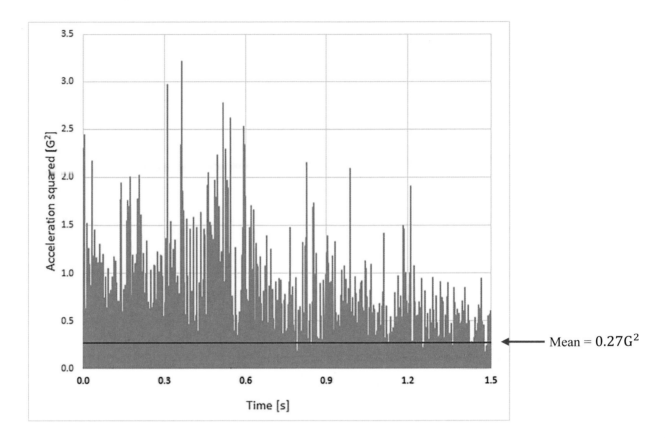

Figure 19-2: Squaring acceleration time history function (Figure 19-1) produces a function with a non-zero mean.

As shown in Figure 19-2, calculating the mean square value gives $G_{RMS}^2 = 0.27G^2$. The square root of the mean square value gives $G_{RMS} = 0.51G$.

The square root of the mean value is the Root-Mean-Square (RMS) acceleration and has units of $[G]$. The same applies to RMS displacement, velocity, stress, etc.

In random vibration, the magnitudes of acceleration, velocity, displacement, etc. all follow a normal distribution. The RMS value corresponds to one standard deviation σ characterizing the normal distribution. To explain this, we refer again to Figure 19-2. The acceleration, as characterized by the given acceleration time history, has a 68% probability of remaining between -0.51G and +0.51G. Consequently, it has a 32% probability of being less than -0.51G or more than 0.51G.

Acceleration Power Spectral Density

Let us assume that the acceleration time history in Figure 19-1 is a stationary random process where probability numbers characterizing this process do not change with time. In this case, the acceleration time history can be used to calculate the Acceleration Power Spectral Density (PSD) curve (the variation of any property with respect to frequency is called "spectrum").

The overall G_{RMS}^2 of random vibrations shown in Figure 19-2 is $0.27G^2$. However, random vibrations are composed of a large number of frequencies. Let us say we wish to investigate G_{RMS}^2 individually for a number of frequencies in the range from 0 to 2000Hz. Therefore, we divide the 0-2000Hz range into 20 sections (bins), each 100Hz wide, and calculate G_{RMS}^2 characterizing each section by filtering out all frequencies falling outside of the section (Figure 19-3).

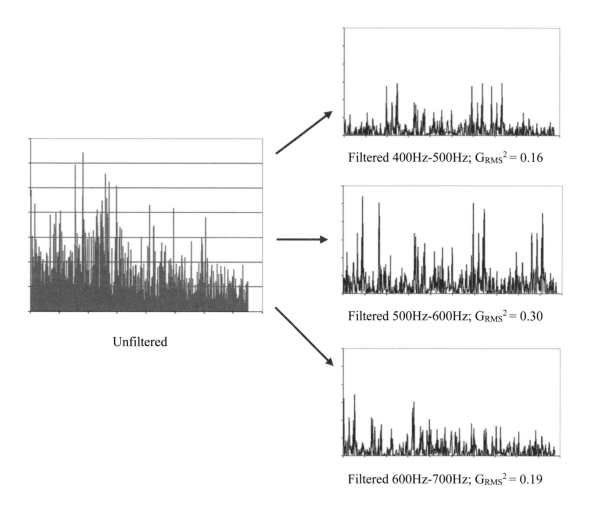

Filtered 400Hz-500Hz; $G_{RMS}^2 = 0.16$

Filtered 500Hz-600Hz; $G_{RMS}^2 = 0.30$

Unfiltered

Filtered 600Hz-700Hz; $G_{RMS}^2 = 0.19$

Figure 19-3: G_{RMS}^2 calculated individually for specified frequency ranges.

The graph on the left shows squared acceleration time history from Figure 19-2. Only three frequency ranges (sections) are illustrated here for brevity.

Having found G_{RMS}^2 values obtained for each frequency range, we can now calculate individual "densities" of G_{RMS}^2 in each section by dividing G_{RMS}^2 in each section by the width of the section. Results obtained for all sections may be plotted as a function of the frequency in the center of each section. This function is the Acceleration Power Spectral Density (Figure 19-4).

Band pass filter	Band center	G_{RMS}^2	Bandwidth	Acceleration PSD
Hz	Hz	G^2	Hz	G^2/Hz
400 – 500	450	0.16	100	0.0016
500 - 600	550	0.30	100	0.0030
600 - 700	650	0.19	100	0.0019

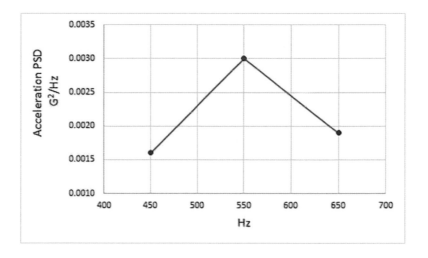

Figure 19-4: The Acceleration Power Spectral Density function (PSD).

Three points of the Acceleration PSD curve have been calculated by dividing G_{RMS}^2 in each section (each frequency range) by the width of the section.

The Acceleration Power Spectral Density (PSD) allows for a compression of data and is commonly used to characterize a random process. Mechanical vibrations are commonly described by the Acceleration Power Spectral Density, which is easily generated by testing equipment. Design specifications and test results of devices subjected to random vibration are typically given in the form Acceleration PSD.

Analysis of random vibration of a hard drive head

Having completed this short introduction to random vibration, we begin a random vibration analysis of a hard drive head. Open part HD HEAD and create a **Frequency** study with properties shown in Figure 19-5. Call the study *01Modal*.

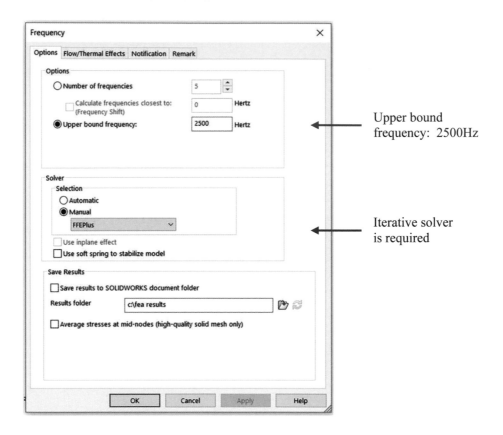

Upper bound frequency: 2500Hz

Iterative solver is required

Figure 19-5: Properties of study Modal.

All frequencies in the range of 0-2500Hz will be calculated.

Define restraints as shown in Figure 19-6.

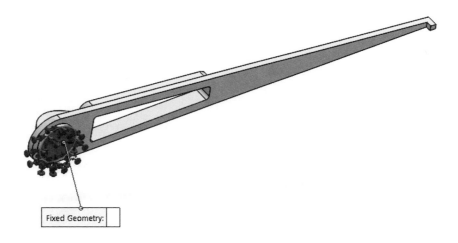

Figure 19-6: Fixed restraints applied to HD HEAD model.

Place Workflow Sensitive sensor at the tip.

Apply a **Mesh Control** as shown in Figure 19-7 and mesh with the default element size.

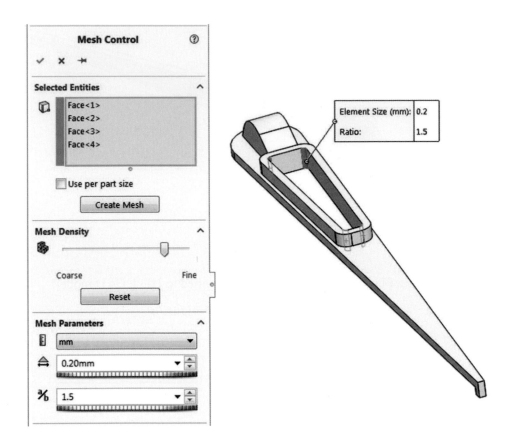

Figure 19-7: Mesh controls (0.20mm) applied to four fillets.

Use the default global element size.

If displacement results were our only objective, the default mesh would be acceptable for both Frequency Analysis and subsequent Random Analysis. However, since in the next study we intend to analyze displacements and stresses, mesh controls are required to ensure correct element shape and size in the area of stress concentrations. Prior to this exercise, analyses were run without mesh controls to find where mesh controls are required.

Solve the *Modal* study and review the results shown in Figure 19-8. Notice that there are four modes of vibration within the requested range of frequencies 0 - 2500Hz.

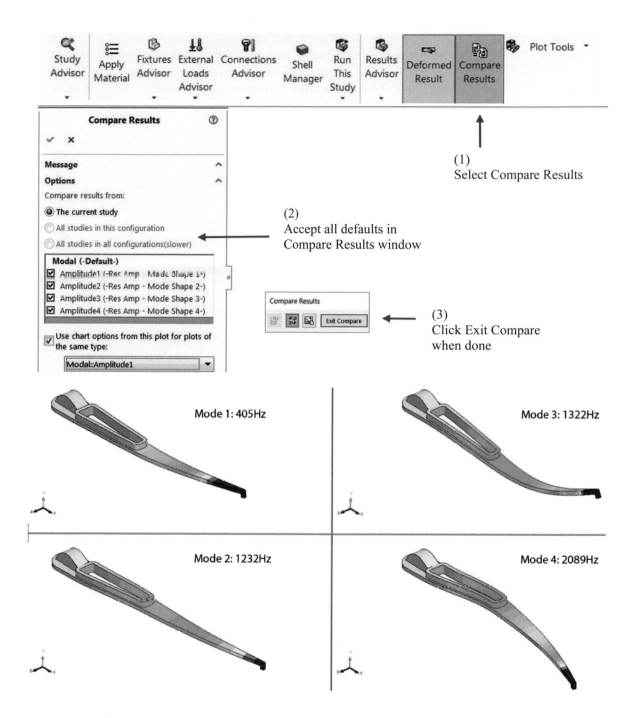

Figure 19-8: Modes of vibration within the range of 0 – 2500Hz.

Vibrations in mode 1 and 3 take place in the XY plane; vibrations in mode 2 and 4 take place in the XZ plane. All four plots can be shown together using Compare Results function from the Command Manager. To use Compare Results follow the steps shown above.

Proceeding to the analysis of Random Vibration, we could create a new **Dynamic** study independent from the completed **Frequency** study, this time with the **Random** option selected. However, a **Frequency** analysis would then have to be repeated within a **Dynamic Random** study. To avoid this repetition, we can copy the results of the **Frequency** study into a **Dynamic Random** study as shown in Figure 19-9.

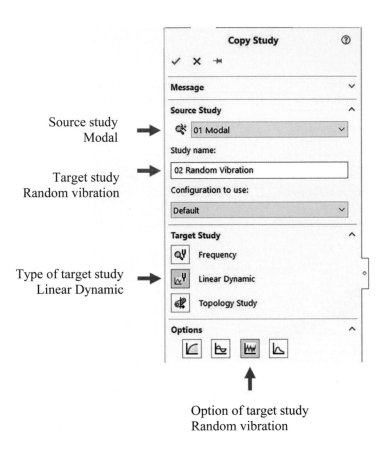

Figure 19-9: Results of a Frequency study can be copied to a new Dynamic study.

Copying the Frequency study into a Dynamic Random study also copies the mesh information and restraints.

The properties of the **Random Vibration** study are shown in Figure 19-10.

Random Vibration Options

Frequency Options

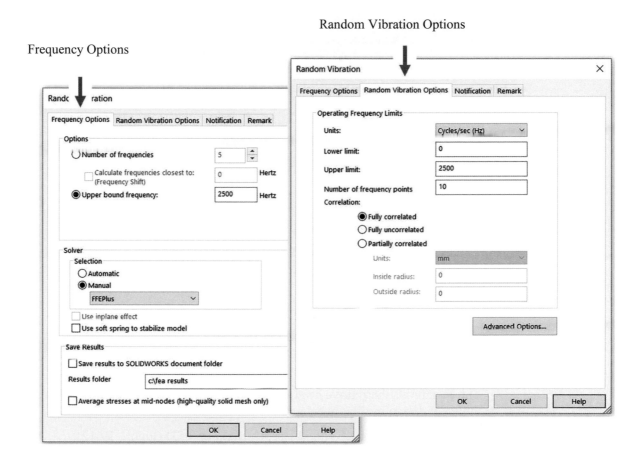

Figure 19-10: Properties of the Random Vibration study.

The Frequency Options specify all modes in the range of 0-2500Hz to be considered in the analysis of random vibration.

In the Random Vibration options, specify the Upper limit as 2500Hz to investigate responses to Random Vibration in the frequency range from 0 to 2500Hz.

Define **Global Damping** as shown in Figure 19-11.

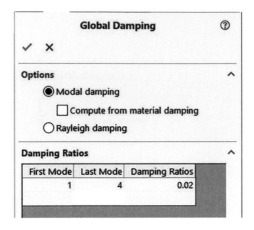

Figure 19-11: Global damping definition.

Global damping is defined as 2% of critical damping. The number of modes (4) corresponds to the number of modes in the frequency range 0-2500Hz, as is specified in Figure 19-10.

Assigning the same damping ratio to all modes represents a simplified and conservative approach. In most cases damping for higher modes will be higher than for lower modes.

The load on the hard drive head comes from random excitation of the base in the global Y direction. This acceleration PSD has been obtained from testing.

Define Uniform Base Excitation as Acceleration PSD (Figure 19-12).

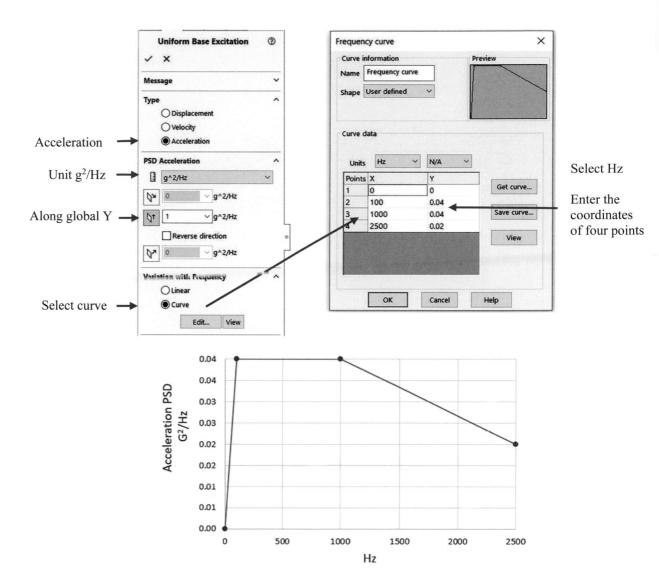

Acceleration

Unit g²/Hz

Along global Y

Select curve

Select Hz

Enter the coordinates of four points

Click View in Uniform Base Excitation window to review the PSD curve

Figure 19-12: Uniform Base Excitation defined as Acceleration PSD in the global Y direction acting on all restraints present in the model (here only one restraint is present).

This PSD graph has been formatted in Excel.

Define a **Sensor** as shown in Figure 19-13.

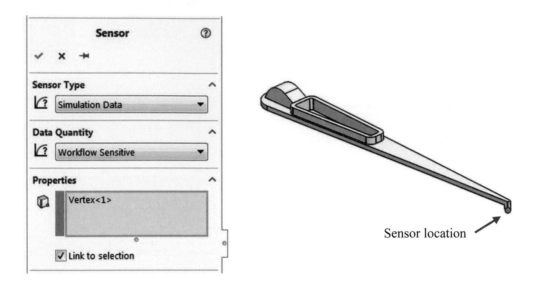

Figure 19-13: Sensor location.

Select the vertex at the tip of the head. Model comes with sensor already defined.

Define **Result Options** as was shown in Figure 18-10. Obtain the solution and analyze the RMS displacement results and the PSD displacement results.

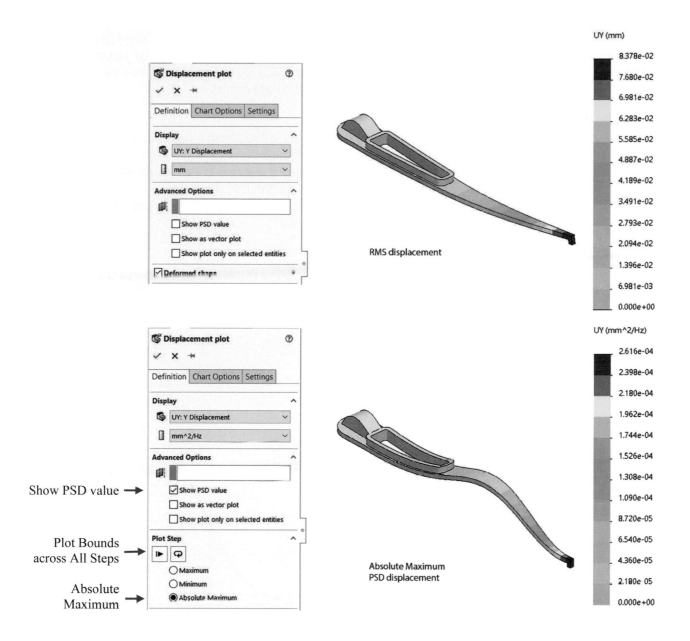

PSD displacements are displayed in units of mm²/Hz for the selected frequency, Minimum, Maximum or Absolute Maximum as used here. The Absolute Maximum of PSD displacement corresponds to the first mode frequency 405Hz. See Figure 19-17 for explanations of the PSD results.

It is important to understand the meaning of results in a Random Vibration analysis. The displacement results in Figure 19-14, top, are the RMS displacements. The maximum RMS displacement is 0.084mm meaning that the magnitude of displacement has a 68% probability of remaining under 0.084mm. The probability of the maximum displacement magnitude exceeding 0.084mm is of course 32%.

Remembering that the probability of a given displacement is defined by a normal distribution for which σ = 0.084mm, we can calculate the probability of displacement magnitude exceeding any defined value (Figure 19-15).

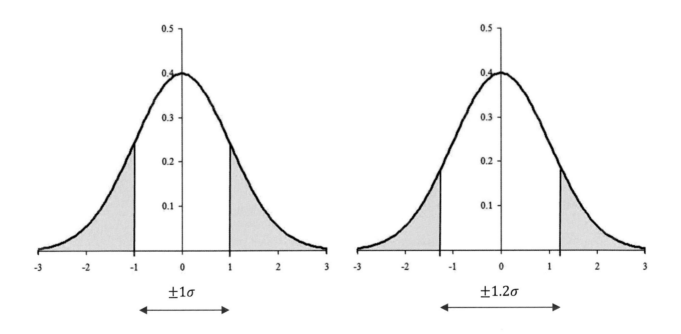

Figure 19-15: The total area under the normalized Gauss curve is 1. The probability of displacement magnitude exceeding ±σ (left) and ±1.2σ (right) is equal to the corresponding shaded areas.

The probability of the displacement magnitude exceeding 1*RMS displacement (here 0.084mm) is given by the area outside ± 1σ which is 32% (left).

The probability of displacement magnitude exceeding 1.2*RMS displacement (0.10mm) is given by the area outside ± 1.2σ which is 23% (right).

The interpretation of results shown in Figure 19-15 applies to all results of a Random Vibration analysis. The RMS SX stress result is shown in Figure 19-16.

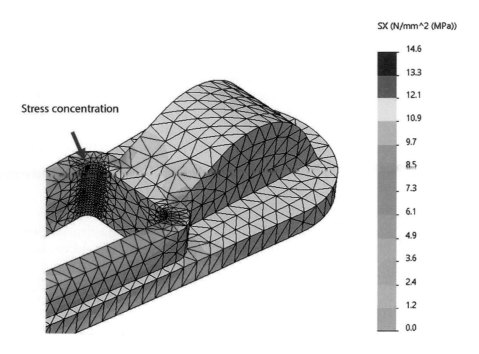

Figure 19-16: RMS SX stress results.

The maximum SX stress has a 68% probability of remaining below 14.5MPa.

Notice that stress singularities present in the model geometry (sharp re-entrant edges) do not show as stress concentrations because of the large element size used for meshing.

Refer to Figure 19-16 and compare the size of elements to the size of the stress concentration. Even with mesh controls applied, the mesh is at best marginal to model stress concentrations. Repeat the analysis with a more aggressive mesh control.

The results of the **Dynamic Random** analysis presented as RMS values provide one result for the entire frequency range of excitation. All results (displacement, stresses, etc.) can also be presented as PSD values (Figure 19-14 bottom), which are different for each excitation frequency. Examine the PSD options of displacements and stress result plots and notice that displacement results are given in mm^2/Hz, and stress results are given in MPa^2/Hz. These units are a consequence of the base excitation being defined as acceleration squared per frequency range, in our case G_{RMS}^2/Hz.

The most informative way to review PSD results is to graph them over the frequency range. Create a graph showing PSD displacement in the selected location shown in Figure 19-13. Define a Y Displacement **Response Graph** as shown in Figure 19-17.

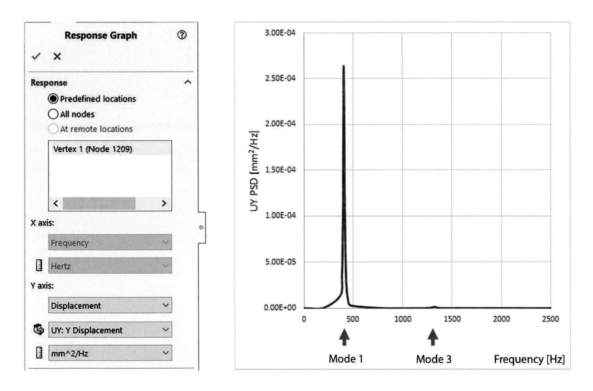

Figure 19-17: PSD displacement as a function of excitation frequency.

Mode 1 and mode 3 are excited. Mode 3 is barely visible on the graph.

Upon examination of the graph in Figure 19-17, we find that the applied base acceleration excites mode 1 and mode 3 because these modes occur in the XY plane. Mode 2 and mode 4 occur in the XZ plane which is orthogonal to the direction of base acceleration and are therefore not excited.

Repeat this exercise with a base excitation where the direction is not orthogonal to any global direction to see that all four modes will be excited. In Random Vibrations study, the directions of excitation are aligned in the global coordinate system; therefore, this will require rotating the model relative to the global coordinate system. This analysis is presented in the book "Vibration Analysis with SOLIDWORKS Simulation."

Models in this chapter

Model	Configuration	Study Name	Study Type
HD HEAD.sldprt	*Default*	*01 Modal*	Frequency
		02 Random Vibration	Random Vibration

Notes:

20: Topology Optimization

Topics covered

❑ Definition of Topology Optimization

❑ Design space

❑ Goals and constraints

❑ Topology Optimization criteria

❑ Examples of Topology Optimization

What is a Topology Optimization?

Topology Optimization finds where material should be placed within the available design space. Topology Optimization starts with a design space which represents the maximum allowed size for a component. It considers loads, restraints and manufacturing constraints. Topology Optimization works with the given optimization objective (for example reduce mass by 50%); it follows an iterative process that seeks the best material distribution within the design space to maximize stiffness to mass ratio, minimize displacement, etc.

We will introduce the concept of Topology Optimization using an example of a bridge. The bridge has been designed originally as a hollow prismatic tube. It is loaded with a uniformly distributed force applied to the bottom inside face (Figure 20-1); the bridge is supported at both ends (Figure 20-2).

Open part BRIDGE and examine the bridge in its original shape.

Design Space

We want to know where to place material within the **Design Space** which is the space occupied by the tube. There is not enough material to fill out the entire Design Space; we have less material to work with because we want to reduce the mass. The objective is to place the material in such a way as to make the best use of what we have.

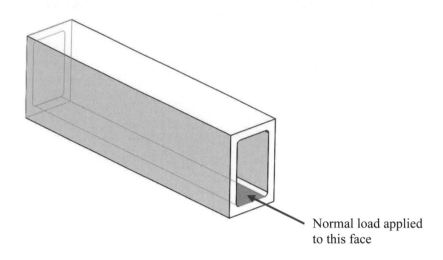

Normal load applied
to this face

Figure 20-1: BRIDGE in its original shape as a hollow rectangular tube. The tube is the Design Space. This is the maximum allowed volume of the BRIDGE.

Normal load is uniformly distributed over the face highlighted in blue.

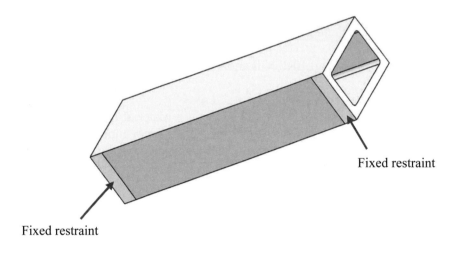

Fixed restraint

Fixed restraint

Figure 20-2: Fixed restraints are applied to two faces highlighted in yellow.

Rotation about supports is not allowed.

Create **Topology Study** called *50% mass reduction* as shown in Figure 20-3.

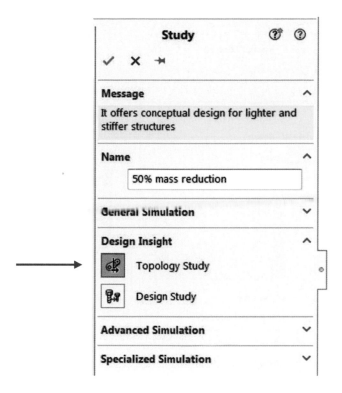

Figure 20-3: Definition of Topology Study.

Read the comment highlighted in yellow.

Define a uniformly distributed load to the face as shown in Figure 20-4. Define fixed restraints to two bottom faces highlighted in yellow (Figure 20-2).

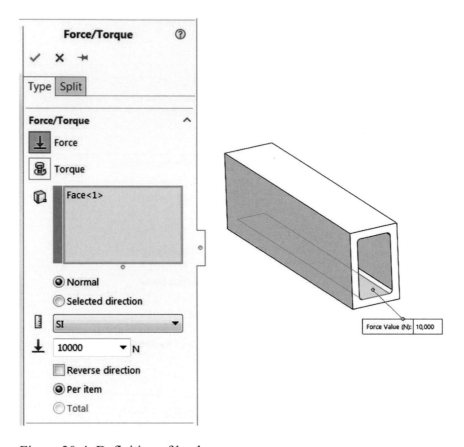

Figure 20-4: Definition of load.

The load magnitude does not matter in this qualitative study.

Goals and Constraints

Right-click **Goals and Constraints** in the study window to display the pop-up window listing three available optimization criteria; select the default **Best Stiffness to Weight ratio** (Figure 20-5).

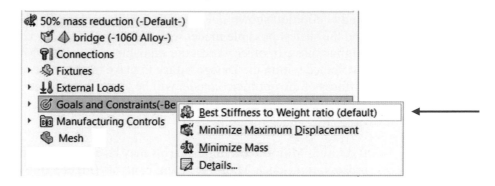

Figure 20-5: Selection of the optimization goal; select Best Stiffness to Weight ratio.

Notice the element order symbol showing the first order elements. First order elements are default in Topology Optimization studies.

Any of the three goals shown in Figure 20-5 may be used in Topology Study:

- Best Stiffness to Weight ratio
- Minimize Maximum Displacement
- Minimize Mass

Selection of the optimization goal opens the **Goals and Constraints** window shown in Figure 20-6.

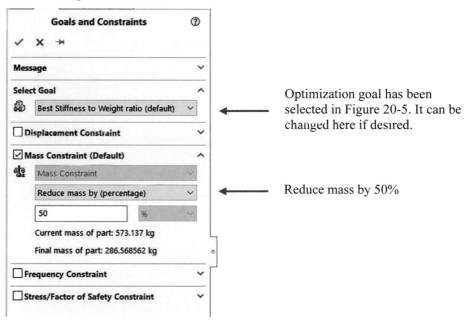

Optimization goal has been selected in Figure 20-5. It can be changed here if desired.

Reduce mass by 50%

Figure 20-6: Selection of the optimization goal.

The goal is to reduce mass by 50%. Mass before optimization is 573kg.

Target Mass after optimization is 287kg.

Study definition shown in Figure 20-6 will direct the Topology Optimization to find the stiffest possible model with mass approximately equal to 50% of the original mass. In other words, the available mass of 0.5*573kg = 287kg will be distributed within the **Design Space** to make the best use of material. Notice that stiffness of the optimized model will be lower than stiffness of the model before the optimization, but the ratio of stiffness to mass will be higher.

Manufacturing Controls

If desired, **Manufacturing Controls** may be added to control the optimization process; right-click **Manufacturing controls** (Figure 20-5) to display a list of **Manufacturing Controls.**

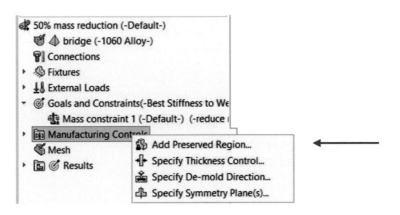

Figure 20-7: Selection of Manufacturing Controls.

Select Add Preserved Region.

Selecting **Add Preserved Region** opens window shown in Figure 20-8.

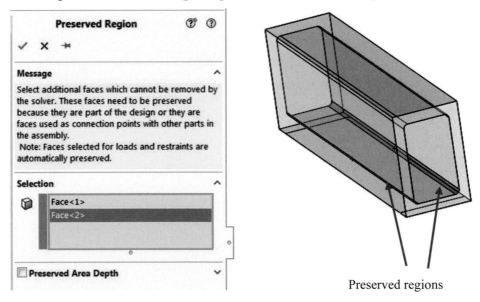

Figure 20-8: Definition of Preserved Region.

Select two fillets on both sides of the face where load is defined. Do not select Preserved Area Depth in this exercise.

Entities where loads and restraints are defined are treated automatically as **Preserved Regions**. All **Preserved Regions** will not be changed in **Topology Optimization** study.

Refer again to Figure 20-7 and select **Specify Thickness Control** to open window shown in Figure 20-9.

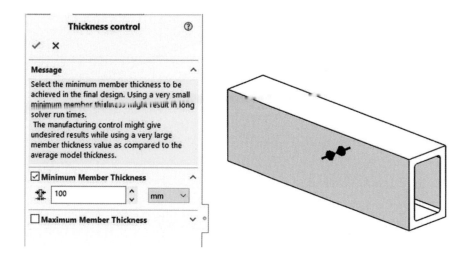

Figure 20-9: Definition of Thickness control; specify the minimum thickness 100mm.

Symbol of thickness control is shown.

Refer one more time to Figure 20-7 and select **Symmetry control** to open window shown in Figure 20-10. Select two reference planes shown in Figure 20-10 to define Quarter Symmetry.

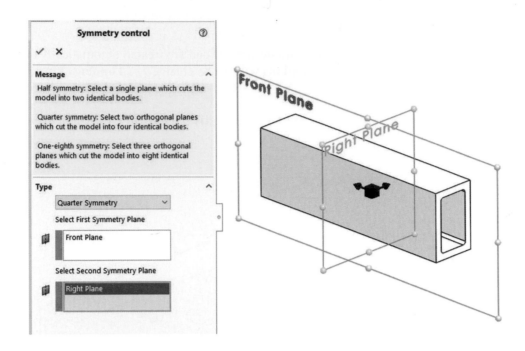

Figure 20-10: Definition of Symmetry control; select Quarter Symmetry.

Symbol of Quarter Symmetry control is shown.

Mesh

Create Mesh using default mesh settings in the **Topology Optimization** study. Review the mesh settings and notice that the Draft Quality Mesh is used, meaning that the mesh uses the first order elements.

Convergence of Goals and Constraints

Run study *50% mass reduction* and watch convergence graphs of **Goal: Best Stiffness** and of **Constraint: Mass** graphs during the solution process. You may also examine those graphs after the solutions ends; right-click *Results* folder and follow steps in Figure 20-11.

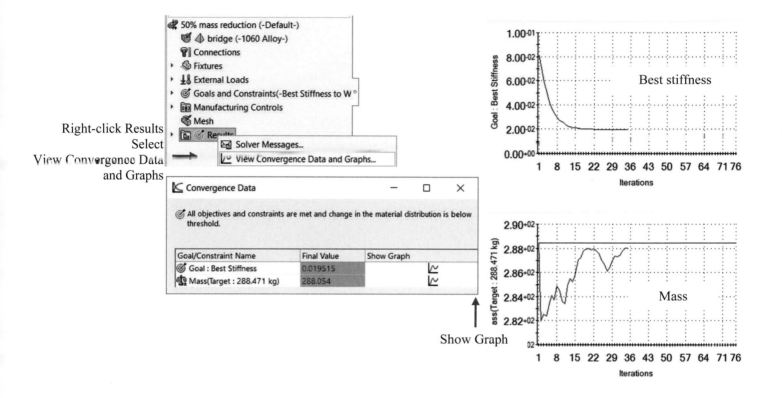

Figure 20-11: The progress of Topology Optimization can be visualized by selecting View Convergence Data and Graphs.

Convergence graph of Goal: Best Stiffness is shown in progress. Convergence graph of Constraint: Mass can be shown the same way.

Material Mass plot is a default plot created by **Topology Study**; it is shown in Figure 20-12.

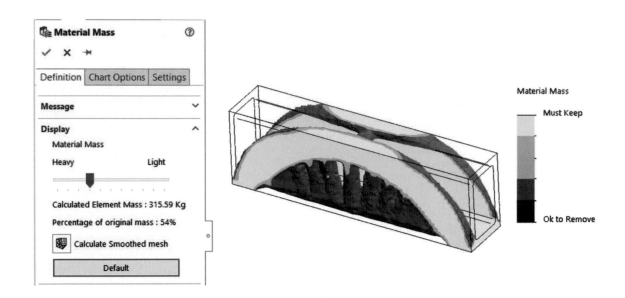

Figure 20-12: Material Mass plots, shows the model with mass closely matching the Mass Constraint. Colors indicate material that must be kept and material that may be removed.

Select Default to display Calculated Element Mass and Percentage of original mass.

The mass match is not exact; 50% mass reduction was specified in study definition; the plot shows model with 54% mass.

In **Material Mass** window set **Material Mass** slider in the middle between **Heavy** and **Light** and select **Calculate Smoothed Mesh**; this will display the plot shown in Figure 20-13.

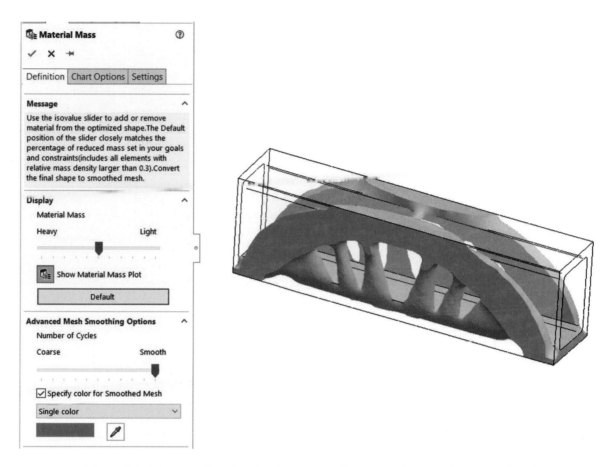

Figure 20-13: Smoothed Mesh plot shows elements with relative mass density larger than 30%. The wireframe indicates the Design Space.

You may toggle between Material Mass plot (Figure 20-12) and Smoothed Mesh plot (this plot) by selecting Show Material Mass plot or Calculate Smoothed mesh. A custom color for Smoothed Mesh has been specified for this plot.

Use **Material Mass** window (Figures 20-12 and 20-13) to experiment with different settings of the **Material Mass** slider that may be set anywhere on the scale **Heavy-Light**, and with different settings of the **Number of Cycles** slider that may be set anywhere on the scale from **Coarse** to **Smooth**.

Copy study *50% mass reduction* into a study named *65% mass reduction* and into a study named 80% *mass reduction*. Modify settings of two new studies as shown in Figure 20-12.

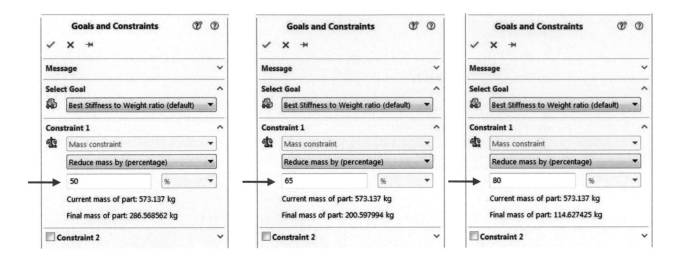

Study *50% mass reduction*   Study *65% mass reduction*   Study *80% mass reduction*

Figure 20-14: Summary of Goals and Constraints definition in three Topology Optimization studies.

Study 50% mass reduction has been already completed; it is shown here for reference only.

Run studies *65% mass reduction* and *80% mass reduction,* then summarize results as shown in Figure 20-15.

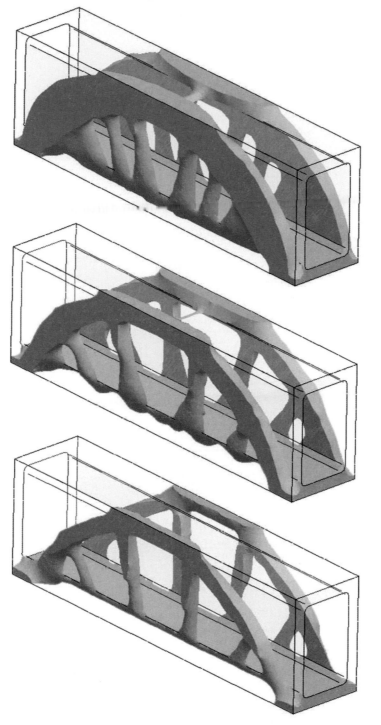

Study *50% mass reduction*

Study *65% mass reduction*

Study *80% mass reduction*

Figure 20-15: Smoothed mesh in three studies.

All meshes have been generated with slider in Smooth position in Advanced Mesh Smoothing Options.

Besides **Material Mass** plot, the results of **Topology Optimization** study may be analyzed using **Topology Stress Plot**, **Topology Displacement Plot** and **Topology Strain Plot** (Figure 20-16).

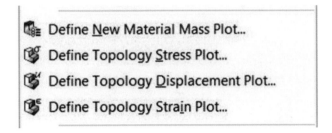

Figure 20-16: Topology plots are available in the Topology Study results.

All these plots offer qualitative results.

A **Topology Stress Plot** as shown in Figure 20-17.

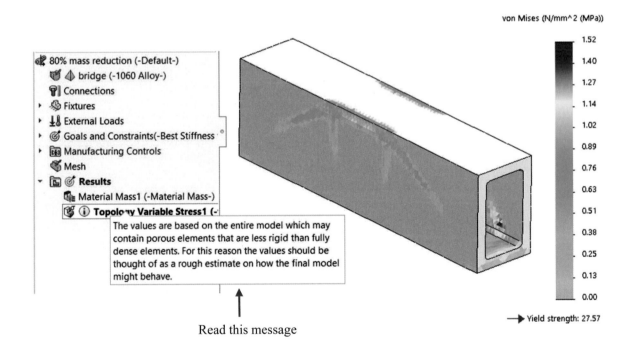

Read this message

Figure 20-17: Topology plots are available in the Topology Study results.

The qualitative nature of topology plots is explained in the pop-up window. Notice that the message should say "less stiff" rather than "less rigid".

The smoothed mesh can be saved as a new model configuration or as a new part. Follow steps in Figure 20-18 to save meshes shown in Figure 20-15 as new parts.

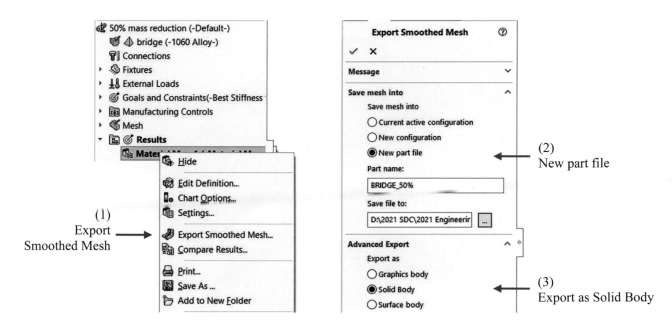

Figure 20-18: Creating a new part from a smoothed mesh.

In Advanced Export you may also select Graphics body or Surface body to reduce the file size.

Models created from smoothed meshes are shown in Figure 20-19. These models are saved as part files: BRIDGE_50%, BRIDGE_65%, BRIDGE_80%.

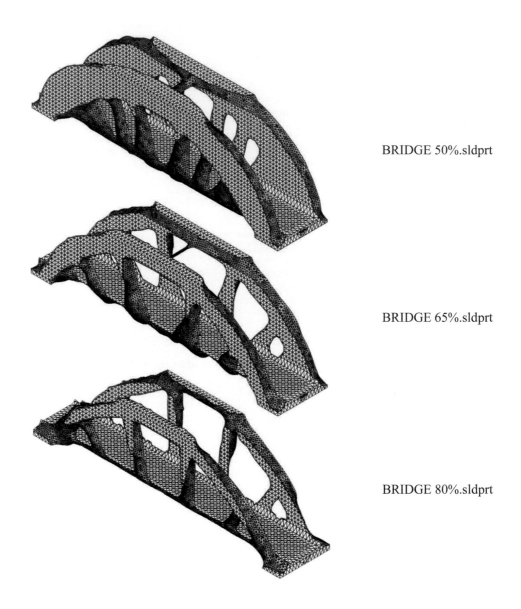

BRIDGE 50%.sldprt

BRIDGE 65%.sldprt

BRIDGE 80%.sldprt

Figure 20-19: Models created from Smoothed Mesh in three completed Topology Optimization studies.

These models look like finite element meshes but are not finite element meshes. Faces are small triangles, and this gives models an appearance of a finite element mesh.

The second example of **Topology Optimization** is presented using part model WATER BASIN (Figure 20-20).

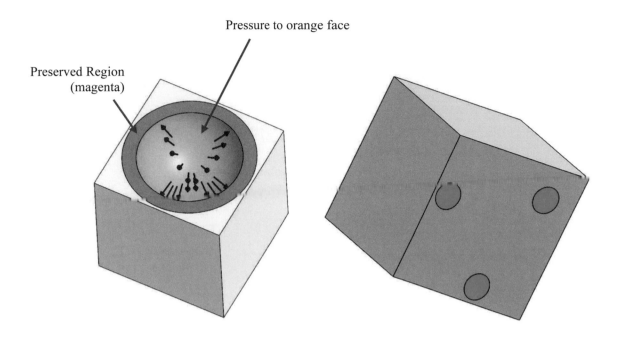

Figure 20-20: WATER BASIN model before Topology Optimization; it is a solid block of ceramic porcelain.

Symbols of restraints are not shown.

To simplify definition of this qualitative study, define water pressure as uniform 1MPa. Define fixed restraints to the three patches on the bottom face. We will not use the gravity load to drive the topology optimization process.

Use **Manufacturing Control** to define the magenta ring on the top face (Figure 20-20) as a **Preserved Region**; and define 100mm as the minimum thickness in **Thickness control**.

To make the best use of the available material **Topology Optimization** will place the available material along the load paths.

Define **Goals and Constraints** as shown in Figure 20-21.

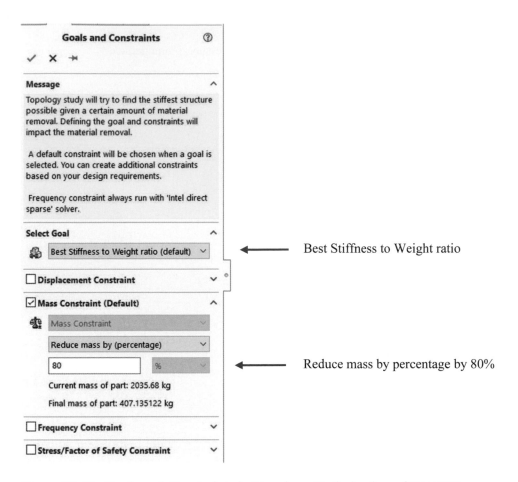

Best Stiffness to Weight ratio

Reduce mass by percentage by 80%

Figure 20-21: Goals and Constraints in Topology Optimization of WATER BASIN model.

80% mass reduction is requested. Mass before optimization is 2036kg.

The expected mass after optimization is 407kg.

The goal of **Topology Optimization** study is to place the available 407kg of ceramic porcelain in such a way as to produce a shape with the highest possible stiffness.

Run the study and review results shown in Figure 20-22.

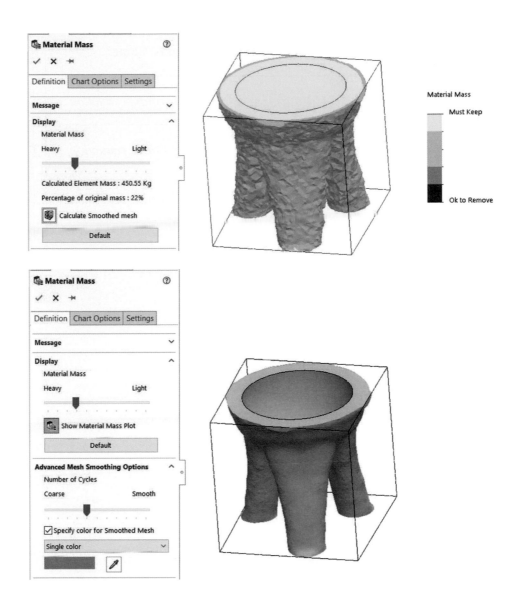

Figure 20-22: Material Mass plot (top) and Smoothed Mesh (bottom).

Mass after the optimization is 451kg; this is 22% of the original mass. Compare this to the optimization target 407kg.

Models in this chapter

Model	Configuration	Study Name	Study Type
BRIDGE.sldprt	*Default*	*50% mass reduction*	Topology Study
		65% mass reduction	Topology Study
		80% mass reduction	Topology Study
BRIDGE_50%.sldprt			
BRIDGE_65%.sldprt			
BRIDGE_80%.sldprt			
WATER BASIN.sldprt	*Default*	*Topology Study 1*	Topology Study

21: Miscellaneous topics – part 1

Topics covered

- Mesh quality
- Solvers and solver options
- Displaying the mesh in result plots
- Automatic reports
- E-drawings
- Non uniform loads
- Frequency analysis with pre-stress
- Interference fit analysis
- Rigid connector
- Pin connector
- Bolt connector
- Remote load/mass
- Weld connector
- Bearing connector
- Cyclic symmetry
- Strongly nonlinear problem
- Submodeling
- Automated detection of stress singularities
- Stress averaging at mid-side nodes
- Terminology issues in Finite Element Analysis

The analysis capabilities of **SOLIDWORKS Simulation** go beyond those we have discussed so far. In this chapter we review a variety of topics that have not been addressed in previous exercises. Some models in this chapter come with partially or fully defined studies.

Mesh quality

The ideal shape of a tetrahedral element is a regular tetrahedron. The aspect ratio of a regular tetrahedron is assumed to be 1. Analogously, an equilateral triangle is the ideal shape for a shell element. During meshing, elements are mapped onto model geometry. This distorts the element shape. The aspect ratio becomes higher when the element departs further from its original shape (Figure 21-1). An aspect ratio that is too high causes element degeneration, which negatively affects the quality of results.

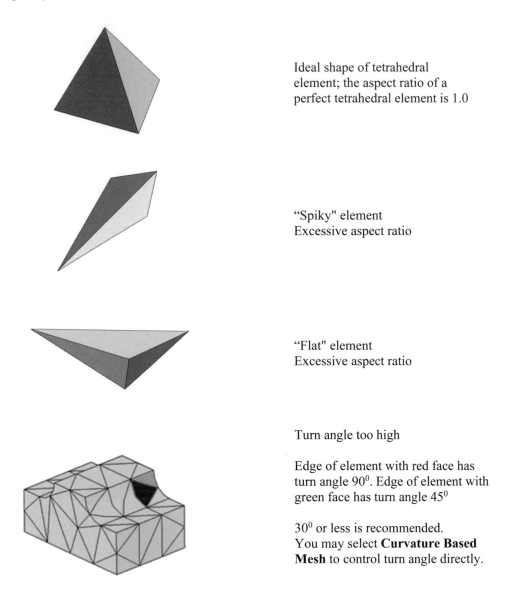

Ideal shape of tetrahedral element; the aspect ratio of a perfect tetrahedral element is 1.0

"Spiky" element
Excessive aspect ratio

"Flat" element
Excessive aspect ratio

Turn angle too high

Edge of element with red face has turn angle 90^0. Edge of element with green face has turn angle 45^0

30^0 or less is recommended. You may select **Curvature Based Mesh** to control turn angle directly.

Figure 21-1: Tetrahedral element shapes: ideal and after mapping.

A tetrahedral element with an ideal shape (top) has an aspect ratio of 1. "Spiky" and "flat" elements shown in this illustration (middle) have excessively high aspect ratios. "Concave" elements (bottom) have excessive turn angles.

While the mesher tries to create elements with aspect ratio close to 1, the nature of geometry sometimes makes it impossible to avoid high aspect ratios (Figure 21-2).

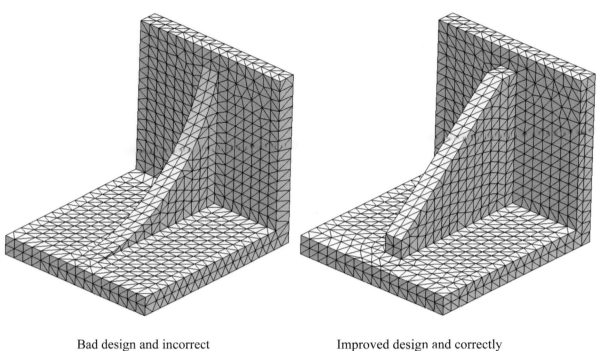

| Bad design and incorrect element shape | Improved design and correctly shaped elements |

Figure 21-2: Mesh of a curved and a straight gusset.

Meshing a curved gusset creates highly distorted elements near the tangent edges (left). Notice that this is also a bad design. Simple modification shown on the right improves the design and improves meshing. In both cases, large elements are used for the clarity of illustration. In most cases a more refined mesh would be required. Stresses along the sharp re-entrant edges will not be valid, regardless of the element size.

Review mesh problems illustrated in Figure 21-2 using the part model GUSSET.

A failure diagnostic can be used to spot problem areas if meshing fails. To run a failure diagnostic, right-click the **Mesh** icon. This opens the associated pop-up window from which **Failure Diagnostic …** is selectable.

With right-click the *Mesh* folder you can **Create Mesh plot** showing the mesh itself or mesh quality measures such as **Aspect Ratio** and **Jacobian**, explained in the table below, and shown in Figure 21-3.

Aspect ratio	The aspect ratio of an element is defined as the ratio between an element's longest edge and the shortest height normalized with respect to a perfect tetrahedron. The aspect ratio check assumes straight edges connecting the four corner nodes. It cannot differentiate between first order elements (straight edges) and second order elements which may have curved edges.
Jacobian	The Jacobian is a measure of the quality of second order elements. The Jacobian of an element, with all mid-side nodes located exactly at the middle of the straight edges, is 1.0. The Jacobian increases as the curvatures of the edges increase. The Jacobian at a point inside the element provides a measure of the degree of distortion of the element at that location. The software calculates the Jacobian at the selected number of points for each element.

Use the model MESH QUALITY 01 to review plots shown in Figure 21-3.

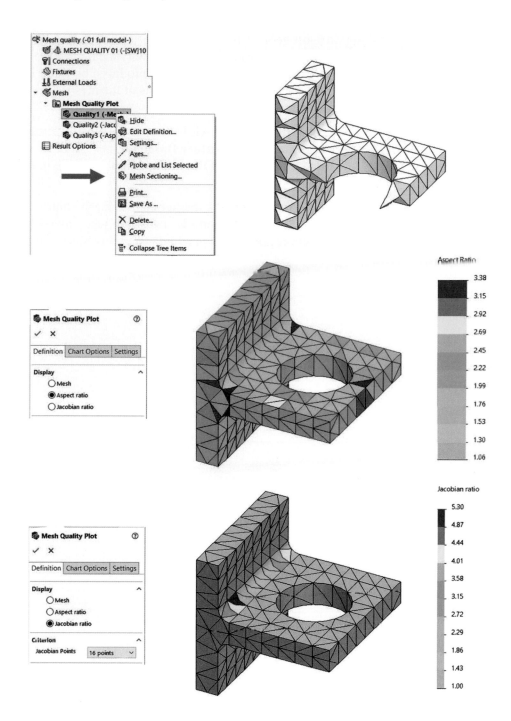

Figure 21-3: Mesh Quality plots: Mesh (top), Aspect ratio (middle) and Jacobian (bottom). For clarity of illustrations a coarse mesh is used for these plots.

Mesh plot is duplicated to demonstrate mesh sectioning. Jacobian plot may be based on evaluation using different number of points as specified in Criteria. The default number is 16.

Thick legend with eight colors is shown in the Aspect Ratio plot. Thin legend with user defined twelve colors is shown in the Jacobian plot.

Meshing difficult geometries may sometimes result in highly distorted elements without any warning. If mesh distortion is only local, then we can simply not look at the results (especially the stress results) produced by those degenerated elements. If distortion affects large portions of the mesh, then even global results cannot be trusted.

We have studied different methods to control the mesh size such as **Mesh Controls**, **Standard** and **Curvature Based** meshes. Another way to control a mesh is to use the **Automatic Transition** option available in the **Standard** mesher.

With **Automatic transition**, the mesher applies mesh controls to small features (Figure 21-4). Do not use **Automatic transition** when meshing large models with many small features and details to avoid generating a very large number of elements.

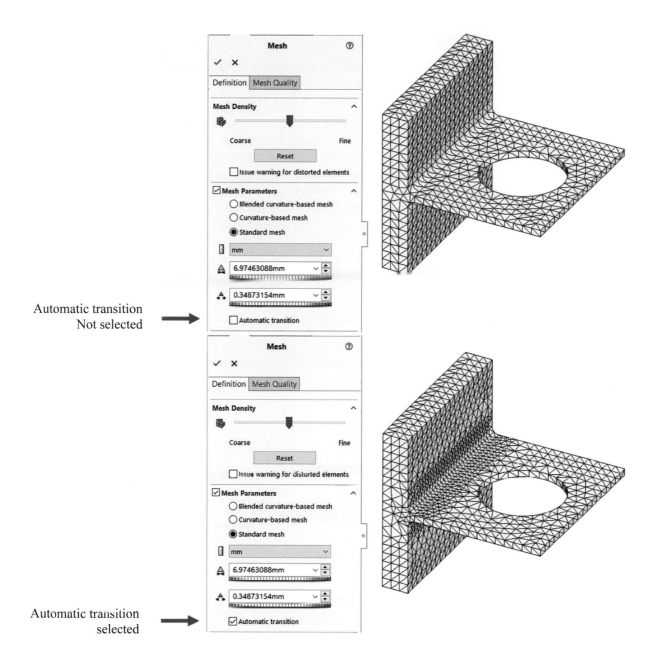

Automatic transition
Not selected

Automatic transition
selected

Figure 21-4: Mesh without Automatic Transition option (top) and mesh with Automatic Transition option (bottom).

Automatic transition has a similar effect to applying mesh controls to the fillets.

Review the effect of **Automatic transition** using the model MESH QUALITY 02.

Solvers and Study Options

In Finite Element Analysis, a problem is represented by a large set of linear algebraic equations that must be solved simultaneously. There are two classes of solution methods: direct and iterative.

Direct methods solve equations using exact numerical techniques, while iterative methods solve equations using approximate techniques. With an iterative method, a solution is approximated iteratively and the associated errors are evaluated. The iterations continue until the errors become acceptable. The **Direct Sparse** solver (usually slower) uses a direct solution technique, and the **FFEPlus** solver uses an iterative technique. **Large Problem Direct Sparse** solver can handle simulation problems that exceed the physical memory of your computer. There are also Intel specific solvers. **Automatic** solver option allows for automatic selection of solver best suited for the type of problem. A solution can therefore be run with five solver options (Figure 21-5).

Automatic or Manual solver selection

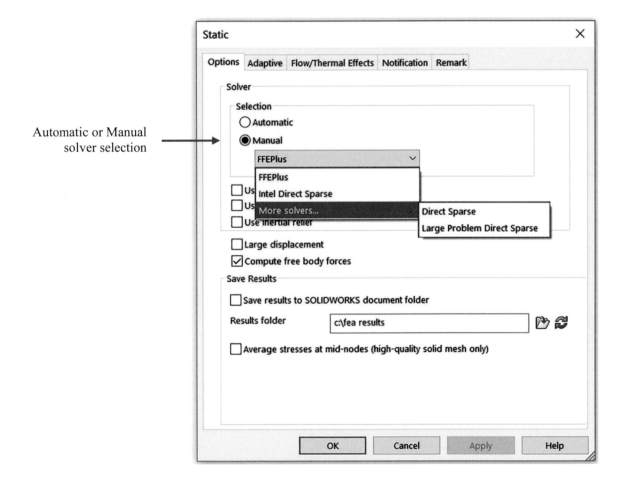

Figure 21-5: Different solvers are available in SOLIDWORKS Simulation.

Using Automatic Solver Selection, the fastest solver available for a given problem is automatically selected. Refer to help for description of solver types.

Different solver types give comparable results if the required options are supported. It is generally recommended to use the **Automatic** option to select the solver automatically based on user specified solver options.

Three solution options are available in Static study:

Option	Purpose
Use in plane effect	In a static analysis, use this option to account for changes in structural stiffness due to the effect of stress stiffening (when stresses are predominantly tensile) or stress softening (when stresses are predominantly compressive).
Use soft springs to stabilize model	Use this option primarily to locate problems with restraints that result in rigid body motion. If the solver runs without this option selected and reports that the model is insufficiently constrained (an error message appears), the problem can be re-run with this option selected (checked). Insufficient restraints can then be detected by animating the displacement results. An alternative to using this option is to run a frequency analysis, identify the modes with zero frequency (these correspond to rigid body modes), and animate them to determine in which direction the model is insufficiently constrained.
Use inertial relief	Use this option if a model is loaded with a balanced load, but no restraints. Due to numerical inaccuracies, the balanced load will report a non-zero resultant. This option can then be used to restore model equilibrium. This option is most often used to balance a model with loads imported from a Motion Simulation.

Problem definition option and analysis of result option:

Option	Purpose
Large displacement	See Chapter 14
Compute free body forces	Enables probing of forces and moments transmitted by nodes.

Displaying mesh in result plots

The default brightness of **Ambient** light, defined in the **SOLIDWORKS Display Manager** in the **Lighting** folder, is usually too dark to display the mesh, especially a high-density mesh. A clear display of the mesh (Figure 21-6) requires increasing the brightness of ambient light. You will also notice that using models with light colors, disabling perspective, shading and other display effects make it easier to work with **Simulation** models.

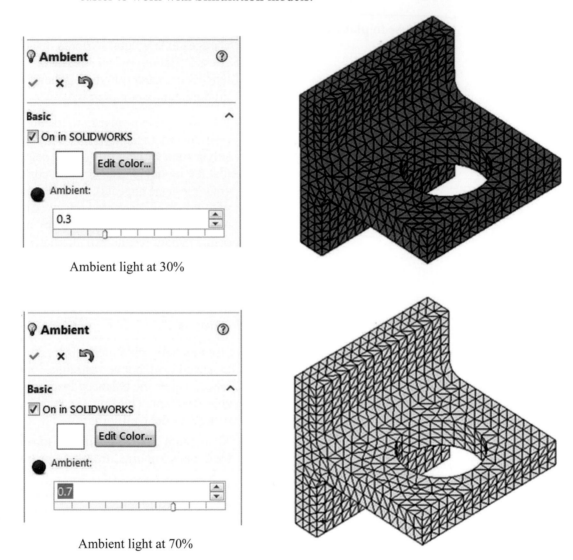

Ambient light at 30%

Ambient light at 70%

<u>Figure 21-6: Mesh display in default ambient light (top) and adjusted ambient light (bottom).</u>

A finite element mesh is displayed with ambient light brightness suitable for a CAD model (top), and with the brightness adjusted for displaying the mesh (bottom). All lights are defined in SOLIDWORKS model.

Many models used in this book use bright colors, bright ambient light, and directional lights.

Automatic reports

SOLIDWORKS Simulation provides automated report creation. After a solution finishes, select Report from the Simulation menu or from the Simulation **Command Manager** to define the report format and the items it will contain. The report is created in a few steps and contains all plots from the result folders (Figure 21-7).

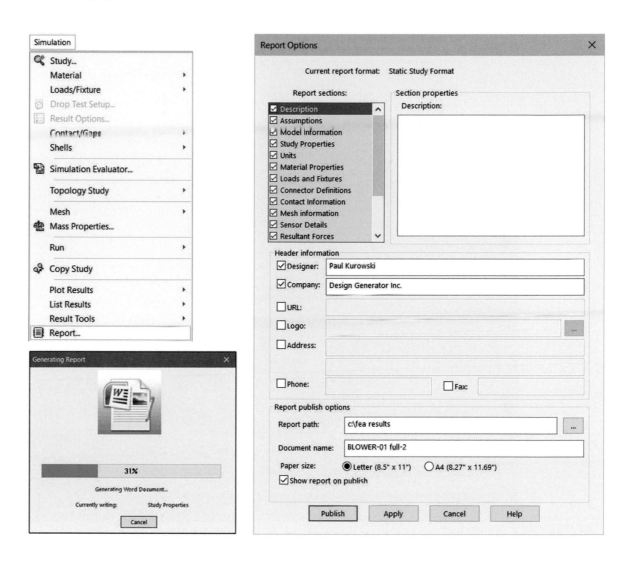

Figure 21-7: The Report is automatically created with the options specified in the Report Options windows.

Click Publish to create a report. Reports in MS Word format are created individually for each study.

eDrawings

Each result plot can be saved in various graphic formats (Bitmap, JPEG, PNG, VRML), as well as in the **SOLIDWORKS** eDrawings format. To save a plot, first show it then right-click it and select **Save as**. Select the desired format from the pop-out menu.

Examine the **eDrawings** format which offers a very convenient 3D way of communicating analysis results to people who do not have **SOLIDWORKS Simulation** or no **SOLIDWORKS** at all (Figure 21-8).

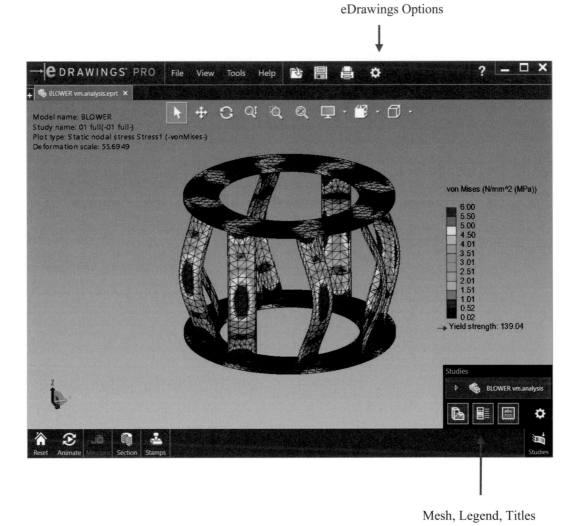

Figure 21-8: Simulation results saved in eDrawing format can be viewed in 3D.

Review options and choices indicated in this illustration.

Non-uniform loads

We will illustrate the use of non-uniformly distributed loads with an example of hydrostatic pressure acting on the walls of a 2m deep tank, in TANK model. Notice that this model uses meters for the unit of length. The pressure magnitude expressed in [N/m^2] follows the equation p = 10000x, with x being the distance from the top of tank (where the coordinate system *cs1* is located). The pressure definition requires selecting the coordinate system and the face where pressure is to be applied. The formula governing pressure distribution can then be entered, as shown in Figure 21-9.

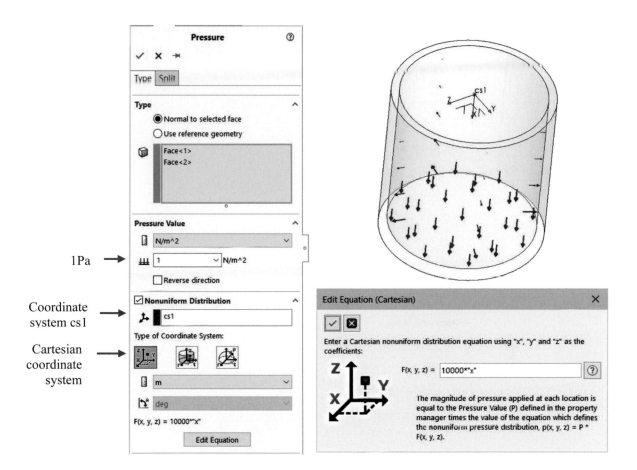

Figure 21-9: A water tank loaded with hydrostatic pressure requires a linearly distributed pressure. The pressure is defined in a Cartesian coordinate system cs1.

Notice that the vector lengths correspond to pressure magnitudes that vary with the x axis of the cs1 coordinate system. This model uses meter as unit of length, and Pascal as unit of pressure.

This illustration uses a transparent model view. Important geometry details such as fillets are not modeled for the clarity of this illustration.

Complete this exercise by applying restraints and material of your choice.

Frequency analysis with pre-stress

A frequency analysis of rotating machinery most often must account for stress stiffening. Stress stiffening is the increase in stiffness due to tensile loads. We will illustrate this concept with the example of a helicopter blade. Since HELICOPTER ROTOR model has four identical blades, switch to the *02 section* configuration, which will work with geometry containing only one blade. The model comes with assigned material properties and two defined **SOLIDWORKS Simulation** studies: *no preload* and *preload*.

A load definition is not required in a **Frequency** analysis. However, if loads are defined, their effect will be considered. The **Direct sparse** solver is the only solver that accounts for the effect of loads in a **Frequency** analysis.

A centrifugal load is defined as shown in Figure 21-10. An axis or a cylindrical face is required as a reference to define a centrifugal load. Review the restraint, which is the same in both studies. Restraints may be represented as **Fixed** restraints applied to hub. Selecting the cylindrical face (Figure 21-10) defines the axis of rotation.

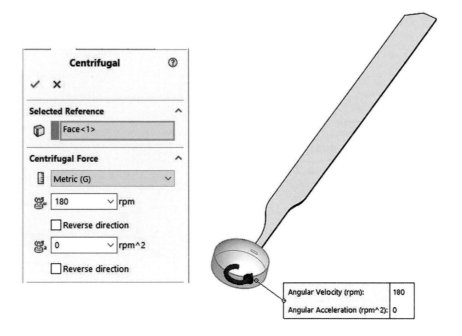

Figure 21-10: The Centrifugal load window with centrifugal force defined.

A centrifugal load resulting from an angular velocity of 180 RPM is applied to the model, simulating the effect of rotation about the axis of the cylindrical face. Angular acceleration can also be defined. Metric units are used to express the angular velocity in RPM.

Solve both studies and compare frequency results (Figure 21-11) with and without the pre-stress effect caused by a centrifugal force. **Frequency** study with pre-load requires **Direct** solver; select **Direct Sparse** or **Intel Direct Sparse** or select **Automatic**.

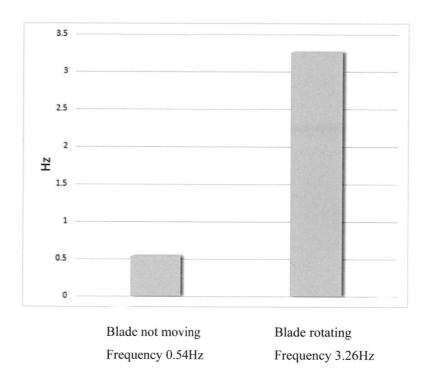

Blade not moving

Frequency 0.54Hz

Blade rotating

Frequency 3.26Hz

Figure 21-11: The frequency of the first mode of vibration without and with the effect of a centrifugal load.

As shown in Figure 21-11, the presence of preload significantly increases the first mode of vibration. In the first mode, the frequency increases 600%.

The opposite effect (decrease in the natural frequency) would be observed if, hypothetically, the blades were subjected to a compressive load.

For more practice, try conducting a frequency analysis of a beam under a compressive load. You may re-use the HELICOPTER ROTOR model. The higher the compressive load, the lower the first natural frequency. Higher frequencies follow the same pattern. The magnitude of the compressive load that causes the first natural frequency to drop to zero is the buckling load. This is where frequency and buckling analyses meet.

Interference fit

Interference fit, called in **Simulation Shrink Fit**, is another type of **Contact** condition. We use it here to analyze stresses developed as a result of the interference (press fit) between two assembly components. Open the SHRINK FIT assembly. The definition of the interference fit condition is shown in Figure 21-12. Review this model for definitions of restraints, supports and contact conditions.

The diameter of the hole (housing) is 20mm and the diameter of the shaft 20.050mm. In the Basic Hole fits system this is force fit 20 H7/u6.

Notice that the contact condition does not include friction; therefore, the inside cylindrical face of the pressed-in component has been restrained in circumferential and axial directions to prevent rigid body motions.

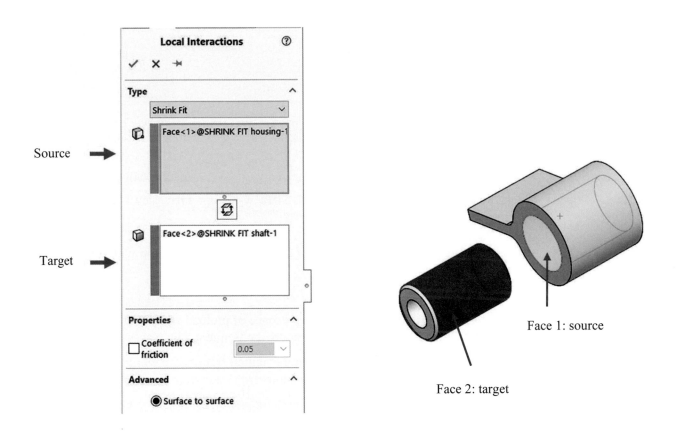

Figure 21-12: Cylindrical Face 2 has a larger diameter than cylindrical Face 1. Solving the model with a Shrink Fit contact condition eliminates this interference.

An exploded view is used to select the interfering faces.

Apply a restraint to the "tail" of the housing and restrain the hole of the shaft as shown in Figure 21-13. This is necessary to eliminate rigid body motions of the shaft.

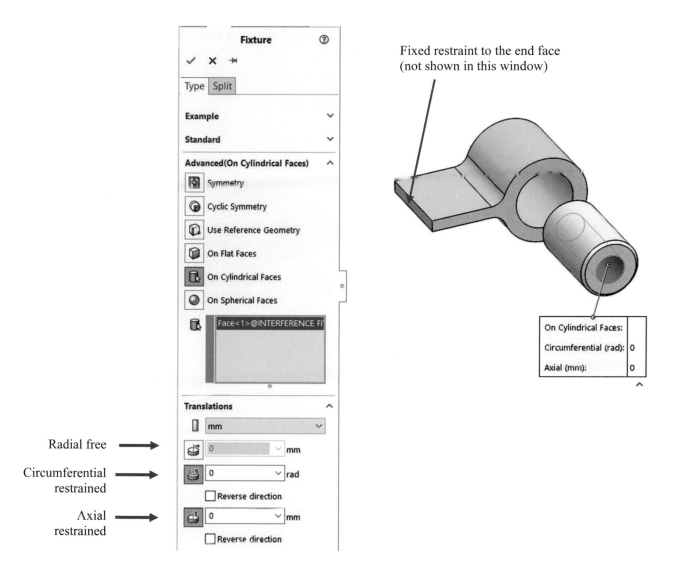

<u>Figure 21-13: Restraints applied to the hole in the shaft are needed to eliminate rigid body motions.</u>

Restraints are defined in the local cylindrical coordinate system using an On Cylindrical Faces restraint. The same could have been defined using Axis1 as the reference geometry.

There are no external loads acting on the model.

Mesh the model with a 1mm element size using a **Curvature based mesh** and obtain the solution. Display the SX stress plot using Axis1 as reference geometry to convert SX stress into radial stress (Figure 21-14). When stresses SX, SY, and SZ are plotted using an axis as a reference, as in Figure 21-14, SX becomes radial stress, SY becomes circumferential stress and SZ becomes axial stress. The symbol in the lower right corner indicates that the results are presented in a local cylindrical coordinate system.

Axis1 defines the cylindrical coordinate system

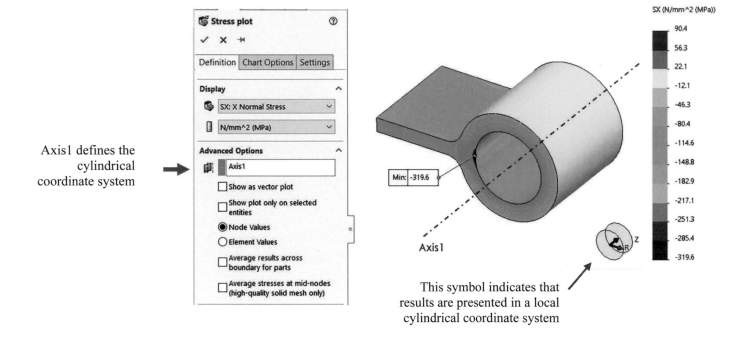

This symbol indicates that results are presented in a local cylindrical coordinate system

Figure 21-14: Radial stresses developed due to the force fit.

Notice that on contacting faces, the magnitude of contact stress is the absolute magnitude of SX stress. The highest magnitude of contact pressure is the numerically lowest number: -320MPa. Detailed analysis of this stress concentration would require a more refined mesh.

Review the options in the **Stress Plot** window in Figure 21-14 and use vector display option to visualize radial direction of SX stress (Figure 21-15).

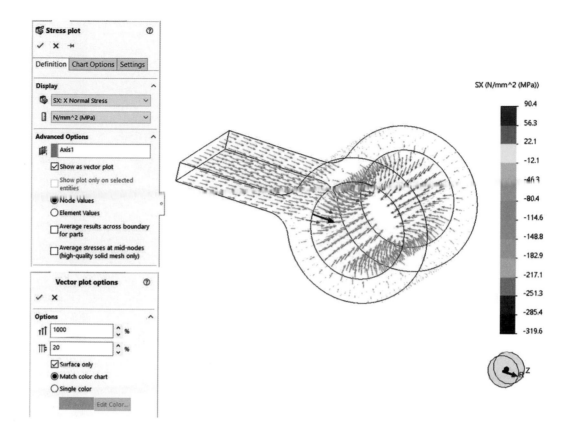

Figure 21-15: Radial stresses SX presented as vectors.

Use the Vector plot options window to adjust the vector plot.

Only the housing component is shown; the shaft is hidden in SOLIDWORKS assembly model.

Using unexploded assembly view, hide Shaft component, return to the fringe plot, and probe radial stresses on contacting face of Housing. Next, show Shaft and hide Housing component and probe radial stress on the face of Shaft. Results of both probes are shown in Figure 21-16.

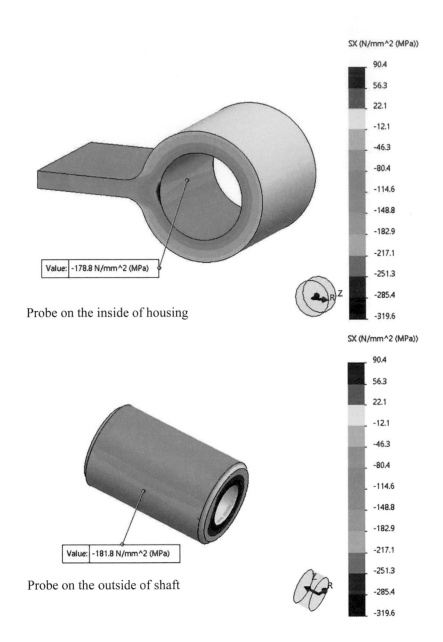

SX (N/mm^2 (MPa))

Value: -178.8 N/mm^2 (MPa)

Probe on the inside of housing

SX (N/mm^2 (MPa))

Value: -181.8 N/mm^2 (MPa)

Probe on the outside of shaft

Figure 21-16: Radial stress SX on the face of the tube (top) approximately equals the radial stress SX on the side of the hole (bottom) because probing has been done in approximately the same location. In both cases SX stress is compressive.

Radial stresses correspond to contact pressure which is the same on both contacting faces due to the equilibrium conditions. In the housing, probing is done on the inner face, which is not visible in the view used in this illustration.

Right-click stress plot and select Probe to invoke Probe window (not shown in this illustration).

Connectors

A connector defines how an entity (vertex, edge, face) is connected to another entity or to the ground. Using connectors simplifies modeling because, in many cases, you can simulate the desired behavior without having to create the detailed geometry or define contact conditions.

SOLIDWORKS Simulation offers several types of connectors listed as **Spring**, **Pin**, **Bolt**, **Bearing**, **Spot Welds**, **Edge Welds**, **Link**, **Rigid Connection** and **Linkage Rod** (Figure 21-17). Selected connectors are briefly introduced in this chapter. For more information, refer to the SOLIDWORKS Simulation help documentation which offers extensive explanations with examples. To learn about Simulation functionality (including connectors), you may also use Advisor found in the SOLIDWORKS task pane.

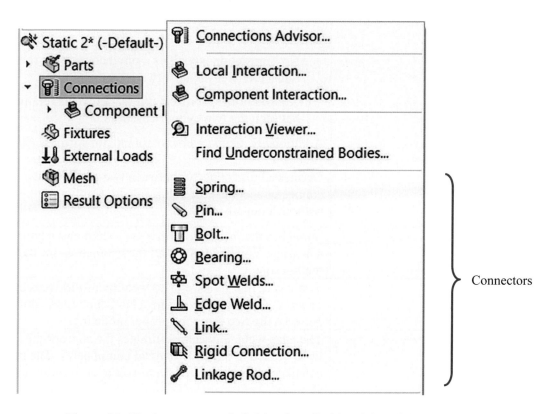

Figure 21-17: A connector definition is called by right-clicking the Connectors folder and selecting the desired connector type.

Connectors available in **SOLIDWORKS Simulation** are summarized in the table below.

TYPE OF CONNECTOR	FUNCTION
Spring	Connects a face of a component to a face of another component by defining total stiffness or stiffness per area. Both normal and shear stiffness can be specified. The two faces must be planar and parallel to each other. Springs are introduced in the common area of the projection of one of the faces onto the other. You can specify a compressive or tensile preload for the spring connector.
Pin	A pin connects cylindrical faces of two components. Two options are available: No Translation: specifies a pin that prevents relative axial translation between the two cylindrical faces. No Rotation: specifies a pin that prevents relative rotation between the two cylindrical faces. Additionally, axial and/or rotational pin stiffness can be defined.
Bolt	Defines a bolt connector between two components. The bolt connector accounts for bolt pre-load. Configurations with and without a nut are available.
Bearing	Simulates the interaction between a shaft and a housing through a bearing. You have to model the geometries for the shaft and the housing.
Spot weld	You can define spot welds to weld two solid faces or two shell faces. You should also define a No Penetration contact condition between the two faces for proper modeling.
Edge weld	The edge weld connector estimates the appropriate size of a weld needed to attach two metal components. The program calculates the appropriate weld size at each mesh node location along the weld seam.
Link	The Link connector ties any two locations in the model by a rigid bar that is hinged at both ends. The distance between the two locations remains unchanged during deformation. The link connector is available for static, buckling, and frequency studies.
Rigid Connection	Defines a rigid link between the selected faces. Faces connected by a rigid link do not translate or rotate in relation to each other.
Linkage rod	

Pin Connector and Linkage Rod Connector

We will review the use of a **Pin** and **Linkage Rod** connectors using assembly CRANE. The assembly in *01 full* configuration requires three **Pin Connectors** as shown in Figure 21-18.

Selected faces in Pin Connector definition window →

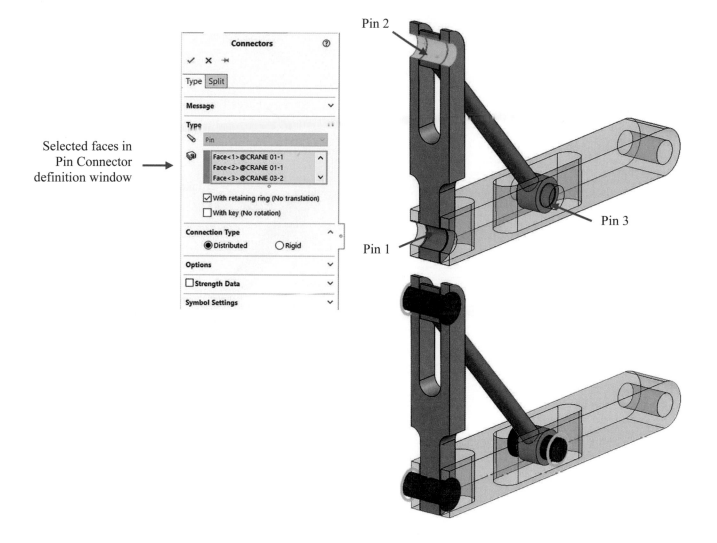

Figure 21-18: The Pin connector connects three faces of two components (top). The blue pins are the pin connector symbols.

Both illustrations use section views and transparency to show pin connector symbols. You need to select the Strength Data box to perform a pass/no pass pin check.

Definition of **Linkage Rod** connector is shown in Figure 21-19. This is done in configuration *02 simplified* where one assembly component is replaced by **Linkage Rod** connector.

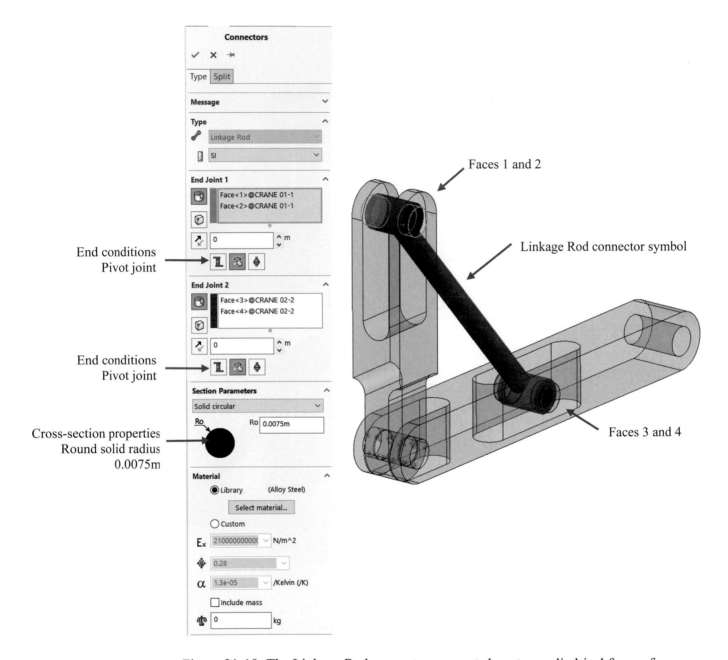

<u>Figure 21-19: The Linkage Rod connector connects here two cylindrical faces of two components.</u>

Rigid joint, Pivot joint and Spherical Joint end conditions may be selected individually for each end of Linkage Rod connector.

Linkage Rod symbol is a realistic representation of a rod with two pivots. It is shown in a custom color.

Both assembly components are shown in transparent view.

Notice that the torsional stiffness of the **Pin** connector shown in Figure 21-18 is specified as 0 (this is the default value). This means that **Pin** connectors allow for rotation between the two components. All degrees of freedom on the selected faces are coupled (must be the same) except for circumferential translations, which are disjoined. Axial and torsional stiffness is specified in Advanced Options. **Pin** mass may be specified in **Pin** connector definition.

Practice using **Pin** connectors using the assembly STAND. Our objective is to find the first mode of vibration of the assembly. This model has little relevance to real life devices but offers a good opportunity to study **Pin** connectors and contact conditions.

Create Frequency study called *Modal 01* and delete **Global Interaction** condition. In the absence of **Global Interaction** condition, all touching faces are treated as free. Consequently, the four links are not bonded to the top or to the bottom plate. Connect them with **Pin** Connectors as shown in Figure 21-20.

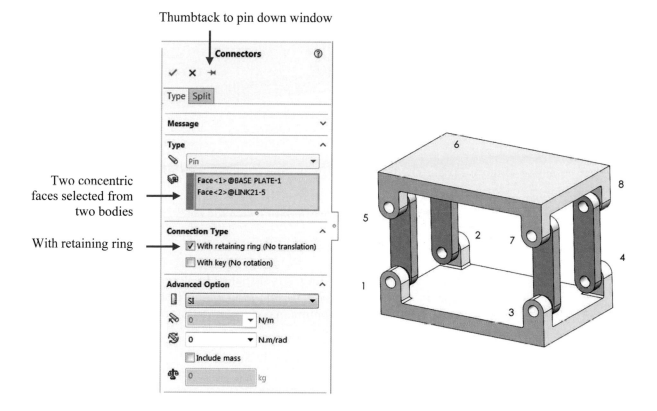

Figure 21-20: Definition of one of eight Pin connectors connecting legs with the top and bottom plate. Connectors are numbered 1, …,8.

You may pin down the Connectors window to create several pin connectors in one step. All connectors will be placed in a separate folder.

Notice that "pin" as in "Pin down the window" has nothing to do with "pin" as in the Pin connector.

Apply **Fixed** restraint to the bottom face of the bottom place and solve *Modal 01* study. Review the results and notice that the frequency of the first mode is close to zero; this is a **Rigid Body Mode**. Model with all **Pin** connectors allowing rotations is a four bar linkage mechanism. It can collapse in the direction indicated by animation of the first mode (Figure 21-21).

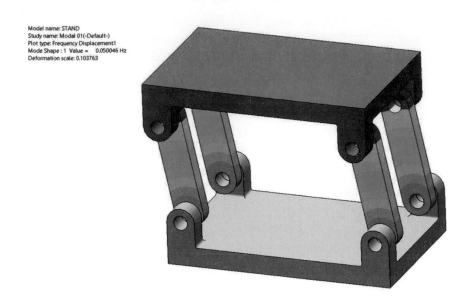

Model name: STAND
Study name: Modal 01(-Default-)
Plot type: Frequency Displacement1
Mode Shape : 1 Value =   0.050046 Hz
Deformation scale: 0.103763

Figure 21-21: The first mode of vibration in model with all Pin connectors allowing rotation.

Animate this mode to see the direction in which the model will collapse as a mechanism.

To eliminate the **Rigid Body Mode** it is sufficient to eliminate rotation in just one **Pin connector**. Copy study *Modal 01* into *Modal 02*. To eliminate rotation in **Pin connector 1**, select **With key** option in connector definition window (Figure 21-20).

Figure 21-22 shows the first four modes of vibration with rotation in **Pin connector 1** eliminated.

Mode 1: 121Hz

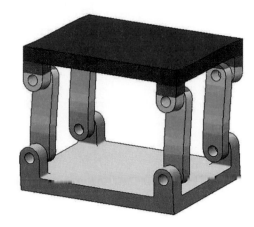

Mode 2: 238Hz

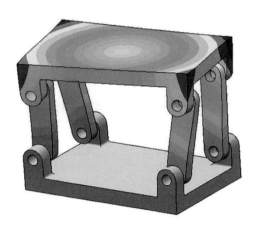

Mode 3: 325Hz

Mode 4: 651Hz

Figure 21-22: The first four modes of vibration of STAND with rotations in Pin connector 1 eliminated.

Examine closely deformations and notice bending in the link where rotation in Pin connector is eliminated.

Bolt connector

To review **Bolt** connector as well as load type called **Remote Load**, open the assembly PIPES in the *01 long* configuration and go to **Simulation** study *01 long*. The model comes with six bolt connectors already defined. Right-click one of the **Bolt Connector** icons and select **Edit Definition** to open the **Connectors** windows (Figure 21-23).

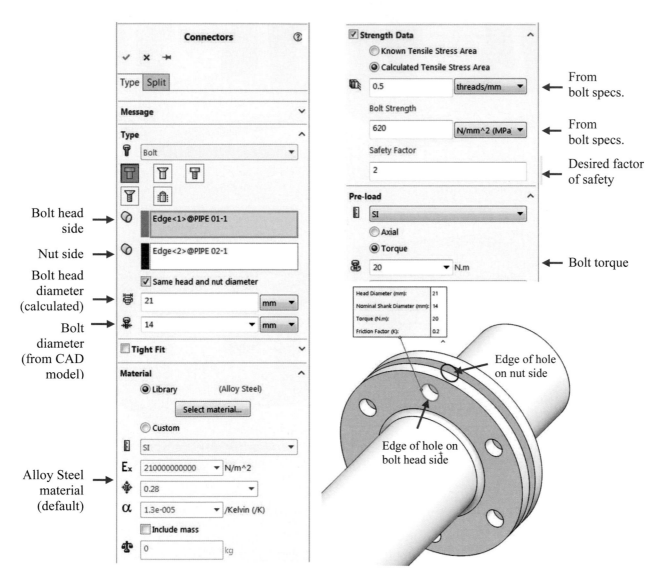

Figure 21-23: One of six bolt connectors in the PIPES assembly model in *02 short* configuration. A no penetration contact condition must be defined between the touching faces of the two flanges. To transmit the shear force in the absence of friction between the flanges, a Tight Fit must be defined.

The pre-load definition indicates that each bolt is loaded with 20Nm of torque; the Friction Factor is 0.2. Strength data is required only if bolt check is to be conducted.

Defining the **Bolt Connector** offers several options. In this example, we model the bolt with a nut. The bolt is made out of Alloy Steel, has a loose fit, and a diameter of 14mm. Automatically calculated diameters of the bolt head and nut are accepted. The bolt is preloaded with a torque of 20Nm. Review **Advanced Options** before proceeding.

There is one set of contacting faces in the model and a contact condition must be defined, so you may define **No Penetration** as a **Global Contact**, **Component Contact** or a **Contact Set**.

Define six bolt connectors, loads and restraints as shown in Figure 21-24.

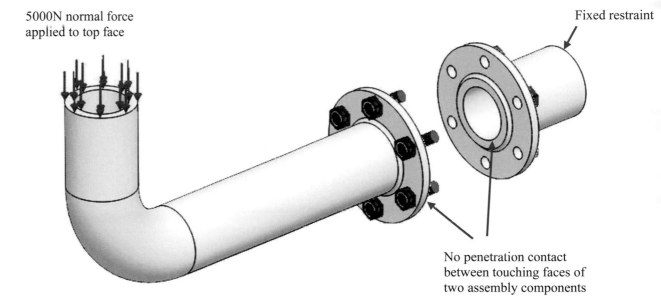

5000N normal force applied to top face

Fixed restraint

No penetration contact between touching faces of two assembly components

Figure 21-24: Loads, restraints, connectors and contact conditions defined in PIPES model.

Exploded view is used; restraints symbols are not shown.

Apply mesh control 1.5mm to four rounds present in the assembly and mesh the model with 5mm mesh size. Upon solution, review the bolt forces which are available by right-clicking on the *Results* folder and selecting **Define Pin/Bolt Check Force**. (Figure 21-25).

Click OK to open Pin/Bolt Check Plot
window shown below

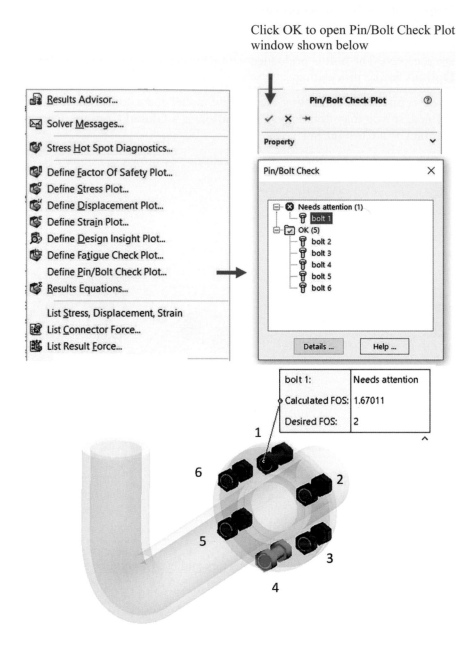

Figure 21-25: The pop-up menu activated by right-clicking the Results folder
(left). The Pin/Bolt Check window (top right) shows failed bolts in red. Bolt 1
needs attention; Bolts 2, 3, 4, 5, 6 are OK.

This window is accompanied by a plot locating passed and failed bolts.

Review von Mises stress results (Figure 21-26). Maximum von Mises stress magnitude is above the yield strength indicating that the assembly is overloaded.

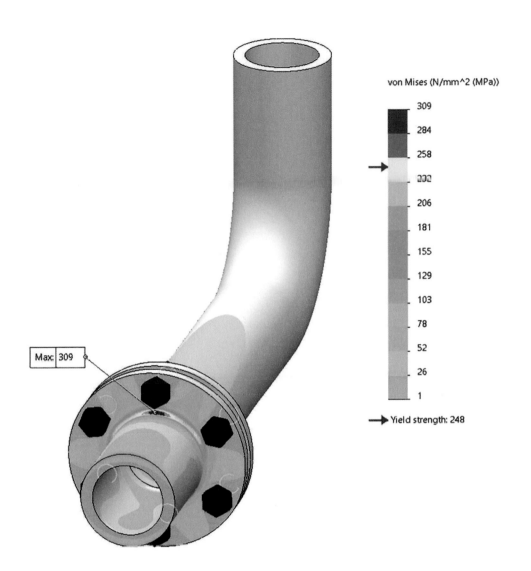

Figure 21-26: Von Mises stress results; stress concentrations are located in fillets.

If stress above yield does not disqualify the design right away, it certainly requires an analysis with a non-linear material to investigate the extent of the yield zone.

Remote Load/Mass

Stay with the PIPES assembly model and switch to the *02 short* configuration and go to **Simulation** study *02 short*. We use this configuration to demonstrate a **Remote load.** It allows to reduce the model size by cutting the pipe and applying a load as if the eliminated portion of the model were still present (Figure 21-27).

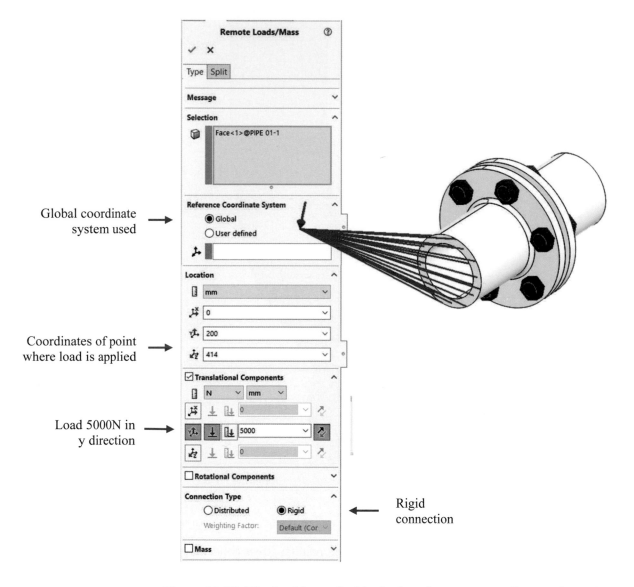

Figure 21-27: The load is applied in the location corresponding to Figure 21-24. The point of the load application is rigidly connected to the end face of the pipe in the *02short* configuration.

The load is transferred from the point of application to the face by means of rigid connectors.

There are two types of **Remote Loads**: **Rigid** and **Distributed**. The choice between these two types depends on the stiffness of the suppressed/eliminated component as compared to the rest of the model. A thick wall pipe, as in our case, stiffens the end face of the short flange where it is attached, so we specify a **Rigid** connection.

For example, a thin-walled tube would just transfer the load to the face of the link and therefore the **Distributed** connection type would be a better choice. **Remote Loads/Mass** can also be used to define a remote mass. **Distributed** connection and **Mass** options are not used in PIPES exercise.

Remote Loads/Mass also offers a convenient way to apply a moment to a solid element model. Using **Remote load**, moment loads are automatically translated into equivalent force loads.

Apply a fixed restraint to the opposite end of the model in configuration *02 short*. Apply a mesh control of 1.5mm to all rounds and mesh the model with 5mm mesh size (Figure 21-28).

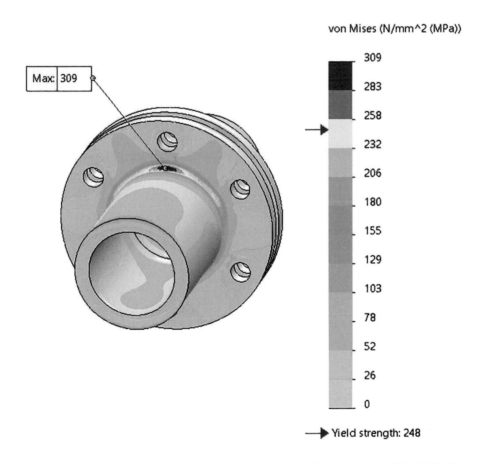

Figure 21-28: Von Mises stress plot produced by the simplified *02 short* configuration model with remote load.

Bolt connectors are not shown.

Repeat this exercise using h adaptive solution.

Edge Weld connector

Open the assembly TUBE WELDMENT (Figure 21-29). The assembly consists of a square hollow tube modeled as a solid, endplates modeled as solids, and a hanger modeled as a surface. The hanger is connected to the tube by welds. We need to check if the welds are "strong enough."

We use an **Edge Weld** connector to connect the plate to the tube and calculate the weld loads. **Edge Welds** model connections along a line where the weld would be located. Notice that the actual weld is not modeled, and welds are represented by lines connecting the surface to the solid. Loads transmitted by those lines are calculated and these loads are then used to assess if the specified weld is adequate.

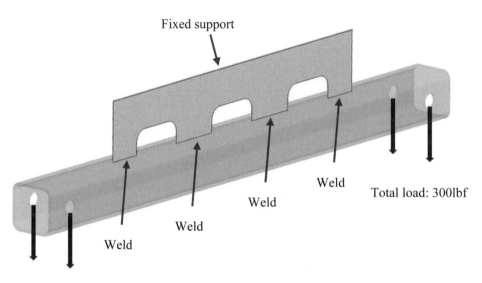

Figure 21-29: TUBE WELDMENT consists of a square tube and a plate connected by welds.

The hanger thickness is 0.25", and the weld is a double sided 3/16" fillet weld. Loads are pointing down and are applied to split faces on the cylindrical surfaces of the holes. See Figure 21-30 for load details.

Tube is shown in transparent view to show all four holes where load is applied.

The load applied to the model is explained in Figure 21-30.

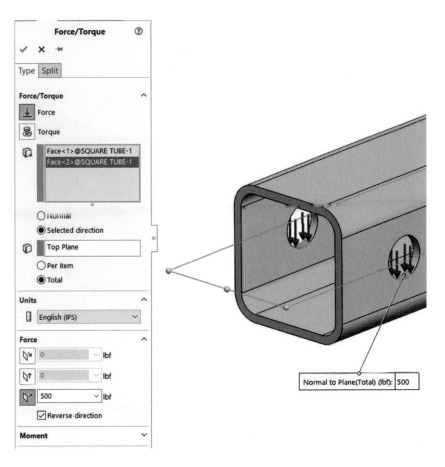

Figure 21-30: Force load applied to the left endplate.

Force is distributed uniformly over two split faces. Apply 300lbf force to the other end.

Four **Edge Weld** connectors must be defined in the model. Figure 21-31 shows the steps necessary to define them.

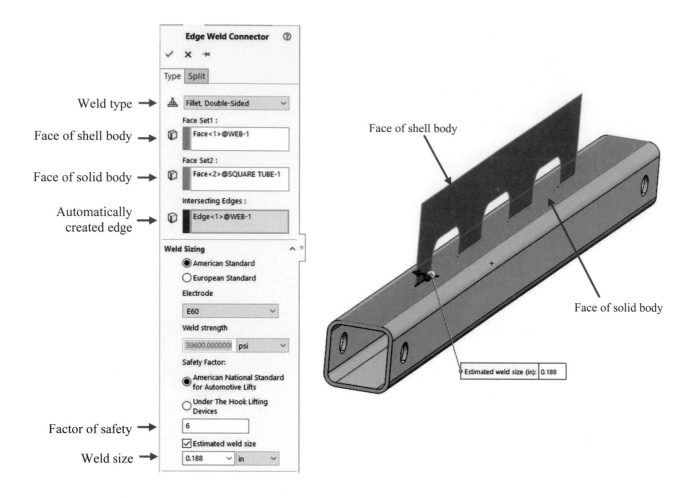

Weld type

Face of shell body

Face of solid body

Automatically created edge

Factor of safety

Weld size

Face of shell body

Face of solid body

<u>Figure 21-31: Edge weld definition.</u>

The definition includes the weld type and size: it is a double-sided fillet weld 0.188" in size. It also includes the electrode material type, here E60. The above window creates all four weld connectors.

Mesh the assembly with Curvature based mesh element size 0.125" to place two elements across the tube wall thickness. Run the analysis and display the **Weld Check Plot** following the steps shown in Figure 21-32.

Click OK to open Weld Check Plot
window shown below

Define Weld
Check Plot

Review List Weld
Results

Weld 4
needs attention

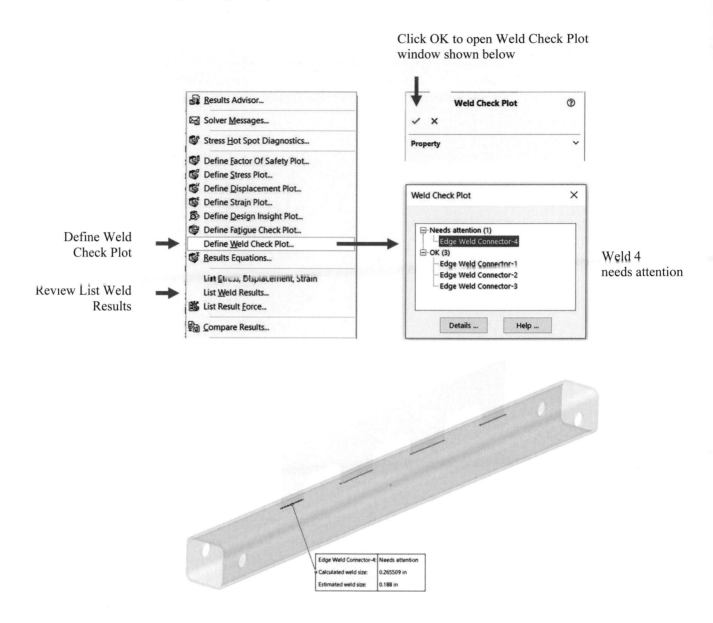

Figure 21-32: The Weld Check Plot window indicates that a weld failed the check.

The plot is accompanied by a Weld Check Plot window listing passed and failed welds.

Complete the analysis of the TUBE WELDMENT by reviewing **List Weld Results**. Review von Mises stress results and compare them to the yield strength of the assembly components.

Bearing connector

A bearing connector models support offered by a bearing and allows for angular rotation of the supported shaft. In the BEARING SUPPORT assembly model, a shaft loaded with a radial force is supported by two bearings (Figure 21-33).

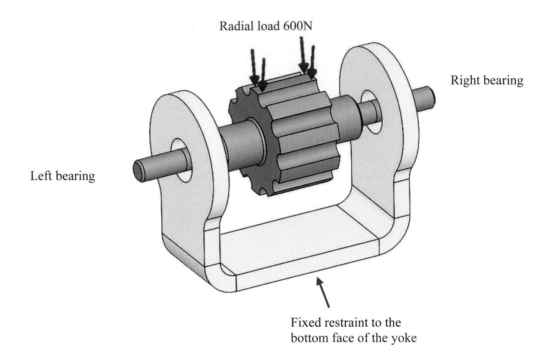

Figure 21-33: Shaft supported by two bearings.

Each bearing support is modeled by a Bearing connector. Notice that bearings are not modeled explicitly. The bearing connector connects the shaft to the bearing housings. Examine split lines on both ends of the shaft. Load is applied to the split face on the rotor, and restraint is applied to the bottom face of the yoke.

Notice that the yoke supporting the rotor has been created as a sheet metal part. By default, **Simulation** would mesh it with shell elements (Figure 21-34 left). To force meshing with solid elements, follow the steps explained in Figure 21-34. Mesh using default mesh settings. Notice that mesh would not be adequate for stress analysis due to sharp re-entrant edges causing stress singularities as well as due to high angular distortions of many elements. We limit this exercise to analysis of displacements.

Right-click
Surface body Yoke
Select
Treat as Solid

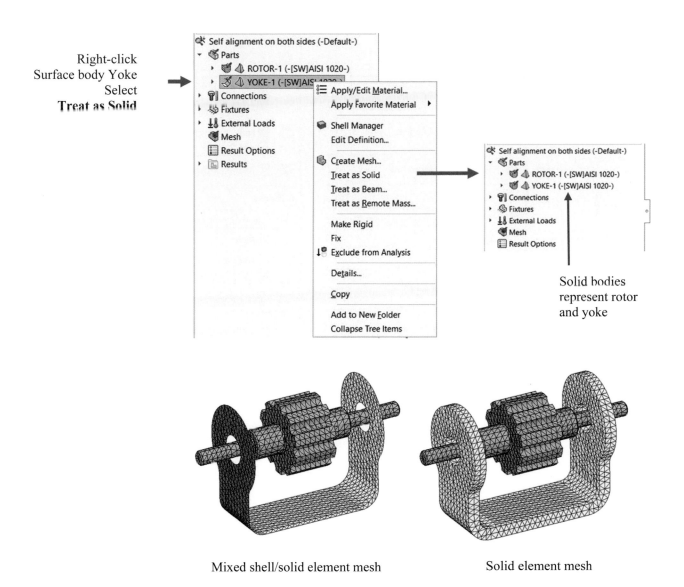

Solid bodies
represent rotor
and yoke

Mixed shell/solid element mesh

Solid element mesh

Figure 21-34: Converting the shell element mesh to a solid element mesh.

Open the Parts folder and right-click the yoke part. Select Treat as Solid from the pop-up menu.

The definition of **Bearing connectors** is shown in Figure 21-35.

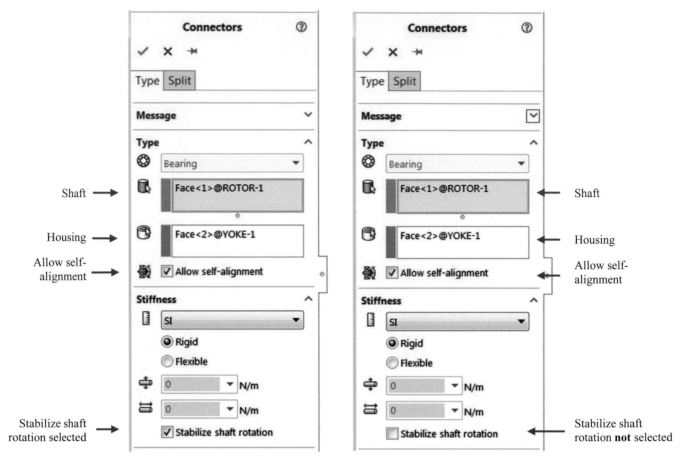

Left bearing connector definition   Right bearing connector definition

Figure 21-35: Definition of Bearing connectors.

The "Stabilize shaft rotation" option is selected in the left bearing connector. Self-alignment is allowed on both sides.

The definition of the connectors on the left and right sides differs because we need to eliminate the rotation of the shaft about the z axis. This would result in a rigid body motion.

Both bearings are modeled as rigid. **Allow self-alignment** is selected meaning that these are spherical bearings, and the deflecting shaft can rotate about the center of the imaginary bearing.

Deformed shape of the shaft is shown in Figure 21-36.

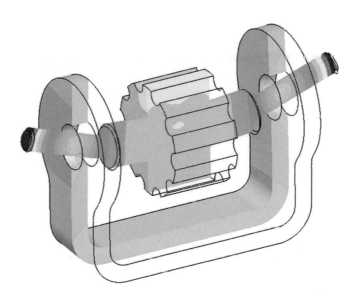

Figure 21-36: Deformed shapes of the shaft with supports defined as Bearing connectors.

A user defined scale of deformation 800:1 is used to show this plot. Section plot is used in this illustration.

Cyclic symmetry

The use of **Cyclic symmetry** simplifies the analysis of a model with cyclic geometry. Generally, a cyclic pattern may be linear or angular but in SOLIDWORKS Simulation **Cyclic symmetry** applies to angular patterns only. **Cyclic symmetry** allows simplifying a model to one repetitive segment. The geometry, restraints, and loads must be the same for all segments made in the model. This means that the geometry as well as the loads and restraints must be characterized by the same angular pattern. Machine components suitable for analysis with **Cyclic symmetry** include turbines, fans, flywheels, motor rotors, etc.

Open the BLOWER part which is an idealized representation of a centrifugal blower; it has three configurations: *01 full, 02 half, 03 section*.

Next, switch to *03 section* configuration and review 1/8th angular section (Figure 21-37). *03 section* configuration represents 1/16th of the model and requires both **Cyclic symmetry** and **Symmetry** boundary conditions.

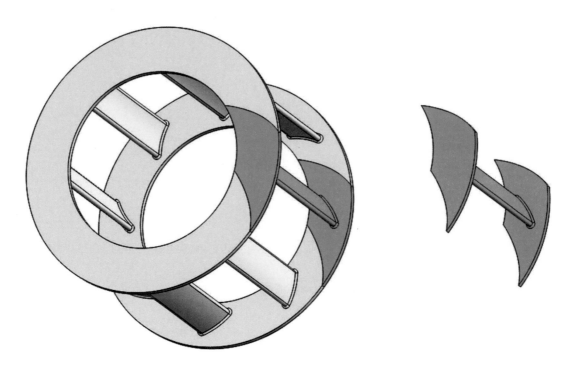

Figure 21-37: Complete model, and 1/8th section for analysis with Cyclic symmetry (right).

The 1/8 section has been created with arbitrary curvilinear cut; any cut that defines a section with a cyclic symmetry can be used.

Definition of **Cyclic symmetry** makes **Simulation** enforce equal displacements at corresponding locations on two faces. The faces where **Cyclic symmetry** is enforced do not have to be of any particular shape. The only requirement is that the segment is a part of a repetitive geometry; that requirement itself makes the faces identical in shape.

We'll use *03 section* configuration first; this requires **Cyclic symmetry** and **Symmetry** as shown in Figure 21-38. Next, we'll use *02 half* configuration requiring **Symmetry** only.

The definitions of **Cyclic symmetry** and **Symmetry** are explained in Figure 21-38.

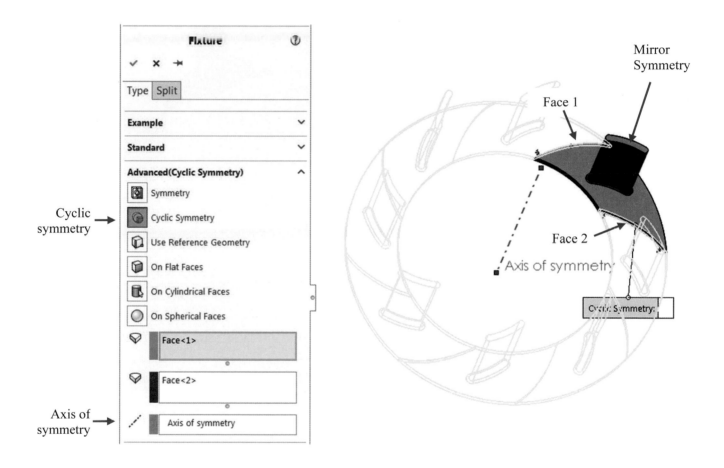

Figure 21-38: Definition of Cyclic Symmetry. Also shown is the face where Symmetry boundary condition must be defined.

The definition of Cyclic symmetry requires a reference axis aligned with the central hole.

Define a **Fixed** restraint to the central hole; define **Centrifugal** load 3000RPM. Apply Mesh control 0.4mm to all fillets and mesh the model with standard mesh size 2mm and obtain solution.

Next, switch to *02 half* configuration, define the same supports (except, of course, for **Cyclic symmetry**), load and mesh parameters.

Compare the results to see that both configurations provided the same results (Figure 21-39).

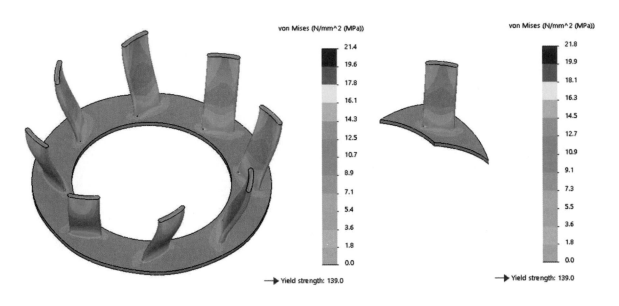

Configuration *02 half*   Configuration *03 section*

Figure 21-39: Von Mises stress results in the half model (left) and in one half of the 1/8th section of the model (right).

The differences in results are caused by different discretization and solution errors in the above two models.

Review closely von Mises stress results and notice that the size of stress concentrations at the base of blades is small as compared to the element size even though an aggressive mesh refinement is used. Further mesh refinement would result in a very large model if we used a complete model but is quite easy when working with one angular section. Repeat analysis in *03 section* configuration using sufficiently small elements in the fillets.

Finally, solve BLOWER in *01 full* configuration.

Strongly nonlinear problem

In the introduction, we said that not all **Simulation** capabilities such as nonlinear analysis will be covered. Indeed, nonlinear buckling analysis and large strain analysis have not been discussed so far. It is the author's belief that the topics presented in the book have provided you with an understanding of finite element tools and methods, as well as preparing you to tackle those more complex problems. To give you a taste of things to come, we will review two examples which present the unexplored capabilities of **Simulation**, in the analysis of strongly nonlinear problems.

Open the CLAMP assembly and review the material properties of the clamp made from Nylon 6/10 and a tube which has custom material properties of a silicon-like material: $E = 1MPa$, $\varepsilon = 0.45$. This model is intended to analyze deformation of both components as the clamp moves over the tube from its initial to its final position (Figure 21-40).

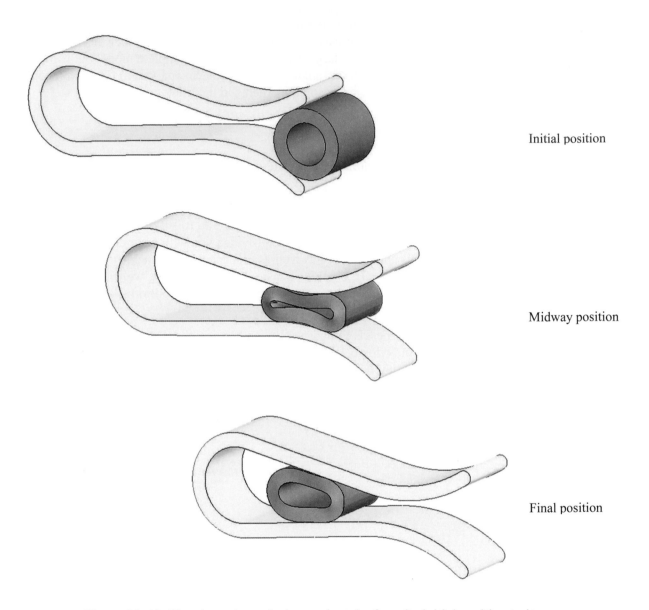

Initial position

Midway position

Final position

Figure 21-40: The clamp is pushed over the tube from its initial position to its final position.

Both tube and clamp experience deformation.

Go to study *01 NL*, review the restraints, and notice that the clamp restraint is defined as a prescribed displacement to the split face. This is what forces the clamp to move over the tube which has a **Fixed** restraint applied to the split face. There are no loads defined in the model. Also review the *Local Interactions* to see which faces participate in no penetration contact.

Solution of the CLAMP assembly model requires **large displacements** formulation. Strong nonlinearities present in this model cause significant numerical difficulties if the default solver settings are used. Review Figure 21-41 to see that, besides **Large displacement formulation**, **Advanced Options** in solver settings are used for a successful solution. Both components use linear material models.

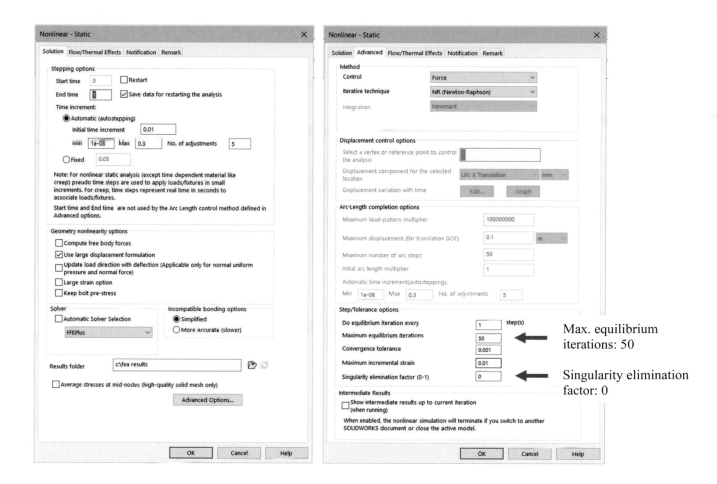

Figure 21-41: Strong nonlinearities in CLAMP require custom settings of Step/Tolerance options.

Advanced settings window is opened by selecting Advanced Option in Solution tab. Advanced Solution settings often must be adjusted following an unsuccessful solution.

Apply **Mesh Control** 1.5mm to assembly component TUBE and use **Curvature Based** mesh with default element size to mesh the model. Solve the model with the options shown in Figure 21-41; be prepared for a long solution time.

Figure 21-42 shows strain results for time step 14 at 0.44s which corresponds to the displacement of the CLAMP of 0.44*50mm = 22mm.

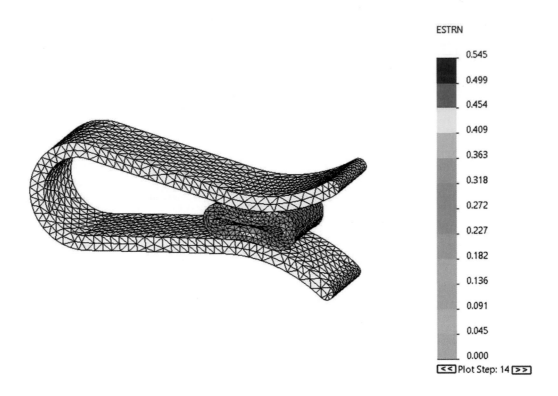

Figure 21-42: Strain plot at step 14. Step 14 is close to where TUBE experiences the highest deformation.

The maximum strain in the rubber tube is 55%.

Large displacements cause high element distortion seen in Figure 21-43.

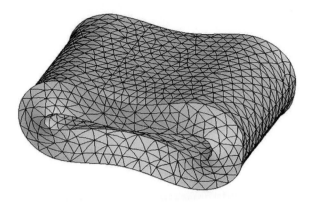

Figure 21-43: The deformed shape of TUBE at step 14 when it experiences its highest deformation.

Notice the "bulging out" of the tube due to a high Poisson's ratio of the tube material.

You may want to repeat the above analysis treating displacements as small (see study *02 LIN*). After solving, review the displacement plot with a 1:1 scale of deformation to observe an incorrect solution.

A similar problem is presented in assembly TWEEZERS shown in Figure 21-44.

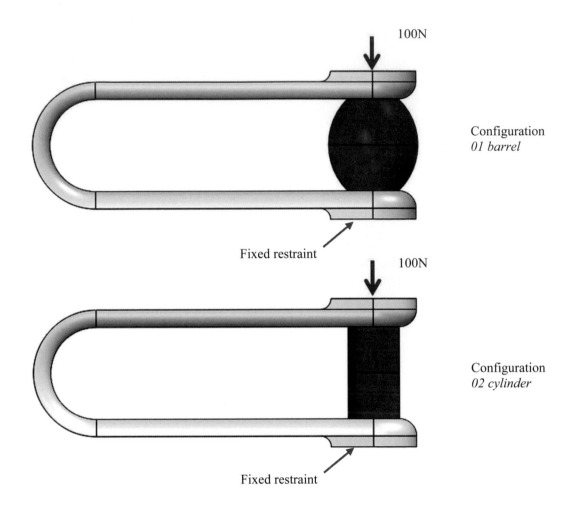

Figure 21-44: Rubber piece squeezed in a clamp.

Notice the initial curvature of the rubber piece in the 01barrel configuration.

Make sure that the model is in *01 barrel* configuration and define a **Nonlinear** study called *01 barrel*. In the study properties select geometry nonlinearity options:

- Use large displacement formulation
- Update load direction with deflection

Define all other settings as shown in Figure 21-41. Use a linear load time curve.

Define one contact set between the upper portion of the rubber piece and the rounds of the upper arm of the clamp. Define the second contact set between the corresponding lower part of the rubber piece and the lower arm. You may review the contact definition because the model comes with study *01 barrel* defined. Define the load and restraint as shown in Figure 21-44 and mesh the assembly with an element size of 4mm with no mesh controls.

Simulation successfully completes the solution producing a deformed shape as shown in Figure 21-45.

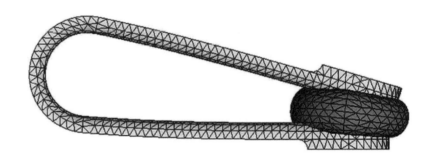

Figure 21-45: Deformed shape of the model in the *01 barrel* configuration.

The deformed rubber piece contacts round edges of the clamp. Therefore, we need contact conditions in this problem. This mesh is shown using an edited color.

Switch to the *02 cylinder* configuration and set up a nonlinear study identical to the previous one. Expecting that the rubber cylinder will "bulge out," define one contact set between the upper portion of the rubber piece and the rounds of the upper arm of the clamp. Define the second contact set between the corresponding lower part of the rubber piece and the lower arm. You may again review contact definitions because the model comes with study *02 cylinder* defined.

This time the solution terminates while executing step 14 producing the error message shown in Figure 21-46. The last successfully completed load step is step 13.

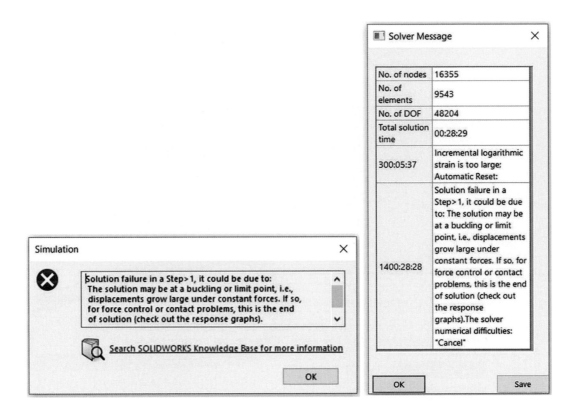

Figure 21-46: Error message produced in step 14 of the solution.

The error message is displayed during solution (left). This error message must be acknowledged before the solver stops. The same message can be read later in the Solver Message window (right).

The last successfully completed step was step 13, which corresponds to a solution time of 0.72s as can be read in the properties of the displacement plot (Figure 21-47). Since the load has a linear time history with a total time of 1s, the last successfully performed step corresponds to 80% of the applied 100N load.

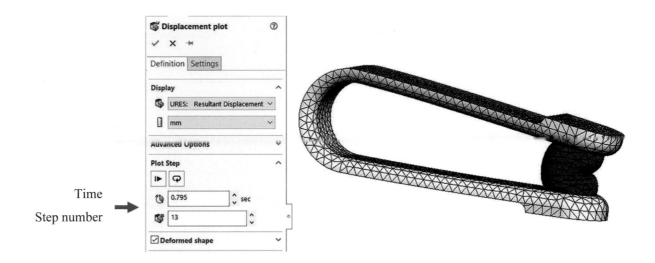

Time

Step number

Figure 21-47: Deformed shape under the load 80N.

Zoom in to investigate the deformed shape of the elements.

The large strain in the rubber cylinder visible in Figure 21-47 led to severe deformation of elements and was the reason for solution failure in step 14.

Submodeling

Submodeling is a technique used when working with large models where the nature of analysis requires a highly refined mesh of a portion of the model, while the rest may be meshed with a coarser mesh. We will demonstrate **Submodeling** using a model named U BEAM. This model consists of three solid bodies. When **Static** study *01 full model* is created, these three solid bodies must be bonded using Incompatible mesh (meshed independently). This is the requirement of submodeling.

The division of the U BEAM into three solid bodies and meshing independently is done to prepare the model for this **Submodeling** exercise.

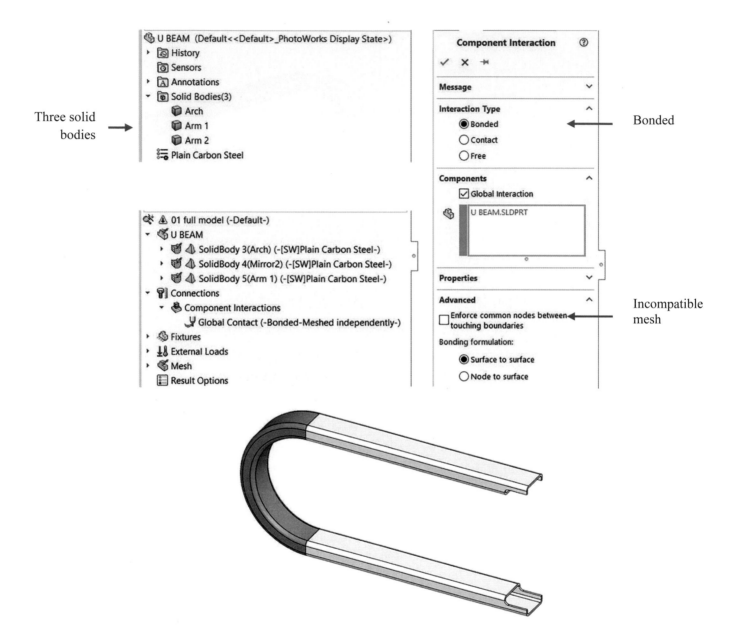

Define Static study *01full model*. Select **Bonded** as Interaction Type in **Component Interaction**. Do NOT select Enforce common nodes between touching boundaries. This will create incompatibility between the three bodies. The bodies will be meshed independently. This is necessary when submodeling is intended (Figure 21-48).

Three solid bodies →

Bonded

Incompatible mesh

Figure 21-48: The Feature Manager window of U BEAM model composed of three solid bodies (top) and Static study *01 full model* window (bottom). Also shown is Component Contact window specifying Incompatible mesh.

Review the Solid Bodies in the Feature Manager Design Tree to identify their locations in the model.

The loads and restraints are shown in Figure 21-49.

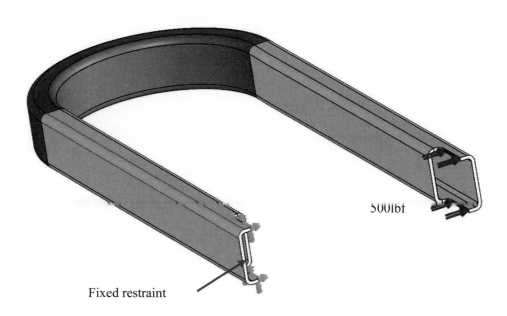

500lbf

Fixed restraint

Figure 21-49: Loads and restraints on U BEAM.

500lbf normal load is uniformly distributed over two red faces at the end of Arm 1. Fixed restraint is applied to the other two rad faces at the end of Arm 2.

Run the **Static** study *01 full model* using a default mesh size. Review von Mises stress results as shown in Figure 21-50 and notice that the irregular pattern of fringes in the stress plot indicates the need of mesh refinement. Also examine the displacement results not shown here.

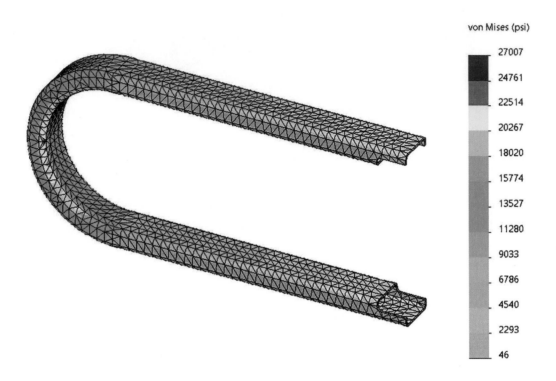

von Mises (psi)

27007
24761
22514
20267
18020
15774
13527
11280
9033
6786
4540
2293
46

Figure 21-50: Von Mises stress results produced by a linear analysis.

Notice the "spotty" stress pattern indicating the need for a more refined mesh. The maximum stress location coincides with the connection between two bodies. This is caused by incompatible mesh.

The results of the *01 full model* study indicate the need for a more refined mesh in the curved portion of the tube.

We could refine the mesh either globally (less efficient) or locally (more efficient) but expecting the need for several runs during design process, we will use **Submodeling**, where *Arch* solid body is isolated from the model and the results of the study *01 full model* are used only to provide displacement boundary conditions applied to this isolated model of *Arch* solid body.

To create a **Submodeling** study based on the *01 full model* study, follow the steps shown in Figure 21-51.

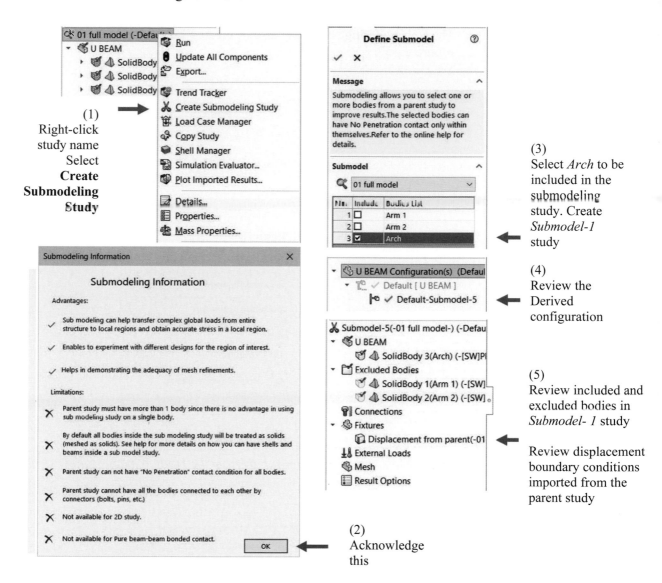

Figure 21-51: Creating and reviewing Submodeling study.

Every Submodeling study is based on a parent study.

In this example, Submodel-5 study is based on the 01 full model study.

Creating a **Submodeling** study adds a derived configuration to the **SOLIDWORKS** model; the study name is created automatically (Figure 21-51). You may rename it later.

The **Submodeling** study may also be created from the Study window as shown in Figure 21-52.

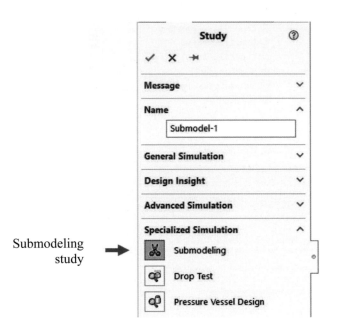

Figure 21-52: Messages related to the Submodeling study.

Having selected Submodeling study from the Study Window, follow steps shown in Figure 21-51.

In the U BEAM model hide solid bodies *Arm 1* and *Arm 2*. The study *Submodel-1* is now ready for meshing; mesh it with an element size of 0.1" (Figure 21-53).

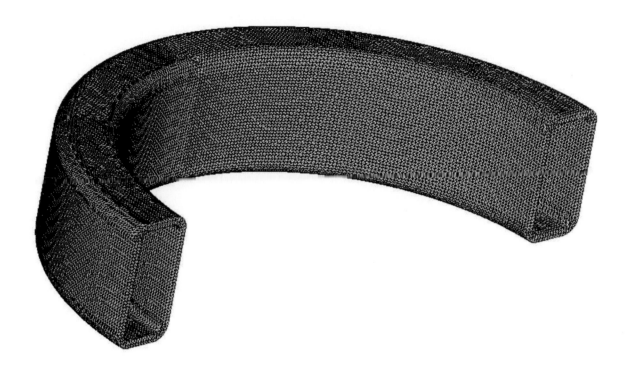

Figure 21-53: Fine mesh of the submodel.

Using this small element size to mesh the parent model would mean a long solution time.

Run the **Submodeling** study where *Arch* is isolated from the full model and subjected to displacement boundary conditions imported from the parent study. Von Mises stress results are shown in Figure 21-54.

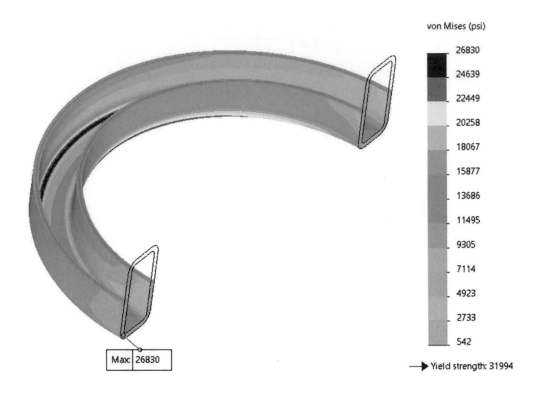

Figure 21-54: Von Mises stress results of the *Submodel-1* study.

Notice an artificial stress concentration (maximum stress annotation) caused by displacement boundary conditions imported from the parent study. This plot uses Section Clipping to show stress concentration inside the tube.

You may now study the effect of different mesh sizes on the results without having to re-run the full model. This can be done by duplicating the **Submodeling** study.

Automated detection of stress singularities

In Chapter 3 stress singularities in the sharp re-entrant edge of L BRACKET were detected during a convergence process. Consecutively smaller elements were used, and divergence of stresses was observed.

Simulation offers an automated way of detecting stress singularities, available only in **Static** studies. We'll demonstrate it using NOTCH model. Von Mises stress plot of NOTCH model is shown in Figure 21-55. A default mesh was used to solve the model and no attempt was made to treat the expected stress concentration. Based on what we learned in Chapter 3, we recognized that the notch produces stress singularity where stress results are invalid. We'll now demonstrate an automated detection of stress singularities.

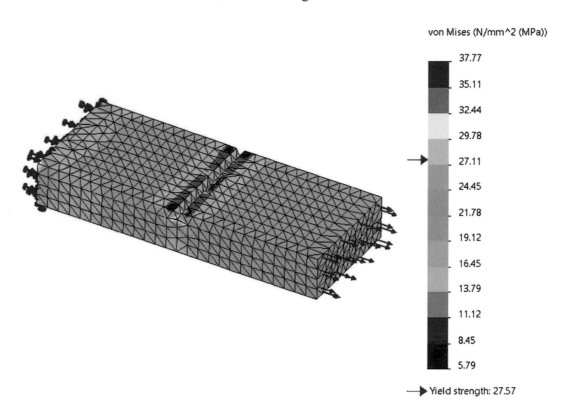

Figure 21-55: Von Mises stress results of a static study of NOTCH model.

Model is subjected to tensile stress 20MPa; a default mesh is used. Sharp notch produces a stress concentration. Fixed restraints symbols are shown in red color.

Right-click the Results folder and select **Stress Hot Spot Diagnostic** to open **Stress Hot Spot** window (Figure 21-56).

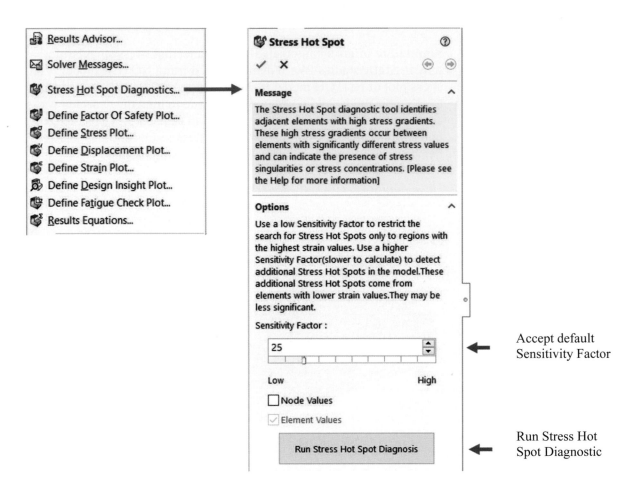

Figure 21-56: Hot Spot detection – part 1.

Read Message and Options; click Run Stress Hot Spot Diagnostic.

Selecting **Run Stress Hot Spot Diagnostic** in **Stress Hot Spot** window (Figure 21-56) produces message shown in Figure 21-57.

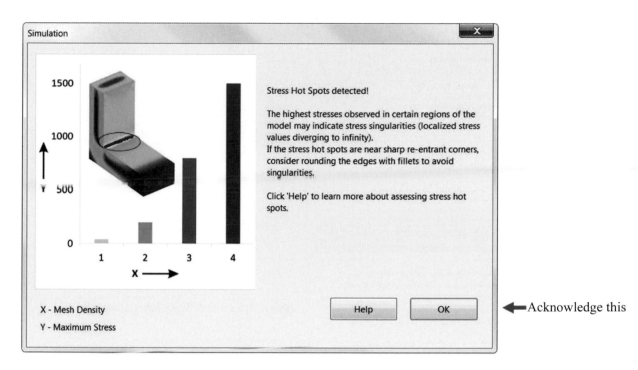

Figure 21-57: Hot Spot detection – part 2.

Stress singularity is called here "Hot Spot". Click OK to continue the diagnostic process of stress singularities.

Window in Figure 21-57 informs that a stress singularity has been detected. To proceed with investigation of stress singularity in the model click OK; this opens window shown in Figure 21-58.

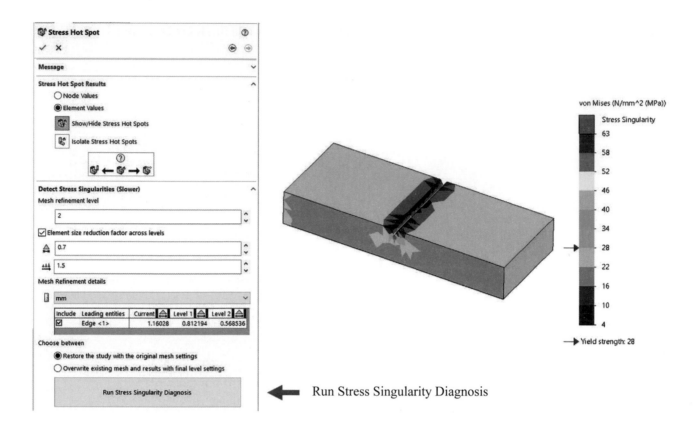

Run Stress Singularity Diagnosis

Figure 21-58: Hot Spot detection – part 3.

Stress Hot Spot window is displayed after the message in Figure 21-57 has been acknowledged. Stress singularity area is shown in grey color on the top of the color legend.

Stress Hot Spot window in Figure 21-58 offers **Run Stress Singularity Diagnosis** option. If selected, this option will apply mesh controls, execute a convergence process, and offer a review of convergence graphs; all similar to what was done in Chapter 3. To complete this exercise, investigate different mesh refinement levels and options **Restore the study with the original mesh settings** and **Overwrite existing mesh and results with final level settings**.

Von Mises stress plot shown in Figure 21-58 will differ depending what options are selected in **Stress Hot Spot** window.

Stress Averaging at Mid-side Nodes

Throughout this book we were using nodal stress plots. Nodal stress plots are created by extrapolating stresses from Gauss points to corner nodes and averaging stresses reported by different element on the shared nodes; this process was explained in Chapter 3.

Simulation also offers a different method of stress averaging called **Stress Averaging at Mid-side Nodes**. Using this method, stresses at the mid-side nodes are calculated by averaging the stress values at the associated corner nodes.

Stress Averaging at Mid-side Nodes can be activated by selecting **Average stresses at mid-nodes** (Figure 21-59). This option works only for second order (high quality) elements.

Figure 21-59 Stress Averaging at Mid-side Nodes.

If Average stresses at mid-nodes is selected, then nodal stress plots use Stress Averaging at Mid-side Nodes method.

This selection of **Stress Averaging at Mid-side Nodes** is available in Stress plot window (Figure 21-59) and also in study **Properties** window and in **Default Options, Results** window.

Figure 21-60 shows von Mises stress plots without and with **Stress Averaging at Mid-side Nodes**. The model shown is HOLLOW PLATE, study *tensile load 02* (Chapter 2). **Stress Averaging at Mid-side Nodes** method may improve the calculation of stresses for tetrahedral elements with high aspect ratios.

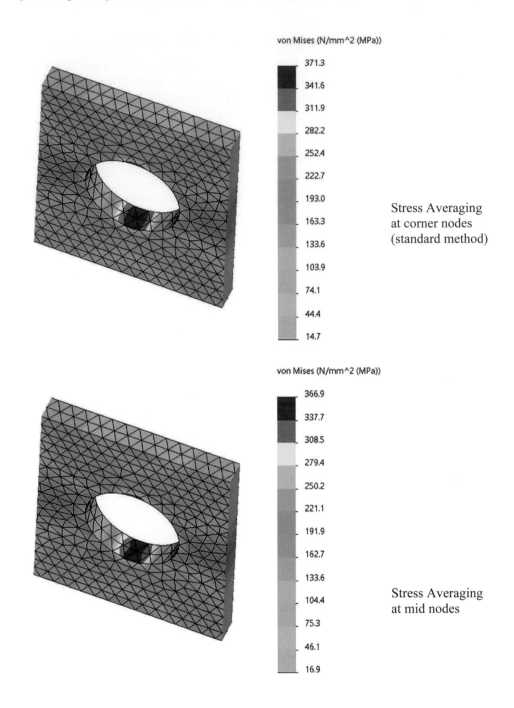

Stress Averaging
at corner nodes
(standard method)

Stress Averaging
at mid nodes

Figure 21-60 Stress Averaging at corner nodes (top) and stress averaging at mid-side nodes (bottom).

A default mesh used in both illustrations. Section Clipping is used.

Terminology issues in Finite Element Analysis

Finite Element Analysis (FEA), or Finite Element Method as mathematicians call it, is one of many numerical techniques of solving partial differential equations that describe, among others, the structural and thermal problems presented in this book. FEA has seen rapid development during the last few decades and it has displaced other numerical techniques into niche applications assuming a dominant position in the market of engineering analysis tools. Still, FEA is a relatively new engineering tool that has evolved from being an exclusive tool for highly trained analysts, to the present day where it has become an everyday tool of design engineers. Deeply rooted in mathematics and often developed independently by competitive commercial firms, FEA shows discrepancies in the development of terminology, which has not yet been unified across the industry.

Users of different FEA programs may use different terminology for similar problems or use the same term describing different things. Constraints, restraints, supports and fixtures may all mean the same for some people while others will understand them differently. Many FEA users will argue that loads and boundary conditions are different entities; while others will say that loads are just one type of boundary condition because they are applied to the boundary of a model (loads external to the model are in fact boundary conditions, volume loads are not). Make sure you understand what is meant by each term you use and do not be afraid to ask exactly what is meant when an element "locks" or what a "nonconforming hexahedral element" is when you hear such a term. Many of those terms come from legacy sources and have long lost their relevance in modern programs such as **SOLIDWORKS Simulation**.

While volumes could and in fact should be written about FEA terminology, here we will only review terminology issues that apply to names of analysis types used by **SOLIDWORKS Simulation**. As you know, the following studies are available: **Static, Frequency, Buckling, Thermal, Drop test, Fatigue, Nonlinear, Linear Dynamic** and **Pressure Vessel Design**. Don't take each name literally as a short description of the analysis capabilities of each study. Instead treat them just as labels; here is why:

Static

This can be a linear static analysis or a nonlinear static analysis. However, a nonlinear analysis is limited to large displacements and/or contact. In a nonlinear analysis conducted under a **Static** study, the user has no control over the load time history which must be linear ("ramping-up" the load at a uniform pace). Nonlinear material is not available.

Frequency

A common name for this type of analysis is modal analysis as you will find in every textbook on vibration analysis. Modal analysis finds natural frequencies and the associated shapes of vibration. A combination of frequency and shape is called a mode of vibration. Modal analysis does not find displacements, strains, or stresses.

Topology Study

A Topology study performs nonparametric topology optimization of parts. Starting with a maximum design space (which represents the maximum allowed size for a component) and considering all applied loads, fixtures, and manufacturing constraints, the topology optimization seeks a new material layout, within the boundaries of the maximum allowed geometry, by redistributing the material. The optimized component satisfies all the required mechanical and manufacturing requirements.

Design Study

This is used to optimize the model with respect to criteria such as the lowest mass, the maximum natural frequency, etc. Specified design variables (model parameters) are assigned ranges of allowable variation and the optimum solution is found that does not violate constraints such as stress.

Thermal

A thermal analysis can be executed as a **Steady State thermal** analysis or a **Transient Thermal** analysis and is utilized to find temperatures, temperature gradients, and heat flux. Notice that thermal stresses are not calculated in a thermal analysis; they are calculated in a **Static** or **Nonlinear** analysis using the temperature results from a **Thermal** analysis.

Buckling

This is a linear buckling analysis which finds buckling load factors and the associated buckling shapes. The name "Eigenvalue based buckling analysis" is sometimes used. Linear buckling analysis does not say how far a structure will buckle or if it will survive buckling. To solve these questions, you must use a nonlinear buckling analysis which is available in **Simulation** under **Nonlinear** analysis.

Fatigue

A fatigue analysis uses results of a Static analysis to calculate fatigue life under periodic loads.

Nonlinear

A **Nonlinear** analysis will do everything that a **Static** analysis can do and much more, but at a higher computational cost. All types of nonlinear behaviors can be analyzed including nonlinear buckling and nonlinear materials. **Simulation** features an extensive library of nonlinear materials available in a **Nonlinear** study. Beware of the common misconception that a **Nonlinear** analysis is used only for nonlinear materials. In this book we have presented many examples where other types of nonlinear behaviors were present. Additionally, a **Nonlinear** analysis can be executed as static or dynamic. And so it is more general than a **Linear Dynamic** analysis.

Linear Dynamic

This should really be called "Linear Vibration" analysis. Remember that FEA is a tool of structural analysis and as such, deals with elastic bodies. Any motion of elastic bodies can only take the form of vibration about the position of equilibrium. Linear Dynamic (Vibration) analysis is based on the Modal Superposition method and this makes it very numerically efficient, but less general than Nonlinear Dynamic (Vibration) analysis. **Linear Dynamic** analysis has four sub-categories in **Simulation**: **Modal Time History, Harmonic, Random Vibration Analysis, and Response Spectrum Analysis**.

Modal Time History

Vibration analysis textbooks call this a Time Response analysis (the term Dynamic Time is also used). This analysis is intended for problems where the load is an explicit function of time.

Harmonic

Vibration analysis textbooks call this Frequency response (the terms Steady State Harmonic analysis and Dynamic Frequency analysis are also used). This analysis is intended for problems where load is a function of frequency which in turn is a function of time. It is assumed that frequency changes very slowly (if at all), hence the alternative name: Steady State Harmonic analysis.

Random Vibration Analysis

Here, loads are given as a Power Spectral Density (PSD) of displacements, velocities or accelerations. Results such as RMS and PSD displacements, velocities and accelerations are calculated only in probabilistic terms.

Response Spectrum Analysis

This analysis is intended for excitation loads of longer duration that are non-stationary and therefore cannot be presented as PSD. Instead, the excitation is presented as a Response Spectrum which is useful to analyze events such as earthquakes.

Submodeling

A portion of a model is analyzed separately from the rest. The analyzed portion is subjected to boundary conditions imposed by the rest of the model. This approach allows the use of small elements which would be impossible if the entire model were analyzed.

Drop Test

This is a specialized type of analysis intended for analysis of collision between two bodies. This is a dynamic analysis based on the direct integration method, which is stable but very time consuming.

Pressure Vessel Design

This analysis offers a convenient way of superimposing results of different Static studies as required in the analysis of pressure vessels for compliance with safety codes. Notice that a Pressure Vessel Design study can be used to analyze superimposed results of anything, not only pressure vessels.

Models in this chapter

Models come with studies partially or fully defined.

Model	Configuration	Study Name	Study Type
GUSSET.sldprt	01 bad design	01 bad design	Static
	02 good design	02 good design	Static
MESH QUALITY 01.sldprt	Default	Mesh quality	Static
MESH QUALITY 02.sldprt	Default	01 no automatic transition	Static
		02 automatic transition	Static
TANK.sldprt	Default	Non uniform load	Static
HELICOPTER ROTOR.sldprt	01 four blades		
	02 one blade	no preload	Frequency
		preload	Frequency
SHRINK FIT.sldasm	Default	Static 1	Static
CRANE.sldasm	01 full	01 pin connector	Static
	02 simplified	02 linkage rod	Static
STAND.sldasm	Default	Modal 01	Frequency
		Modal 02	Frequency
PIPES.sldasm	01 long	01 long	Static
	02 short	02 short	Static
		03 h adaptive	Static
TUBE WELDMENT.sldasm	Default	Welds	Static

BEARING SUPPORT.sldasm	*Default*	*Self-alignment on both sides*	Static
BLOWER.sldprt	*01 full*	*01 full*	
	02 half	*02 half*	Static
	03 section	*03 section*	Static
CLAMP.sldasm	*Default*	*01 NL*	Nonlinear
		02 LIN	Static
TWEEZERS.sldasm	*01 barrel*	*01 barrel*	Nonlinear
	02 cylinder	*02 cylinder*	Nonlinear
U BEAM.sldprt	*Default*	*01 full model*	Static
	Default – Submodel 1	*Submodel-1*	Static
NOTCH.SLDPRT	*Default*	*Static 1*	Static

Notes:

22: Miscellaneous topics – part 2

In this chapter we present several problems that reinforce and expand concepts and modeling techniques discussed throughout this book. Detailed descriptions of all analysis steps are not given but a general analysis outline and objectives are introduced in each case.

2D Plate

* 2D Plane Stress analysis
* Stress singularity
* Displacement singularity
* Classification of singularities

Open 2D BEAM part which models a thin beam. The beam is subjected to 10000N load along the top edge and supported by two point supports as shown in Figure 22-1. Plate thickness is 10mm, material is 1060Alloy.

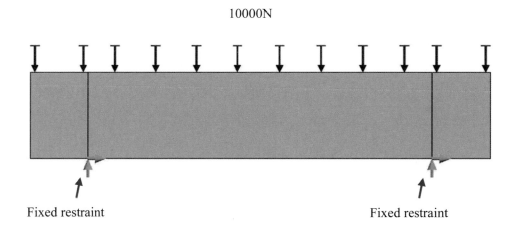

Figure 22-1: Loads and restraints in 2D BEAM model.

The fixed restraint to a node eliminates two degrees of freedom and effectively defines a hinge support. The load is evenly distributed along the top edge.

We will use this model to demonstrate displacement and stress singularities caused by point supports. Create a **Static** study *01 2D* using **Plane Stress** simplification, define a model thickness of 10mm and apply loads and restraints as shown in Figure 22-1. Solve the problem using a default mesh. Next, copy study *01 2D* into *02 2D* and define a **Mesh Control** of 0.3mm to both points where the restraints are defined; obtain the solution. Copy study *02 2D* into *03 2D* and define a **Mesh Control** of 0.05mm; obtain the solution.

Displacement and stress results of the studies are summarized in Figure 22-2.

Study	Mesh control	Max. displacement	Max. von Mises stress
	mm	mm	MPa
01 2D	none	0.079	321
02 2D	0.3	0.091	3118
03 2D	0.05	0.102	18286

Figure 22-2: The maximum displacement and the maximum stress in three studies.

You may use different settings of mesh controls.

The results demonstrate stress divergence caused by stress singularities at the point supports. This should not come as a surprise. What may be less obvious is that displacements are also divergent; with mesh refinement they tend to infinity albeit much slower than stresses. The rate of stress divergence is exponential while the rate of displacement divergence is logarithmic.

To demonstrate this slow divergence of displacement we must use very small elements, and this is easily done using a 2D model.

Observe the pattern of displacement and notice that point supports "cut into" the model (Figure 22-3). Indeed, a restraint defined on an entity with zero area cannot provide a support and the only reason why the model doesn't "sink through" the supports is the discretization error.

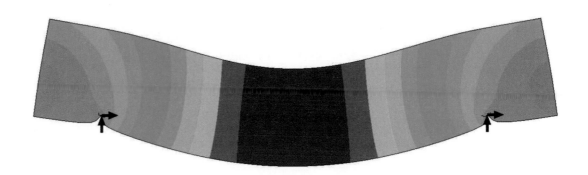

Figure 22-3: Deformation of 2D model supported by two points.

Locations of point supports are indicated by Fixtures symbols.

We now may expand the definition of singularity originally introduced in Chapter 3.

Singularity Type 1: Infinite stress, finite energy

Examples: sharp re-entrant edge, point load, line load, connection between materials with different elastic properties

Singularity Type 2: Infinite stress, infinite energy

Examples: point support, line support

Idler pulley

- Symmetry boundary conditions in solid element model
- Symmetry boundary conditions in shell element model
- Analysis of reaction forces
- Shell element mesh alignment
- Limitations of shell elements

Review two models: ALUMINUM PULLEY and STEEL PULLEY shown in Figure 22-4. The aluminum pulley is manufactured by die casting, and the steel pulley is stamped.

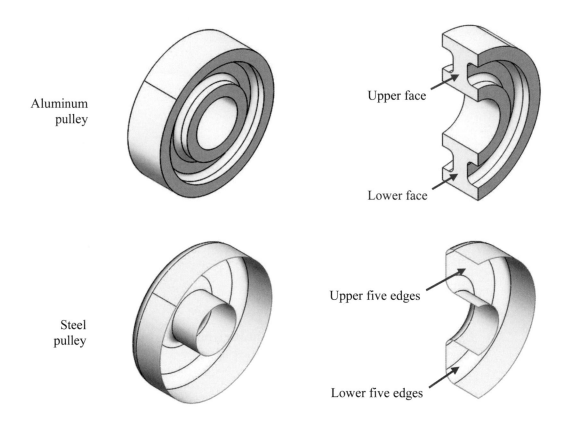

Figure 22-4: ALUMINUM PULLEY (top) and STEEL PULLEY (bottom) models.

Both models have two configurations: 01 full and 02 half. The thickness of the steel pulley is 1.5mm.

Symmetry boundary conditions will be applied to two faces of the ALUMINUM PULLEY and to the total of ten edges of the STEEL PULLEY.

Both pulleys are idler pulleys in a car serpentine belt drive; they do not transmit any moment; they serve only to change belt direction. For this reason, the pulleys are loaded only with a uniformly distributed pressure which we apply to the split faces created for this specific reason. The size of split faces corresponds to a 90° belt wrap angle. The outside diameter of both pulleys is the same: 140mm.

The pressure load and support are shown in Figure 22-5.

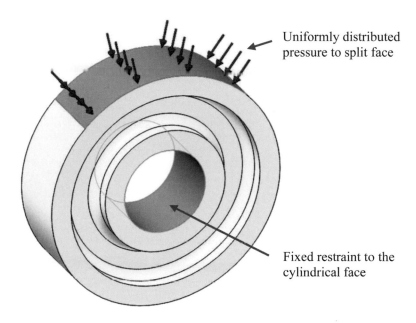

Uniformly distributed pressure to split face

Fixed restraint to the cylindrical face

Figure 22-5: Pressure load and restraint applied to the die cast aluminum pulley; by using a fixed restraint we assume the pulley is supported by a solid shaft.

The aluminum pulley is shown here; loads and restraints on the steel pulley are the same as those applied to the aluminum pulley.

We don't know the magnitude of pressure; what we know is that it causes a 1000N reaction in the bearing. We'll apply a unit pressure of 1MPa, measure the reaction and scale the pressure accordingly. Open ALUMINUM PULLEY, switch to *02 half* configuration, create a **Static** study *Static 1* and apply **Symmetry Boundary Conditions** as shown in Figure 22-6.

Roller/Slider

Figure 22-6: Four different ways of defining Symmetry Boundary Conditions for ALUMINUM PULLEY.

Any of the above four methods of defining the Symmetry Boundary Conditions may be used for ALUMINUM PULLEY. They all have the effect of restraining displacements of two faces in the plane of symmetry in the direction normal to those faces.

Apply a **Fixed** restraint to the hole and run the solution using a default mesh. Review reaction forces on the face where **Fixed** restraints have been defined.

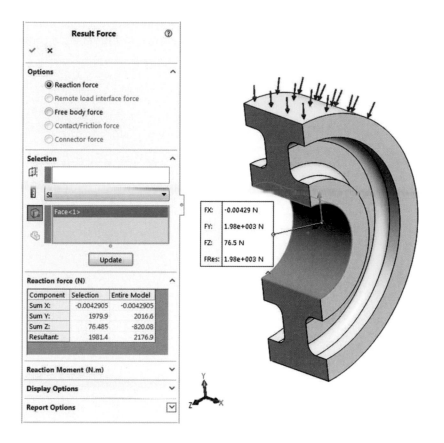

Figure 22-7: Review of reaction forces on the face where pulley is supported by a rigid shaft. Reaction force in Y direction is 1980N.

Confirm that all moment reactions are zero; this is because solid elements do not have rotational degrees of freedom.

The resultant force in Y direction is 1000N but results should be 500N because we are working with half of the model. Therefore, we need to scale the applied pressure by 500/1980 = 0.2525 meaning that the applied pressure should be 0.25MPa if rounded off to two decimal places.

Before we re-run the analysis with the adjusted pressure, let us have a look at the default mesh used for analysis. A quick review of the default mesh shown in Figure 22-8 (top) reveals high angular distortions in all four fillets where the turn angle is 90°. There are many ways to correct this; here we will use **Curvature based mesh** with parameters shown in Figure 22-8 (bottom).

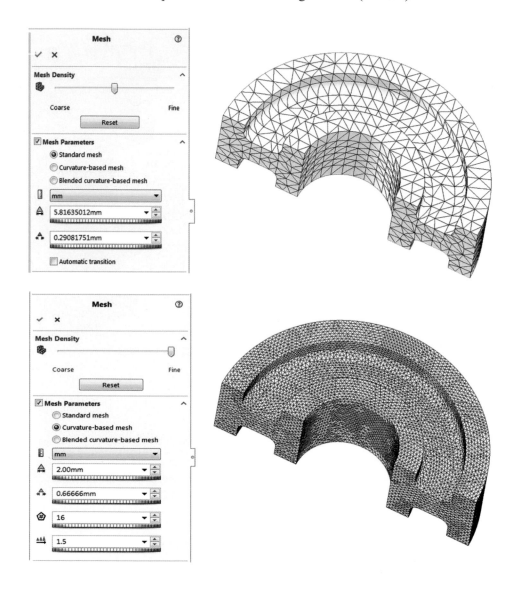

Figure 22-8: Original (default) mesh with high element distortion in all four fillets (top) and the Curvature based mesh (bottom).

The default mesh has elements with 90° turn angle situated along all four fillets. To improve the mesh, use Curvature based mesh with 2mm element size and 16 elements in the circle as shown in the mesh window.

Adjust the pressure to 0.25MPa and obtain the solution with the new mesh. Animate the displacement plot and verify that both faces where symmetry boundary conditions have been applied remain flat while they deform (Figure 22-9).

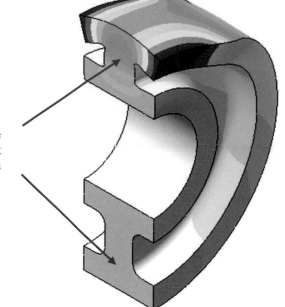

Both faces in the plane of symmetry remain flat while model deforms

Figure 22-9: Animate this displacement plot to see that both faces in the plane of symmetry remain flat.

The flatness is the result of symmetry boundary conditions applied to these two faces.

Next, review the von Mises stress results shown in Figure 22-10.

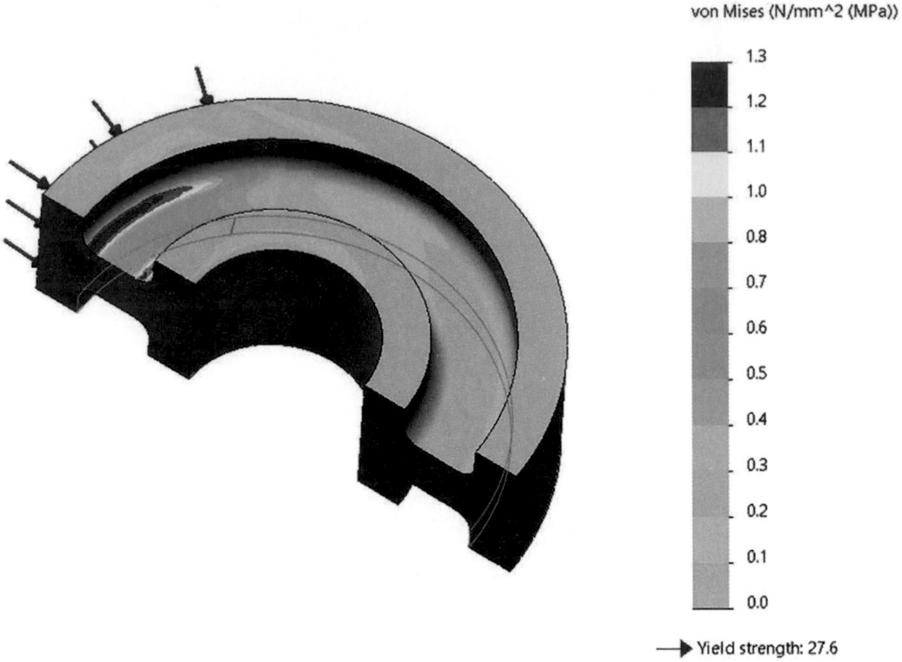

Figure 22-10: Von Mises stress ALUMINUM PULLEY model.

Location of stress concentration is on the side where pressure is applied.

When reviewing stress results, we always want to see fringe patterns that are not affected by the mesh, have no unjustified discontinuities, are not attracted to nodes, etc. Upon inspection of the stress plot in Figure 22-10 we find some minor irregularities of fringes in the area of stress concentration. This indicates the need for further mesh refinement which you may execute by mesh controls and/or by using h adaptive solution.

Now, repeat the above analysis using STEEL PULLEY. There is no need for a review of reaction forces; we will apply the same 0.25MPa pressure. The only difference is the definition of symmetry boundary conditions. Use model in *02 half* configuration and define symmetry boundary conditions as shown in Figure 22-11.

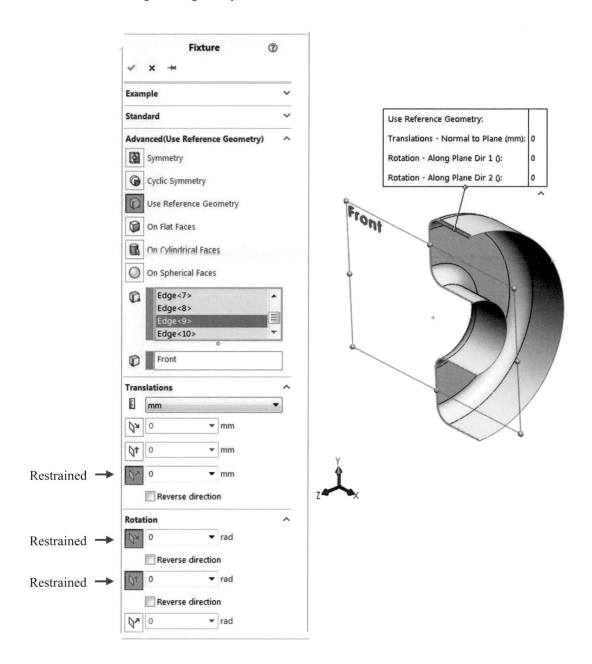

Figure 22-11: Symmetry Boundary Conditions for STEEL PULLEY.

Using Reference Geometry is the only method to define symmetry boundary conditions in the absence of a flat face. Restraints symbols are not shown.

Restraining translational degrees of freedom is not different from the definition of symmetry boundary conditions in ALUMINUM PULLEY. The difference is in rotational degrees of freedom which were irrelevant in ALUMINUM PULLEY; remember that solid elements don't have rotational degrees of freedom. STEEL PULLEY will be meshed with shell elements; therefore, the definition of symmetry boundary conditions must include rotation: both in-plane directions must be suppressed, or else symmetry would be destroyed.

Define a shell thickness of 1.5mm with an offset from the top surface as shown in Figure 22-12.

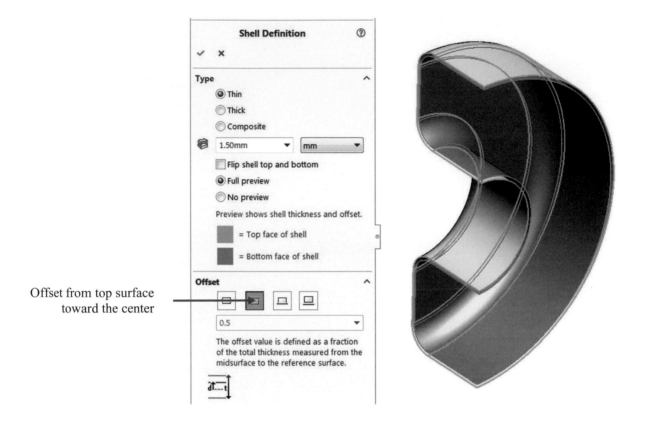

Offset from top surface toward the center

Figure 22-12: Shell element thickness and direction of offset definition for STEEL PULLEY. Offset is toward the center of pulley.

Review the orientation of shell elements we are about to create; the top of shell elements is on the outside and is coincident with the surface. The surface itself is hidden in this illustration.

Mesh the model with an element size of 1mm; this will be a very fine mesh which is required to model stress concentrations on the highly curved faces. The solution time will not be long because shell elements and, consequently, the number of DOF in the model is not large.

We are interested in the maximum tensile stress and the maximum compressive stress in the model. To see the maximum tensile stress, construct a plot of P1 stress on the **Top** of shell elements; this is where the maximum tension is found (Figure 22-13). To confirm this, review an animated stress or displacement and observe the pattern of deformation.

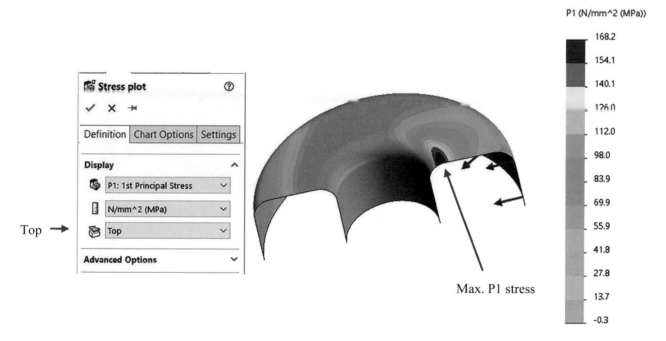

Figure 22-13: Plot of P1 stress on the Top of shell which is the outside of the model. The maximum tensile stress is 168.2MPa.

The maximum tensile stress is 168.2MPa.

To see the maximum compressive stress, construct plot of P3 stress on the Bottom of shell elements (Figure 22-14). Notice that the maximum compressive stress is the numerically lowest of three principal stresses.

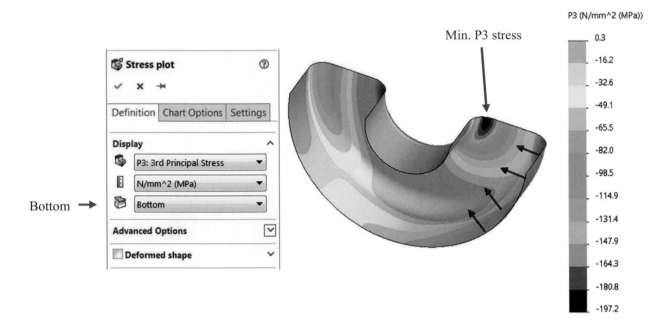

Figure 22-14: Plot of P3 stress on the Bottom of shells which is the inside of the model.

The maximum compressive stress is -197.2MPa.

Compare plots in Figure 22-13 and Figure 22-14 and notice that the maximum tensile stress and the maximum compressive stress are found in the same geometric location; the only difference is definition of location: **Top** or **Bottom** of shell element.

P1 and P3 stresses are vectors; review the above stress plots using vector display.

Figure 22-15 shows von Mises stress plots on the **Top** (outside of the model) and on the **Bottom** (inside of the model).

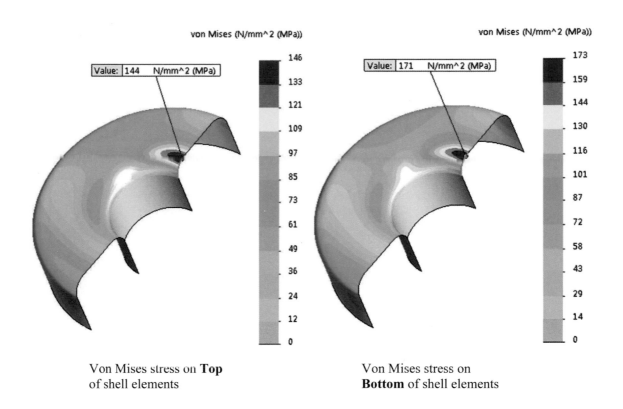

Von Mises stress on **Top**
of shell elements

Von Mises stress on
Bottom of shell elements

Figure 22-15: Von Mises stress on the outside (Top) and on the inside (Bottom) of shell elements.

*Notice that the shell element model is "transparent"; stresses on the inside face (**Bottom**) can be seen even though we are looking at the outside (**Top**) face.*

Difference in von Mises stresses on the outside and on the inside can be shown on one plot if **Render shell thickness in 3D** is selected in **Advanced Options** or if solid elements are used.

The use of solid elements to model stamped steel pulley geometry will be demonstrated using the STEEL PULLEY SOLID model which has been designed to complement the STEEL PULLEY exercise.

Open STEEL PULLEY SOLID part, switch to *02 half* configuration and notice that this model consists of two bodies as shown in Figure 22-16.

Division into two bodies was implemented to facilitate the use of mesh control applied to a body; *Body1* will be meshed with smaller elements than *Body 2*.

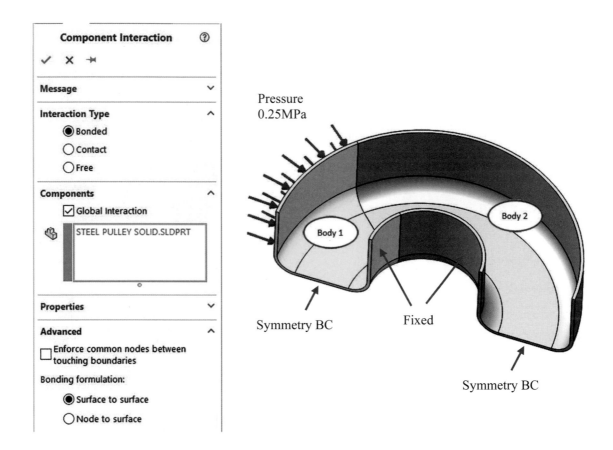

<u>Figure 22-16: STEEL PULLEY SOLID model consists of two bodies to facilitate meshing with small elements.</u>

Body 1 and Body 2 are touching. The option Enforce common nodes between touching boundaries is not selected.

To have *Body 1* and *Body 2* connected, leave **Component Interaction** as **Bonded**, and do NOT select the option **Enforce common nodes between touching boundaries**. This is done to eliminate the transition zone between fine mesh (*Body 1*) and coarse mesh (*Body 2*), to reduce the number of degrees of freedom in the model.

Apply the pressure load, symmetry BC and restraint as shown in Figure 22-16.

Apply a **Mesh Control** of 0.35mm to *Body 2* and mesh the model with an element size of 2.00mm (Figure 22-17).

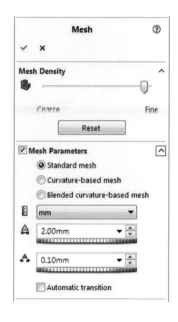

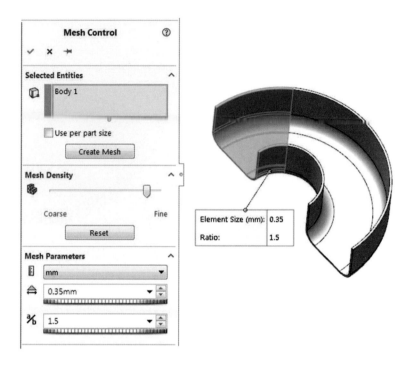

Global Mesh   Mesh Control applied to Body 1
Element size: 2mm   Element size: 0.35mm

Figure 22-17: Mesh settings in STEEL PULLEY SOLID model.

Examine the transition of element size between bodies; a rapid transition is enabled not selecting the option Enforce common nodes between touching boundaries (Figure 22-16).

Obtain the solution and read the solver message to see that despite our efforts to reduce the model size (mesh control to *Body 1* and no mesh transition between *Body 1* and *Body 2*), the problem still has over 4 million DOF.

We now compare results of STEEL PULLEY and STEEL PULLEY SOLID using von Mises stress results (Figure 22-18).

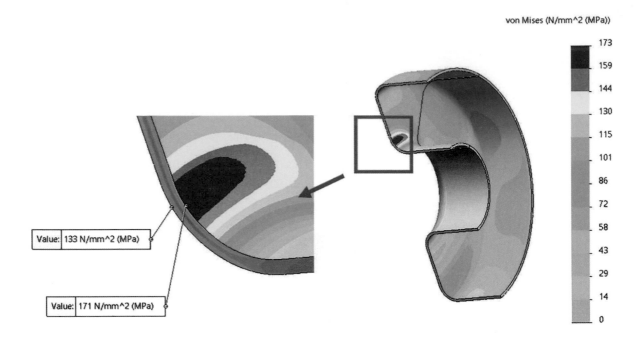

von Mises (N/mm^2 (MPa))

Value: 133 N/mm^2 (MPa)

Value: 171 N/mm^2 (MPa)

Figure 22-18: Von Mises stress results in STEEL PULLEY SOLID model.

Compare this result to that on the Top of the shell element model.

Plank

❑ Nonlinear geometry analysis
❑ Bending stress
❑ Membrane stress
❑ Shell element orientation

We'll use PLANK model which originates from "Vibration Analysis with SOLIDWORKS Simulation" where it is used in a non-linear vibration problem. Here we use it to study different ways of reporting stress results provided by shell elements. Loads and restraints are shown in Figure 22-19.

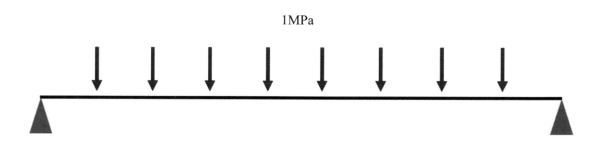

1MPa

Figure 22-19: PLANK model shown in *02 shell* configuration.

Hinge supports cannot move in the horizontal direction.

The PLANK problem is similar to LINK02 from Chapter 14. When loaded and restrained as shown in Figure 22-19, this problem requires a non-linear geometry analysis. This is because, in addition to **Bending stress**, it develops **Membrane stress**.

Make sure the model is in *02 shell* configuration and create a **Static** study *01 static NL* with the **Large displacements** option selected. Define a pressure of 1MPa to the top face.

Define supports to both ends as shown in Figure 22-20.

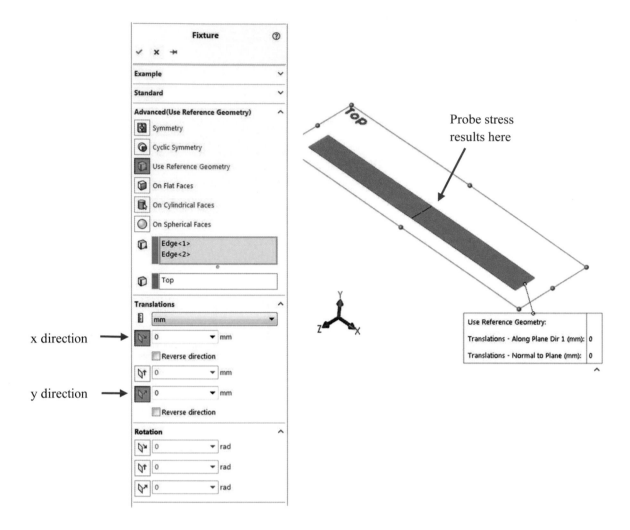

Figure 22-20: Restraints of PLANK model.

Both ends are restrained in X and Y directions as shown above. In addition to this, Rigid Body Movement in the Z direction must be eliminated. This can be easily accomplished with an Immovable restraint applied to any of the four corners. Rotations are not restrained.

A similar restraint could be defined using **Immovable** restraints to both ends but that would prevent the edges from shrinking due to the Poisson's effect.

Use the **Shell Definition** to define a shell element thickness of 1" and **Flip shell top and bottom** to have the top of shell elements on the top face where pressure is applied (Figure 22-21).

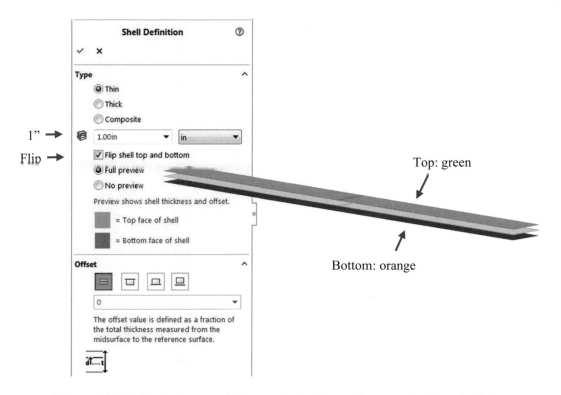

Figure 22-21: Shell element thickness definition and reversal of the shell element normal vector.

Colors do not refer to physical orientation. Colors refer to top and bottom of shell elements that will mesh this surface.

Copy study *01 static NL* into *02 static LIN* and deselect **Large displacement**; the problem becomes linear. **Membrane stresses** will not develop if the two hinges could move closed to each other. Obtain solutions of both studies. Comparison of displacement results of both studies (Figure 22-22) shows the importance of **Membrane stresses** which have a very strong stiffening effect.

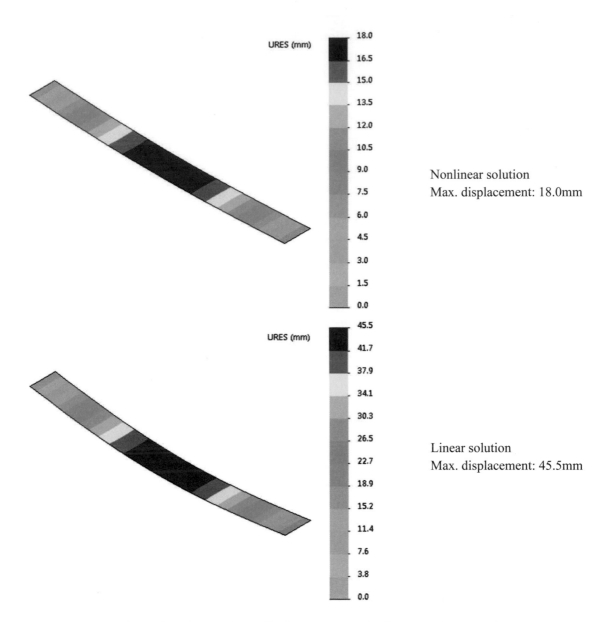

Nonlinear solution
Max. displacement: 18.0mm

Linear solution
Max. displacement: 45.5mm

Figure 22-22: Resultant displacement results from nonlinear study (top) and linear study (bottom).

Neglecting nonlinear effects (membrane stresses) produces 250% error in the resultant displacements.

Having confirmed the importance of **Membrane stress** in the PLANK problem, we now return to the main topic of this exercise: analysis of bending and membrane stresses.

Stay with the *01 Static NL* study and create four SX stress plots: **Top**, **Bottom**, **Membrane** and **Bending**. Probe stress values anywhere along the mid-line shown in Figure 22-20. All results are summarized in Figure 22-23.

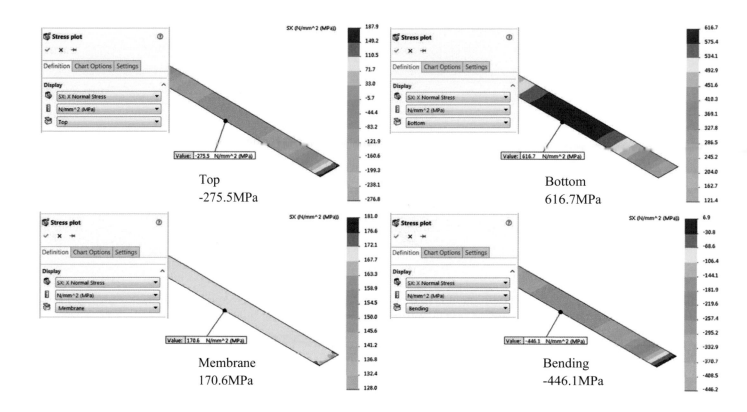

Figure 22-23: SX stress in the mid-span.

Stress on Top is compressive, stress on Bottom is tensile.

The following equations relate **Membrane** and **Bending** stress to **Top** and **Bottom** stresses:

$$SX_{membrane} = (SX_{top} + SX_{bottom})/2 = (-275.5 + 616.7)/2 = 170.6$$

$$SX_{bending} = (SX_{top} - SX_{bottom})/2 = (-275.5 - 616.7)/2 = -446.1$$

The sign of **Bending** stress depends on the orientation of shell elements. Try flipping shell elements and re-run analysis to see that **Bending** stress will be +446.1.

SX stress is linearly distributed across the shell element thickness; this is a part of shell element design. Stresses in any cross section can be separated into **Bending** and **Membrane** as sometimes required by certain design codes, in particular in the field of pressure vessel design. This separation is shown in Figure 22-24 using SX stress results from Figure 22-23.

$SX_{top} = -275.5MPa$

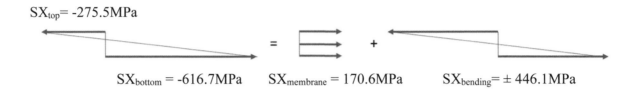

$SX_{bottom} = -616.7MPa$ $SX_{membrane} = 170.6MPa$ $SX_{bending} = \pm 446.1MPa$

Figure 22-24: Separation of SX stress in the mid-span of the PLANK into Membrane and Bending components.

Sign of bending stress depends on the orientation of shell elements.

To complete this exercise, go to study *02 static LIN* and look at plot of $SX_{membrane}$ to confirm that membrane stress does not exist in linear solution.

Torsion bar

- Cyclic symmetry boundary conditions
- Following displacement boundary conditions
- Defeaturing
- Limitations of linear analysis

Open part TORSION BAR and review three configurations: *01 full, 02 full, 03 half.* The first two configurations are shown in Figure 22-25.

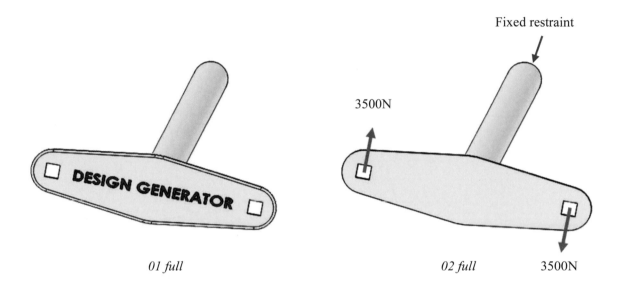

Fixed restraint

3500N

01 full

02 full   3500N

Figure 22-25: Model TORSION BAR in two configurations.

Configuration 02 full shows a couple of forces that apply pure torque to the torsion bar. Also shown is a restraint applied to the end face. Features deemed unimportant for analysis have been suppressed.

The difference between two configurations shown in Figure 22-25 is that *02 full* has small features suppressed. These features (the manufacturer's name, external rounds) have no structural importance and would unnecessarily complicate the FEA model. Remember that defeaturing must be done very carefully so that structurally important features are not removed from the model.

We start this exercise in the *02 full* configuration. Create a **Static** study *full LIN* and apply restraints as shown in Figure 22-25 and loads as shown in Figure 22-26.

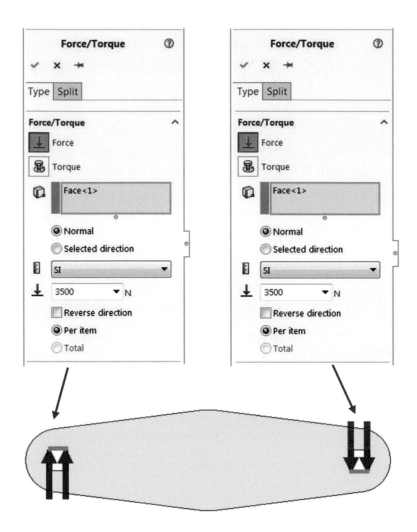

Figure 22-26: Load applied to left hole's top face and load applied to right hole's bottom face.

Load is applied as Normal. Model is shown in the front view. Hole edge where load is applied is highlighted in blue.

Holes are square to simplify definition of a following load in the next study. Square holes have sharp re-entrant edges causing stress singularities. However, we are not interested in stresses around the hole but in the effect of torque load on the torsion bar; therefore, we accept the lack of fillets in both square holes as an acceptable modeling simplification.

Define mesh controls as shown in Figure 22-27.

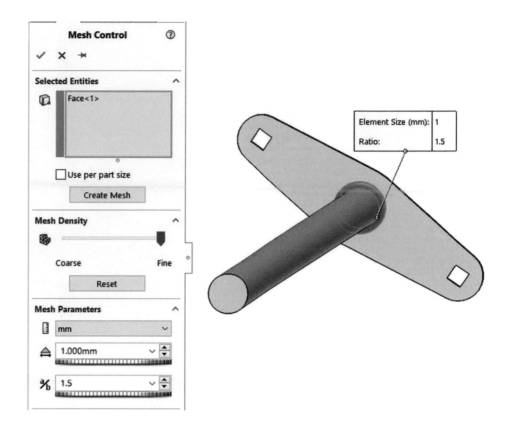

Figure 22-27: Mesh control applied to the fillet.

The fillet is an important structural feature; it cannot be suppressed.

Mesh the model with a 5mm element size and obtain the solution.

Next, create a **Nonlinear** study *full NL* with **Use large displacement formulation** and **Update load direction with deflection** options checked; use defaults for everything else and solve the nonlinear study. Notice that square holes rather than round holes make it easier to define the following load.

Prepare plots of resultant displacements from both studies: *full LIN* and *full NL*; show the undeformed shape superimposed on the deformed shape using a 1:1 scale of deformation. Both plots are shown in Figure 22-28.

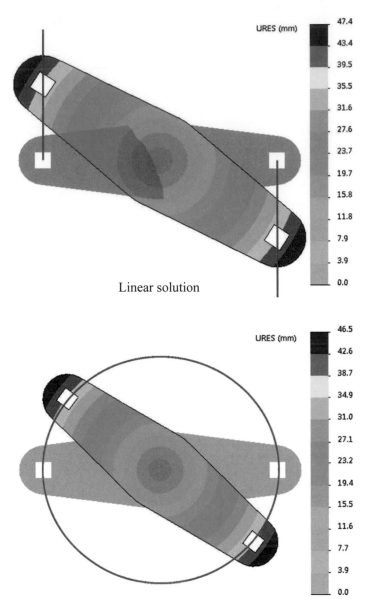

Linear solution

Nonlinear solution with following load

Figure 22-28: Plots of resultant displacements in 1:1 scale of deformation with superimposed undeformed shape: linear solution (top) and nonlinear solution (bottom).

In a linear solution, centers of holes trace straight lines while model deformation is animated. In a nonlinear solution, centers of holes trace an arc.

As plots in Figure 22-28 indicate, the linear solution provides incorrect results. The lug appears to be growing in size and points trace straight lines when the displacement plot is animated. Displacements are large and linear analysis can't be used. A similar problem has been presented using NL002 model in Chapter 14.

We will now use the TORSION BAR model in *03 half* configuration to demonstrate how this torsion problem may be solved taking advantage of the model's **Cyclic Symmetry** (Figure 22-29).

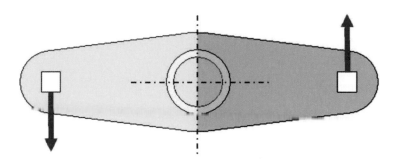

Figure 22-29: Cyclic Symmetry of the TORSION BAR problem; when the left half of the model (yellow) is rotated by 180° about the axis of symmetry it assumes position of the right half (green). This applies not only to geometry but also to load and restraint.

Model is shown in the back view with center lines added.

The symmetry shown in Figure 22-29 is a special case of **Cyclic Symmetry**; it may be called circular symmetry with the multiplier of 2.

Switch to *03 half* configuration and create a **Nonlinear** study *half NL*; select **Use large displacement formulation** and **Update load direction with deflection** in the study properties.

Define the restraint and load, this time to one side only.

Define **Cyclic Symmetry** boundary conditions as shown in Figure 22-30.

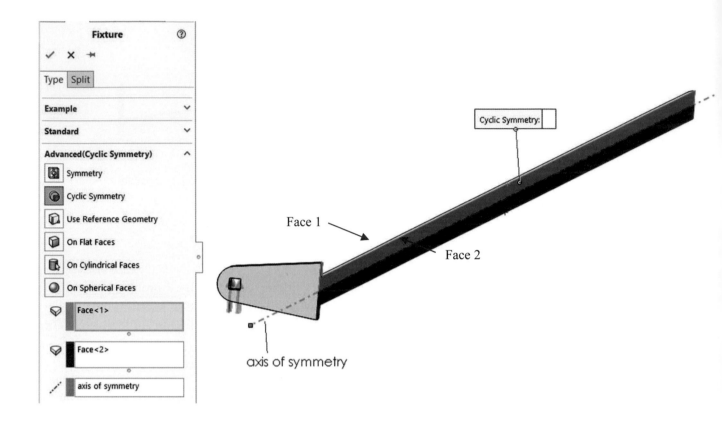

Figure 22-30: Definition of Cyclic Symmetry boundary conditions.

The definition is facilitated by a split line that creates Face 1 and Face 2. Notice Face 1 and Face 2 in the Fixture definition window and in the model view.

Define a mesh control of 1mm to the fillet and mesh the model with an element size of 5mm.

Obtain the solution and review displacements and stress results. A comparison between displacement results of the full model and half model with **Cyclic Symmetry** boundary condition is shown in Figure 22-31.

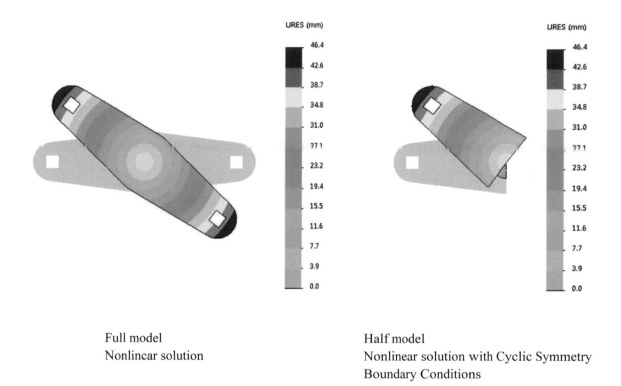

Full model
Nonlinear solution

Half model
Nonlinear solution with Cyclic Symmetry
Boundary Conditions

Figure 22-31: Resultant displacements of full model and half model with Cyclic Symmetry Boundary Conditions.

Both plots are overlaid on the undeformed shape.

A comparison of von Mises stress results in the full model and half model with **Cyclic Symmetry** boundary conditions is shown in Figure 22-32.

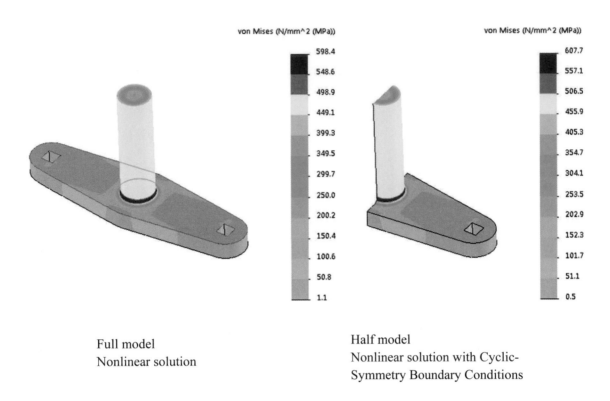

Full model
Nonlinear solution

Half model
Nonlinear solution with Cyclic-Symmetry Boundary Conditions

Figure 22-32: Von Mises stress results in the full model and half model.

Minor differences in stress results are caused by different discretization errors. Plots have been clipped (only a portion of the round bar is shown) to fit better on this page.

Cyclic Symmetry boundary conditions impose the same displacements in corresponding locations of the two faces shown in Figure 22-30; they may be classified as following displacement boundary conditions because the definition of a **Cyclic Symmetry** boundary condition remains valid as faces deform.

To complete the review of TORSION BAR, use configuration *04 full square* and create nonlinear study *full square NL*. Define NYLON 6/10 material; use nonlinear material model Plasticity von Mises type; the yield strength of this material is 139MPa. Define following, anti-symmetric load 1000N to each hole. You may use either automatic time stepping or the fixed time step 0.1s.

Mesh the model with default element size.

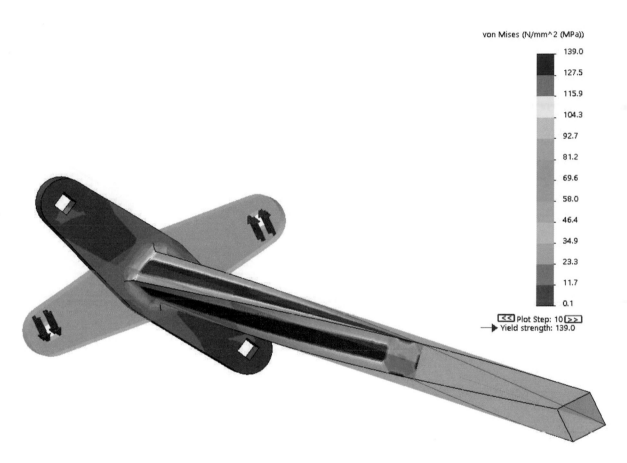

Figure 22-33: Von Mises stress under the maximum load 1000N to each hole.

Section clipping is used to show the extent of yielding in the square shaft.

The square shaft performs about 120° rotation and partially yields under the moment load 120Nm (the distance between holes in 120mm).

Bracket

- Anti-symmetry boundary conditions
- Non-following displacement boundary conditions

Open model BRACKET SYM which is the same as the BRACKET model studied in Chapter 12 except for different configurations. All four configurations are shown in Figure 22-34. The back face is fixed. The bracket is loaded with tractions uniformly distributed over the cylindrical face of the hole; this load produces bending.

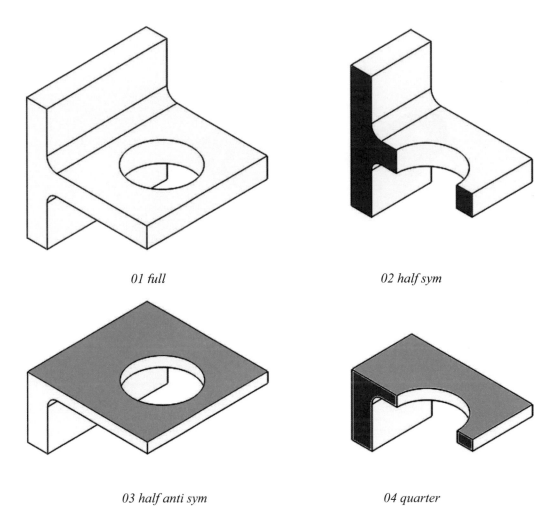

01 full   02 half sym

03 half anti sym   04 quarter

Figure 22-34: BRACKET SYM model in four configurations; configuration *04 quarter* is the only one that will be used for analysis.

Red color indicates faces where symmetry boundary conditions will be applied. Green color indicates faces where anti-symmetry boundary conditions will be applied.

Our objective is to demonstrate the use of symmetry boundary conditions and anti-symmetry boundary conditions in this bending problem. The presence of symmetry in this problem is obvious; the presence of anti-symmetry is explained in Figure 22-35.

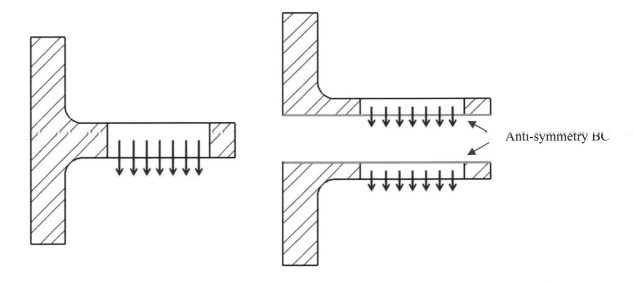

Full model shown in a cross section along the plane of symmetry

Model and load are shown as split into two along the plane of anti-symmetry

Anti-symmetry BC

Figure 22-35: Anti-symmetry boundary conditions apply to faces created by the cut.

Anti-symmetry boundary conditions apply when geometry and restraints are symmetric on both halves of the model while load is the same in magnitude but opposite in its direction. In the illustration on the right two halves are moved apart for clarity of showing load symbols. Edges of faces where anti-symmetry BC are defined are shown in green color.

We may now solve this problem using the *04 quarter* configuration with symmetry boundary conditions applied to red faces and anti-symmetry boundary conditions applied to green faces as shown in Figure 22-34.

Anti-symmetry boundary conditions are, by definition, exactly opposite to symmetry boundary conditions; therefore, we may first define symmetry boundary conditions and then "flip" them into anti-symmetry as shown in Figure 22-36.

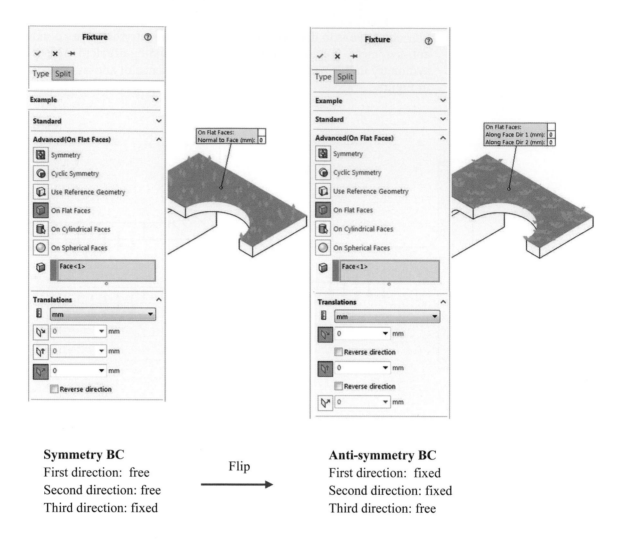

Symmetry BC
First direction: free
Second direction: free
Third direction: fixed

Flip
→

Anti-symmetry BC
First direction: fixed
Second direction: fixed
Third direction: free

Figure 22-36: Anti-symmetry boundary conditions are easiest to define by defining symmetry boundary conditions first, then "flipping" them into anti-symmetry boundary conditions.

This method will work if symmetry boundary conditions are defined either as On Flat faces or Use Reference Geometry because these definitions allow us definition of individual directions. It will not work with Symmetry.

Define symmetry and anti-symmetry boundary conditions, and fixed restrained to the back face. The load magnitude is 10000N but since we are working with ¼ of the model, apply 2500N as shown in Figure 22-37.

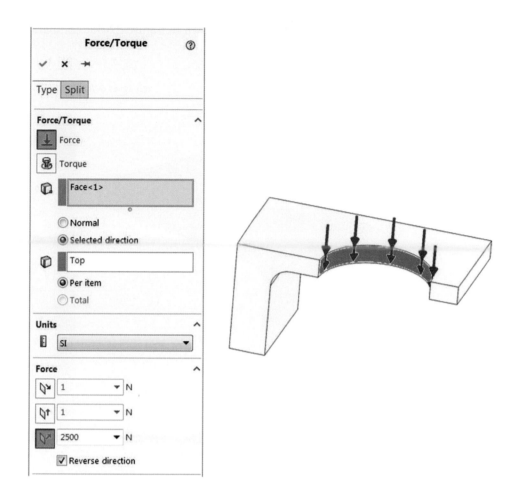

<u>Figure 22-37: Load definition.</u>

The load is uniformly distributed over the cylindrical face of the hole.

The model in *04 quarter* configuration is very simple. We can easily mesh it with small elements without using any mesh controls; use a 2mm global element size to mesh this model.

Von Mises stress results are shown in Figure 22-38; compare them to those shown in Figure 12-6.

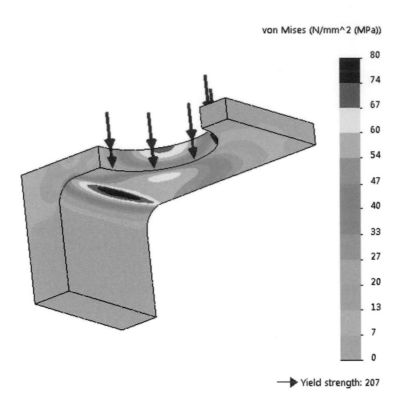

Figure 22-38: Von Mises stress results.

Load symbols are shown to help you orient the model; the face where the anti-symmetry boundary conditions are applied is not visible.

As opposed to **Circular Symmetry** discussed in the previous example, the **Anti-symmetry** boundary conditions do not follow deforming geometry. They may be used only in small displacement problems. Another reason why the **Anti-symmetry** boundary conditions may not be used in a large displacement problem is that the location of neutral bending plane changes with curvature of beam. This would be impossible to model with **Anti-symmetry** boundary conditions even if they did follow the deforming geometry.

Links

❑ Non-linear geometry analysis
❑ Non-following displacement boundary conditions

Open assembly LINKS and review the four configurations shown in Figure 22-39. Two links are connected by a pin and supported by hinges at both ends. The vertical load of 10000N is uniformly distributed between two faces as shown in Figure 22-42.

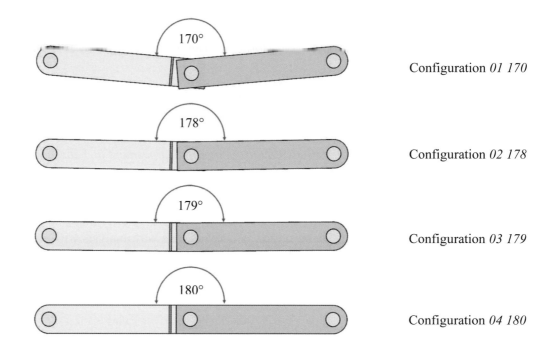

Figure 22-39: LINKS assembly in four configurations.

The difference between configurations is the angle between the two links.

Our objective is to study changes in the assembly stiffness while the load increases from 0 to 10000N. The analysis of stiffness changes during the process of load application requires nonlinear analysis.

In the CAD model, review the **Sensor** definition (Figure 22-40).

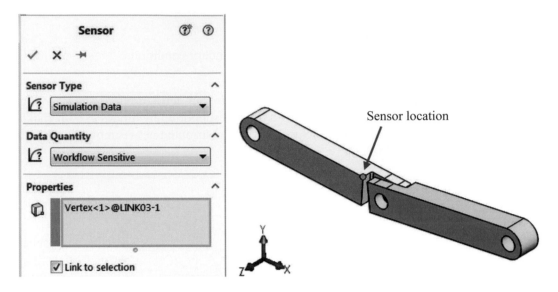

Figure 22-40: Sensor definition and location.

This sensor will be used to plot displacement in Y direction.

We'll start this project in *01 170* configuration. Create a **Nonlinear** study *01 170* and use all defaults in study properties. Verify that **Use large displacement formulation** option is checked; this way analysis will account for geometric nonlinearity if it is present. As always, remember that displacements don't have to be large to change stiffness significantly; this is contrary to what the name **Use large displacement formulation** implies.

Define **Results Options** as shown in Figure 22-41.

Figure 22-41: Results Options definition.

Response Plots will use the Workflow Sensitive sensor shown in Figure 22-40.

Define a **Pin Connector** to make a connection that allows for relative rotation between the two links.

Apply the load as shown in Figure 22-42. Apply **Hinge** Support to the left and the right holes.

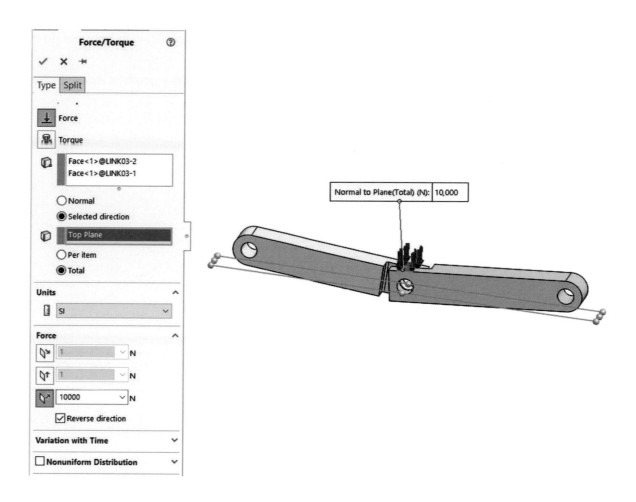

Figure 22-42: Load definition; the total load 10000N is uniformly distributed among two flat faces. Variation with Time does not have to be modified because the load time history is linear.

Loads and restraints symbols are not shown.

Define a following load in the study properties, meaning that load remains perpendicular to faces where it is applied. However, the model won't experience large displacements and the difference between following and non-following load is negligible in this exercise.

Mesh the model with default mesh and notice that elements in fillets are poorly shaped. The effect of poorly shaped elements is only local; therefore, we may accept this mesh if we limit our analysis to displacements only and do not analyze stresses.

Obtain the solution and construct a **Time History** graph as shown in Figure 22-43.

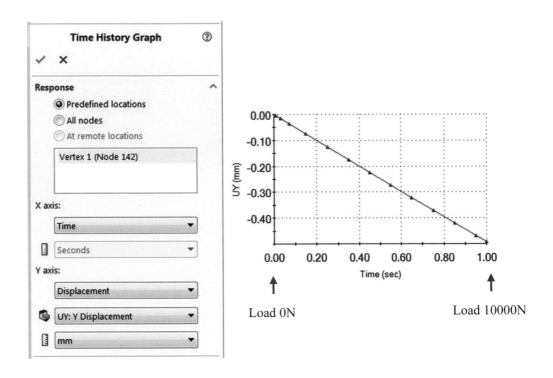

Figure 22-43: Definition of Time History Graph of UY displacement time history.

Predefined locations refer to the vertex where the sensor has been defined.

As shown in Figure 22-43, the load-displacement curve is a straight line; this shows that the model in *01 170* configuration does not exhibit geometric nonlinearities; stiffness is not changing during the process of load application. We could have solved it in a **Static** study with default settings but then we would be unable to see the displacement time history (Figure 22-43).

Repeat the analysis in *02 178* (study *02 178*) and *03 179 (study 03 179)* configurations and notice that the load displacement graphs show curves; this means that stiffness changes during the process of load application (Figure 22-44).

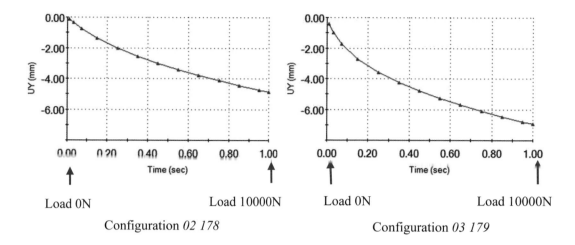

Configuration *02 178*   Configuration *03 179*

Figure 22-44: UY displacement time history of the model in *02 178* and *03 179* configurations.

Scales on the ordinate axes of these graphs have been adjusted to make them the same.

A comparison of UY displacement time history graphs between *02 178* and *03 179* configurations demonstrates that the model in the *03 179* configuration is softer and exhibits stronger nonlinear behavior.

An attempt to solve model in *04 180* configuration produces an error message (Figure 22-45).

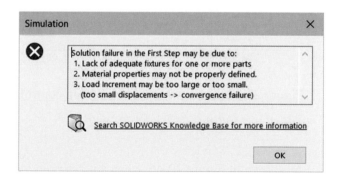

Figure 22-45: Error message produced by an attempt to solve model in *04 180* configurations.

Comments in the windows do not address the real problem.

The reason for this solution failure is that in *04 180* configuration, two links are aligned and have no stiffness in the direction of the applied load. So, what happens when the solution is attempted with default settings shown in Figure 22-46?

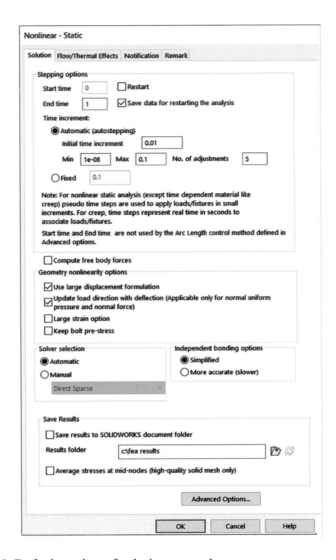

Figure 22-46: Default setting of solution control.

The above settings define force control.

As shown in Figure 22-46, when auto stepping is used, the initial time increment is 0.01s which in our case means that the initial (step 1) load magnitude is 100N. Since the model has no shape stiffness and all (minute) stiffness present in the model comes from discretization errors, that 100N produces a very large displacement. Having detected a very large displacement, the solver attempts to reduce the initial load magnitude until it reaches the maximum allowed number of adjustments, which is 5 in default settings, and an error message is produced (Figure 22-45).

To solve the model in *04 180* configuration we need a different way of solution control. We must use **Displacement Control** available in **Advanced Options** of **Nonlinear** study properties (Figure 22-47). This option requires the **Direct Sparse solver**.

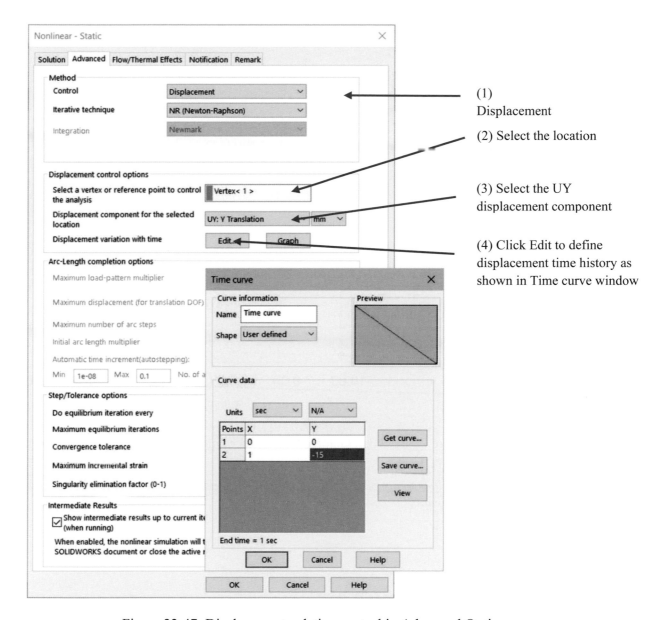

(1) Displacement

(2) Select the location

(3) Select the UY displacement component

(4) Click Edit to define displacement time history as shown in Time curve window

Figure 22-47: Displacement solution control in Advanced Options.

The advanced setting tab must be activated from the Solution tab by clicking Advanced Options.

Using the **Displacement Control** setting shown in Figure 22-47, load increments are adjusted to keep the model in equilibrium after each increase in displacement. This approach overcomes the problem of nonexistent stiffness at the beginning of the solution. A very small initial displacement gives the model some initial stiffness and the solution may proceed until the displacement reaches approximately -15mm as defined in the **Time curve** window (Figure 22-47). A graph showing the UY displacement time history is shown in Figure 22-48.

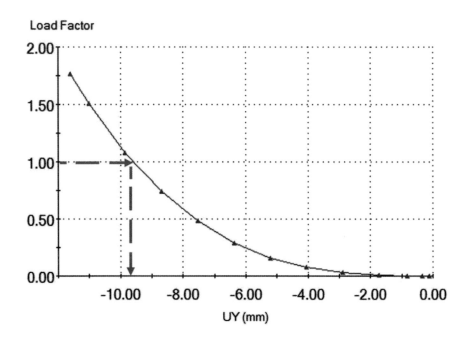

Figure 22-48: Load factor as a function of displacement of the vertex selected in Figure 22-47.

Displacement corresponding to 10000N (load factor equal to 1) may be read from the graph.

Having obtained solutions for all four configurations we may now summarize the results; notice that the ordinate and the abscissa in the graph in Figure 22-48 will have to be reversed to make this graph directly comparable to graphs in Figures 2-43 and Figure 22-44. A summary is shown in Figure 22-49.

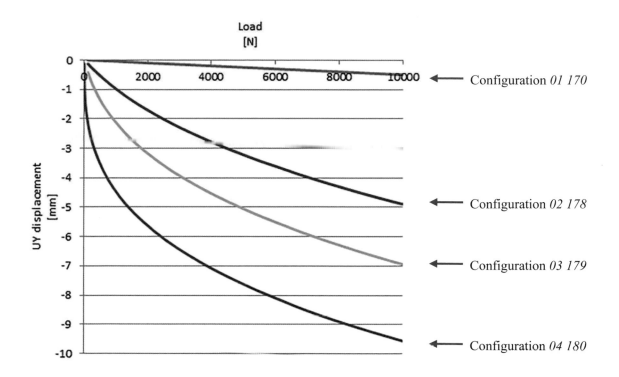

Figure 22-49: Summary of displacement results from all four studies.

Notice that the load-displacement curve for 04 180 configuration is tangent to the ordinate at the origin because the initial stiffness is zero.

The load-displacement curve for *01 170* configuration is a straight line; this proves that the problem is linear and could have been solved using a linear analysis. Configurations *02 178, 03 179, 04 180* require non-linear geometry analysis.

The changing of stiffness is easier to observe if we switch the ordinate and the abscissa and use absolute displacement as shown in Figure 22-50.

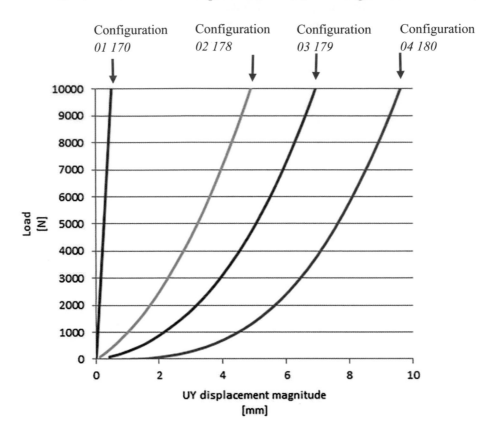

Figure 22-50: Load as a function of displacement.

This way of presenting results concords better with the Displacement control solution method used in Configuration 04 180 where displacement is the directly controlled entity and load follows displacement.

If the angle between the line tangent to the curve and x axis is α, then stiffness in the point of tangency is k = ΔF / Δx = tan α (Figure 22-51).

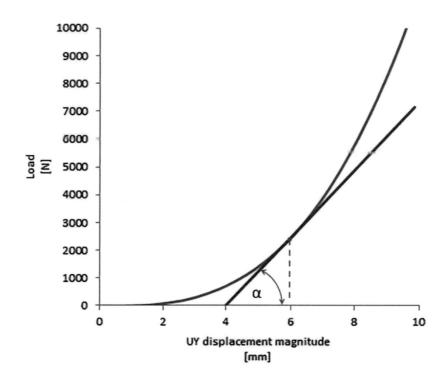

Figure 22-51: Model stiffness for a given displacement.

The curve for 04 180 configuration is shown. As shown in this illustration, tan α is the model stiffness when displacement magnitude equals 6mm.

The graph in Figure 22-51 clearly demonstrates that the initial stiffness in configuration *04 180* is zero.

Repeat studies of your choice with mesh controls added to the fillets to study stress results.

525222

2222

Models in this chapter

All models have studies fully defined and ready to run.

Model	Configuration	Study Name	Study Type
2D BEAM.sldprt	Default	01 2D	Static
		02 2D	Static
		03 2D	Static
ALUMINUM PULLEY.sldprt	01 full		
	02 half	Static 1	Static
STEEL PULLEY.sldprt	01 full		
	02 half	Static 1	Static
STEEL PULLEY SOLID.sldprt	01 full		
	02 half	Static 1	Static
PLANK.sldprt	01 solid		
	02 shell	01 static NL	Static
		02 static LIN	Static
TORSION BAR.sldprt	01 full		
	02 full	full LIN	Static
		full NL	Nonlinear
	03 half	half NL	Nonlinear
	04 full square	full square NL	Nonlinear
BRACKET SYM.sldprt	01 full		
	02 half sym		
	03 half anti sym		
	04 quarter	Static 1	Static
LINKS.sldasm	01 170	01 170	Nonlinear
	02 178	02 178	Nonlinear
	03 179	03 179	Nonlinear
	04 180	04 180	Nonlinear

23: Implementation of FEA into the design process

Topics covered

- Verification and Validation of FEA results
- FEA driven design process
- FEA project management
- FEA project checkpoints
- FEA reports

VERIFICATION AND VALIDATION OF FEA RESULTS

Tools of Computer Aided Engineering (CAE) are now widely used to make design decisions. The reliance on CAE tools such as Finite Element Analysis (FEA) to make design decisions brings about the issue of how relevant results from FEA models are to real life design problems. To make sure that correct decisions are made, FEA results must be verified and validated. The terms "verification" and "validation" are often used interchangeably in casual conversations.

In FEA, verification and validation pertain to different steps in the FEA modeling process. We will define and differentiate between these terms while describing FEA modeling steps. We will expand the discussion started in Chapter 1.

Step 1: Creation of the mathematical model

Every FEA project starts with the creation of the mathematical model. The mathematical model needs information on the geometry of the part or assembly that we analyze for material properties, as well as loads and restraints assigned to that geometry. The definition of the type of analysis along with its simplifying assumptions (for example nonlinear static analysis, linear buckling analysis or transient thermal analysis) completes the creation of the mathematical model (Figure 23-1).

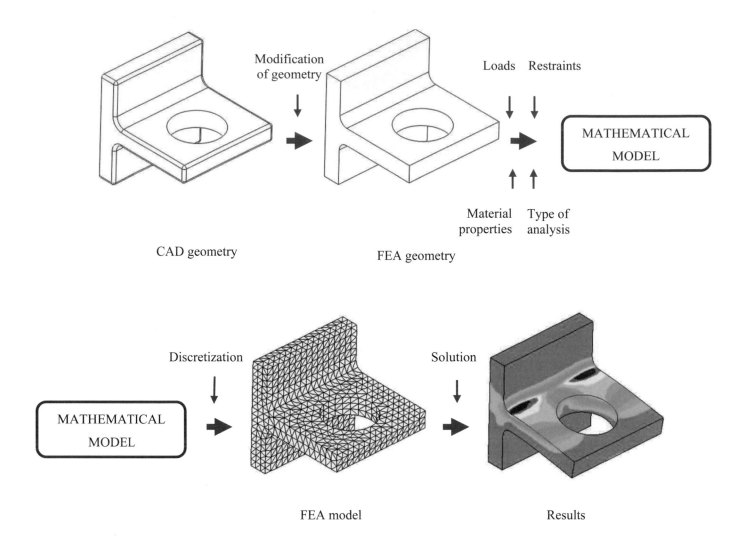

Modification
of geometry

Loads   Restraints

MATHEMATICAL
MODEL

Material   Type of
properties   analysis

CAD geometry

FEA geometry

Discretization

Solution

MATHEMATICAL
MODEL

FEA model

Results

<u>Figure 23-1: Steps in an FEA project.</u>

This is a repetition of Figure 1-2 and Figure 1-3.

All components of a mathematical model definition bring with them inherent simplifying assumptions which affect the results. A correctly formulated mathematical model captures aspects of the real object that are important in analysis. For example, analysis of a compliant link under a static load requires nonlinear formulation due to expected large displacements. Analysis of a cooling process requires transient thermal analysis, and a drop test calls for nonlinear dynamic analysis.

Very serious errors may result if the mathematical model does not capture the physics of the analyzed phenomenon. For example, if we neglect large displacements and use a linear rather than nonlinear, large displacement analysis to calculate beam displacement, we produce nonsensical results as shown in Figure 23-2.

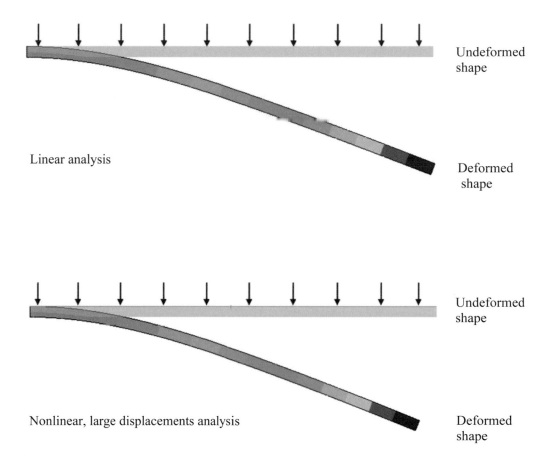

Figure 23-2: A cantilever beam in bending; displacements are shown in 1:1 scale. Incorrect results produced by a linear analysis (top). Correct results produced by a nonlinear, large displacement analysis (bottom).

Undeformed shape is shown with a following pressure load. Notice that the tip of the beam travels along a straight line as deformation progresses in linear analysis.

Review studies in model BEAM for more information.

Similarly, analysis of a flat membrane under pressure requires a nonlinear analysis to account for the change in model stiffness during the deformation process (even though these displacements may be very small). Neglecting this fact and using a linear model formulation leads to a very serious and potentially more dangerous error than that shown in Figure 23-2 because results can look plausible (Figure 23-3).

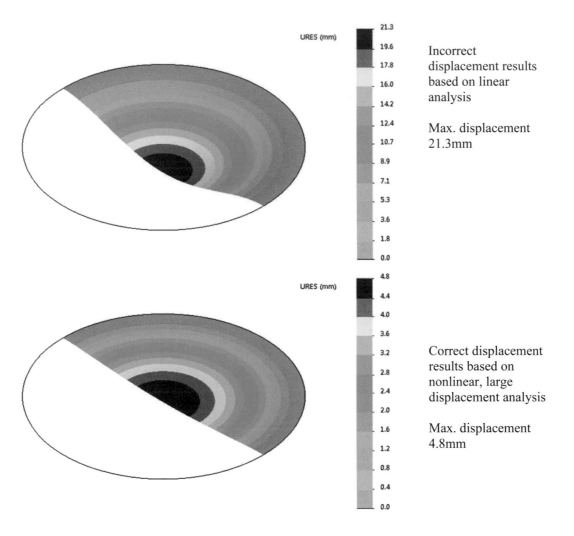

Incorrect displacement results based on linear analysis

Max. displacement 21.3mm

Correct displacement results based on nonlinear, large displacement analysis

Max. displacement 4.8mm

Figure 23-3: A thin plate subjected to a pressure load must be treated as a nonlinear, large displacement problem. Neglecting nonlinear effect of membrane stiffening produces results with 444% error.

Section Clipping is used to show the deformed shape better.

For more information review studies *01 linear* and *02 nonlinear* in model ROUND PLATE NYLON.

A very serious yet common modeling error is using a model with stress singularities to find stress results in those singular locations (Figure 23-4).

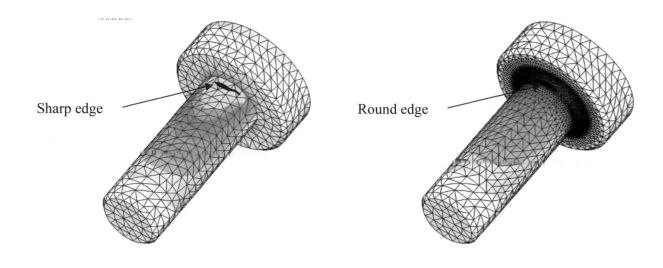

Figure 23-4: Stress singularities caused by a sharp re-entrant edge (top). If stresses at the base of the cantilever are of interest, a fillet must be present in the analyzed geometry, no matter how small the fillet may be (bottom).

Review model BOLT for more information. A similar problem was illustrated in chapter 3 using model L BRACKET.

A mathematical model may also have more "trivial" errors such as incorrect loading or incorrect material properties. In the author's experience, errors are commonly found in the restraints' definitions. For example, applying a rigid support to the entire back face of a support plate rather than to the bolt holes may lead to severe underestimation of displacements and/or stress (Figure 23-5).

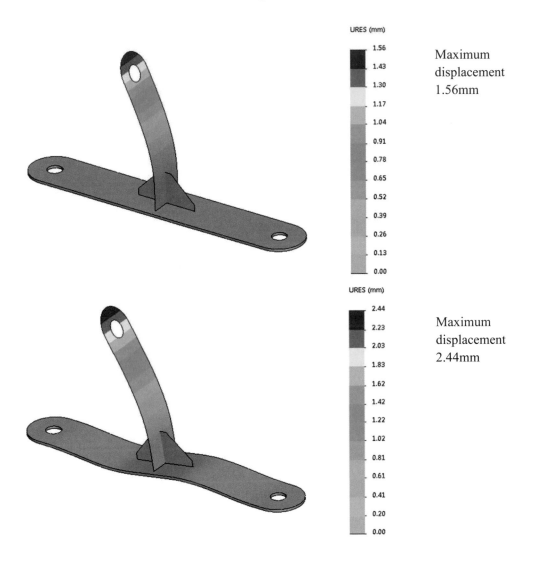

Maximum displacement 1.56mm

Maximum displacement 2.44mm

Figure 23-5: Errors caused by an incorrect restraint definition. Restraints are applied to the entire bottom face (top). Restraints applied to only the holes produce very different displacement results (bottom).

Weld connectors are used to connect components of this assembly. Review model T FLAT assembly for more information.

A mathematical model is never free of all errors. These unavoidable errors are known by a less intimidating term - simplifying assumptions. Every definition in making a mathematical model has some degree of simplifying assumptions which must be justified and are critical to the success of an analysis. It is our responsibility to assure that simplifying assumptions are not made "subconsciously" and that they do not prevent the model from providing trustworthy results.

Mathematical models are seldom simple enough to solve by hand and so we must use numerical techniques to solve them. FEA is one of these numerical techniques which, due to its versatility and ease of use, has dominated the commercial market of engineering analysis software.

Step 2: Creation of an FEA model

As with any other numerical technique, FEA works with a discretized model. Therefore, in preparation for a solution with FEA, the mathematical model must be discretized. In discretization, a continuous mathematical model is split into finite elements in the process commonly known as meshing. While a meshed model is easy to depict graphically, this graphical representation may be confusing because it implies that a mesh is imposed on the model geometry. In fact, there is nothing continuous left in the FEA model. Continuous geometry is replaced by discrete nodes, and the interaction between the nodes is defined by elements connecting these nodes. Finite elements define relations between nodes. It is conceptually important to remember that loads and restraints are also discretized. Discrete loads and discrete restraints are applied to nodes. The model's mass is no longer distributed continuously but rather, it is distributed among nodes. Unfortunately, FEA programs do not have graphical capabilities to show discretization of anything but geometry.

The process of discretization which converts the mathematical model into an FEA model does have problems that add to errors in the mathematical model. Every discretization brings with it discretization errors which may be analyzed (and controlled) in the convergence process where we analyze the effect of element size on the results. There are many "shades" of a convergence process. Most often a mesh is refined, and the results are examined in terms of their sensitivity to the refinement. Many modern FEA programs have capabilities to perform convergence processes automatically.

Discretization of an FEA model leads to the discretization of results. The result's nature depends on the type of elements used. The ability (or the lack of ability) of elements to model the real displacement and stress distribution very strongly impacts the results. A common error is to use too large of elements which are unable to capture local stress concentrations (Figure 23-6).

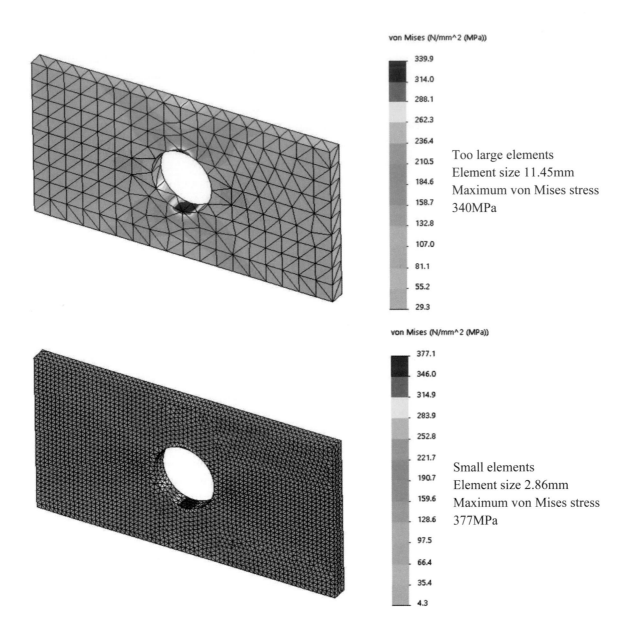

Figure 23-6 Errors caused by incorrect meshing.

You may use model HOLLOW BRACKET from chapter 2 to produce these results.

Problems depicted in Figure 23-6 are easy to catch by a trained eye and can all be rectified by mesh refinement. In fact, understanding the discrete nature of results will prevent the use of inadequate meshes. Even though discretization errors are easily preventable, experience indicates that they still plague analysis results.

Step 3: Solution

Once an FEA model has been created, its solution is just a matter of solving a large number of linear algebraic equations. This can be done by a variety of solvers. The solution introduces numerical errors which are usually very low.

Step 4: Interpretation of results

Finally, results must be analyzed, and a design decision made. Incorrect interpretation of results is a topic for a separate article. Here we just mention a few common errors. Indiscriminate use of von Mises stress as a safety criterion is a common error. Von Mises stress is a valid safety measure for materials showing distinct plasticity on a stress-strain curve. For example, using it to analyze results of a ceramic part is not valid. Another mistake is the incorrect use of element versus nodal stress results, which results from a lack of understanding the difference between these two.

Each of the above steps takes us further away from the reality we are modeling. Errors can be made at each step; some of them are unavoidable such as errors inherent to the method, and others may be grave errors of FEA "malpractice."

We are now in the position to define the terms **verification** and **validation**.

Verification checks if the mathematical model, as submitted to be solved with FEA, has been correctly discretized and solved.

Validation determines if an FEA model correctly represents the reality from the perspective of the intended use of the model. It checks if results correctly describe the real-life behavior of the analyzed object.

The difference between verification and validation is pictured in Figure 23-7.

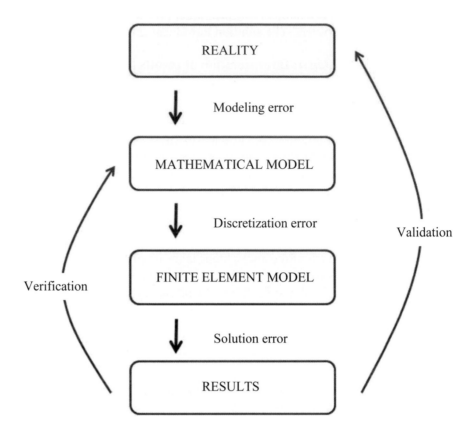

Figure 23-7 Verification and validation of FEA results.

A model with meshing errors would not pass the verification test. For example, having been discretized into too large of elements, the mathematical model would be solved incorrectly. Verification fails if discretization and/or solution errors invalidate results. Convergence analysis will usually reveal problems causing verification test failure and those problems can be treated by mesh refinement or by using higher order elements.

A model with incorrect load definitions would pass the verification test because verification only concerns itself with correctness of solution of the mathematical model, not if the mathematical model itself is correct.

Establishing the correctness of a mathematical model along with the correctness of its solution is the process of validation which should follow verification. Validation will fail because of conceptual errors in the definition of the mathematical model. These conceptual errors are much more dangerous than the errors of discretization. They may escape the modeler's attention, especially since there is no well-defined structured process to reveal conceptual errors. Our only protection is the true understanding of the analyzed problem.

FEA DRIVEN DESIGN PROCESS

We have already stated that FEA should be implemented early in the design process and be executed concurrently with design activities in order to help make more efficient design decisions. This concurrent CAD-FEA process is illustrated in Figure 23-8.

Notice that the design begins in CAD geometry and FEA begins in FEA-specific geometry. Every time FEA is used, the interface line is crossed twice: the first time when modifying CAD geometry to make it suitable for analysis with FEA, and the second time when implementing results.

This significant interfacing effort can be avoided if the new design is started and iterated in FEA-specific geometry. Only after performing a sufficient number of iterations can we switch to CAD geometry by adding all manufacturing specific features. This way, the interfacing effort is reduced to just one switch from FEA to CAD geometry as illustrated in Figure 23-8.

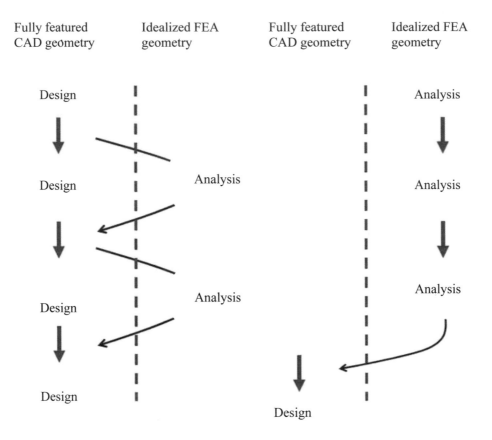

Concurrent CAD-FEA design process

FEA driven design process

Figure 23-8: Concurrent CAD-FEA product development processes (left) and FEA driven product development process (right).

The CAD-FEA design process is developed in CAD-specific geometry, while FEA analysis is conducted in FEA-specific geometry. Interfacing between the two geometries requires substantial effort and is prone to error (left).

CAD-FEA interfacing efforts can be significantly reduced if the differences between CAD geometry and FEA geometry are recognized, and the design process starts with FEA-specific geometry (right).

FEA PROJECT MANAGEMENT

Now let us discuss the steps in an FEA project from a managerial point of view. The steps in an FEA project that require the involvement of management are marked with an asterisk (*).

Do I really need FEA? *

This is the most fundamental question to address before any analysis starts. FEA, just like any other CAE tool, is expensive to conduct and consumes significant company resources to produce results. Therefore, a decision to use FEA should be well justified.

Providing answers to the following questions may help to decide if FEA is worthwhile:

- Can I use previous test results or previous FEA results?
- Is this a standard design, in which case no analysis is necessary?
- Are loads, restraints, and material properties known well enough to make FEA worthwhile?
- Would a simplified analytical model do?
- Does my customer demand FEA?
- Do I have enough time to implement the results of FEA?

Should the analysis be done in house or should it be contracted out? *

Conducting analysis in-house versus using an outside consultant has advantages and disadvantages. Consultants usually produce results faster while analysis performed in house is conducive to establishing company expertise leading to long-term savings.

The following list of questions may help in answering this question:

- How fast do I need to produce results?
- Do I have enough time and resources in-house to complete FEA before design decisions must be made?
- Is in-house expertise available?
- Do I have software that my customer wants me to use?

Establish the scope of the analysis*

Now we decide what type of analysis is required. The following is a list of questions that may help in defining the scope of analysis.

Is this project:

- A standard analysis of a new product from an established product line?
- The last check of a production-ready new design before final testing?
- A quick check of a design in-progress to assist the designer?
- An aid to an R&D project (particular detail of a design, gauge, fixture etc.)?
- A conceptual analysis to support a design at an early stage of development (e.g. R&D project)?
- A simplified analysis (e.g. only a part of the structure) to help make a design decision?

Other questions to consider are:

- Is it possible to perform comparative analyses?
- What is the estimated number of model iterations, load cases, etc.?
- What are applicable criteria to evaluate results?
- How will I know whether the results can be trusted?

Establish a cost-effective modeling approach and define the mathematical model accordingly

Having established the scope of analysis, the FEA model must now be prepared. The best model is of course the simplest one that provides the required results with acceptable accuracy. Therefore, the modeling approach should minimize project cost and duration but should account for the essential characteristics of the analyzed object.

We need to decide on acceptable idealizations of geometry. This decision may involve simplification of CAD geometry by defeaturing or idealization by using surface or wire frame representations. The goal is to produce a meshable geometry properly representing the analyzed problem.

Create a Finite Element model and solve it

The Finite Element model is created by discretization (meshing) of a mathematical model. Although meshing implies that only geometry is discretized, discretization also affects loads and restraints. Meshing and solving are both automated steps, but still require input, which depending on the software used, may include:

- Element type(s) to be used
- Default element size and size tolerance
- Definition of mesh controls (if any)
- Mesher type to be used
- Solver type and options to be used

Review results

FEA results must be critically reviewed prior to using them for making design decisions. This critical review includes:

- Review of assumptions and assessment of results; this is an iterative step that may require several analysis loops to debug the model and to establish confidence in the results
- Studying the overall mode of deformations and animating displacements to ensure that loads and restraints have been defined properly
- Checking for Rigid Body Motions
- Checking for overall stress levels (at least the order of magnitude) using analytical methods in order to verify the applied loads
- Checking for reaction forces and comparing them with free body diagrams
- A review of discretization errors by comparing nodal and element stress results and performing a convergence process
- A review of mesh quality (Aspect Ratio, Jacobian) and an investigation of the impact of element distortions on the data of interest
- Analysis of stress concentrations and the ability of the mesh to model them properly
- A review of results in difficult-to-model locations, such as thin walls, high stress gradients, etc.

Analyze results*

The exact execution of this step depends, of course, on the objective of the analysis.

- Present displacement results (preferably animated)
- Present reaction force results supported by free body diagrams
- Present modal frequencies and associated modes of vibration (if applicable)
- Present stress results and corresponding factors of safety
- Consider modifications to the analyzed structure to eliminate excessive stresses and to improve material utilization and manufacturability
- Discuss results, and repeat iterations until an acceptable solution is found

Complete report*

- Complete a report summarizing the activities performed, including assumptions and conclusions
- Append the completed report with a backup of relevant electronic data

FEA PROJECT CHECKPOINTS

FEA project management requires the involvement of the manager during project execution. The correctness of FEA results cannot be established by only reviewing the analysis of the results. A list of progress checkpoints may help a manager stay in the loop and improve communication with the person performing the analysis. Several checkpoints are suggested in Figure 23-9.

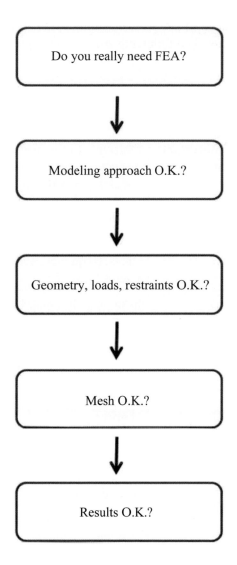

Figure 23-9: Example of checkpoints in an FEA project.

Using the proposed checkpoints, the project is allowed to proceed only after the manager/supervisor has approved each step.

FEA REPORT

Even though each FEA project is unique, the structure of an FEA report follows similar patterns. The following are the major sections of a typical FEA report and their contents.

Section	Content
Executive Summary	Objective of the project, part/assembly number, project number, essential assumptions, results and conclusions, software used, information on where project backup is stored, etc.
Introduction	Description of the problem: Why did the project require FEA? What kind of FEA (static, contact stress, frequency, etc.)? What were the data of interest?
Analysis type	Structural, thermal, linear, nonlinear, types of nonlinearities
Geometry **Material** **Loads** **Restraints**	Description and justification of any defeaturing and/or idealization of geometry Justification of the modeling approach (e.g. solids elements, shells elements, beams elements, 2D elements) Description of the material properties and applicable failure criteria Description of loads and restraints, including free body diagrams
Mesh	Description of the type of elements, global element size, mesh controls, number of elements, number of DOF, type of mesher used Justification of why this particular mesh is adequate to model the data of interest
Solver	Type of solver, adaptive method used (if any)
Analysis of results	Presentation of displacement, strain, stress, temperatures etc. results, including plots and animations Justification of the type of stress used to present results and failure criteria Discussion of errors in the results

Conclusions	Recommendations regarding structural integrity, necessary modifications, further studies needed Recommendations for follow-up testing procedure (e.g., strain-gauge test, fatigue life test) Recommendations on future similar analyses
Project documentation	Full documentation of the design, design drawings, FEA model explanations, and computer back-ups Notice that building in-house expertise requires very good documentation of the project besides the project report itself. Significant time should be allowed to prepare project documentation.
Follow-up	After the completion of tests, append the report with test results Discussion of the correlation between analysis results and test results Discussion of the corrective action taken in case the correlation is unsatisfactory (may involve a revised model and/or tests)

Models in this chapter

All models come with studies fully defined and ready to run.

Model	Configuration	Study Name	Study Type
BEAM.sldprt	*Default*	*01 linear*	Static
		02 nonlinear	Static
ROUND PLATE NYLON.sldprt	*Default*	*01 linear*	Static
		02 nonlinear	Static
BOLT.sldprt	*01 incorrect*	*01 incorrect*	Static
	02 correct	*02 correct*	Static
T FLAT.sldasm	*Default*	*01 back face support*	Static
		02 bolt support	Static

24: Glossary of terms

The following glossary provides short descriptions of selected terms used in this book.

Term	Definition
2D element	Depending on the type, 2D elements are intended for analysis of plane stress, plane strain and axisymmetric problems. Nodes of 2D elements in structural analysis have 2 degrees of freedom.
Aspect ratio	Measure of element distortion (largest/smallest size).
Beam element	A beam element is intended for meshing wire frame geometry. Nodes of beam elements have 6 degrees of freedom.
Boundary Element Method	An alternative to the FEA method of solving field problems, where only the boundary of the solution domain needs to be discretized. Very efficient for analyzing compact 3D shapes, but difficult to use on more "spread out" shapes.
CAD	Computer Aided Design.
Clean-up	Removing and/or repairing geometric features that would prevent the mesher from creating a mesh or would result in an incorrect mesh.
Convergence criterion	A condition that must be satisfied in order for the convergence process to stop. In **SOLIDWORKS Simulation** this applies to studies where an h-adaptive or a p-adaptive solution has been selected.

Term	Definition
Convergence process	This is a process of systematic changes in the mesh in order to see how the data of interest changes with the choice of the mesh and (hopefully) proves that the data is not significantly dependent on the choice of discretization. A convergence process can be performed as h-convergence or p-convergence. An h-convergence process is done by refining the mesh, i.e., by reducing the element size in the mesh and comparing the results before and after mesh refinement. Reduction of element size can be done globally, by refining the mesh everywhere in the model, or locally, by using mesh controls. An h-convergence analysis takes its name from the element characteristic dimension h, which changes from one iteration to the next. A p-convergence analysis does not affect element size. Elements stay the same throughout the entire convergence analysis process. Instead, element order is upgraded from one iteration to the next. A p-convergence analysis is done automatically in an iterative solution until the user-specified convergence criterion is satisfied. Sometimes, the desired accuracy cannot be achieved even with the highest available p-element order. In this case, the user has to refine the p-element mesh manually in a fashion similar to traditional h-convergence, and then re-run the iterative p-convergence solution. This is called a p-h convergence analysis.
Defeaturing	Defeaturing is the process of removing (or suppressing) geometric features in the CAD geometry in order to simplify the finite element mesh or make meshing possible.
Discretization	This defines the process of splitting up a continuous mathematical model into discrete "pieces" called elements. A visible effect of discretization is the finite element mesh. However, model mass, loads and restraints are also discretized.

Discretization error	This type of error affects FEA results because FEA works on an assembly of discrete elements (the mesh) rather than on a continuous structure. The finer the finite element mesh, the lower the discretization error, but the longer the solution time.
Element stress	This refers to stresses at Gauss points of a given element. Stresses at different Gauss points are averaged amongst themselves (but not with stresses reported by other elements) and one value is assigned to the entire element. Element stresses produce a discontinuous stress distribution in the model.
Finite Difference Method	This is an alternative to the FEA method of solving a field problem, where the solution domain is discretized into a grid. The Finite Difference Method is generally less efficient for solving structural and thermal problems but is often used in fluid dynamics problems.
Finite Element	Finite elements are the building blocks of a mesh defined by the position of their nodes and by functions approximating distribution of sought after quantities, such as displacements or temperatures.
Finite Volume Method	This is an alternative to the FEA method of solving a field problem, similar to the Finite Difference Method, and is also often used in fluid dynamics problems.
Frequency analysis	Also called modal analysis, a frequency analysis calculates the natural frequencies of a structure as the associated modes (shapes) of vibration. Modal analysis does not calculate displacements or stresses.
Gaussian points	These points are locations in the element where stresses are calculated. Later, these stress results can be extrapolated to nodes.
h-adaptive solution	An iterative solution which involves mesh refinement. Iterations continue until convergence requirements are satisfied or the maximum number of iterations is reached.

h-element	An h-element is a finite element for which the order does not change during solution. Convergence analysis of the model using h-elements is done by refining the mesh and comparing results (like displacement, stress, etc.) before and after refinement. The name, *h-element*, comes from the element characteristic dimension *h*, which is reduced in consecutive mesh refinements.
Harmonic analysis	Dynamic analysis where excitation is a function of frequency.
Idealization	This refers to making simplifying assumptions in the process of creating a mathematical model of an analyzed structure. Idealization may involve geometry, material properties, loads and restraints. Representing a structure as a surface for shell element meshing or wireframe for beam element meshing are examples of idealization.
Idealization error	This type of error results from the fact that analysis is conducted on an idealized model and not on a real-life object. Geometry, material properties, loads, and restraints are all idealized in models submitted to FEA.
Jacobian	Measure of element curvilinear distortion.
Linear material	This is a type of material where stress is a linear function of strain.
Meshing	This refers to the process of discretizing the model geometry. As a result of meshing, the originally continuous geometry is represented by an assembly of finite elements.
Modal analysis	See Frequency analysis.
Modal Time History analysis	Dynamic analysis where excitation is an explicit function of time.
Modeling error	See Idealization error.
Nodal stresses	These stresses are calculated at nodes by extrapolating stress from Gauss points and then averaging stresses (coming from different elements) at nodes. Nodal stresses, by virtue of averaging, produce continuous stress distributions in the model.
Numerical error	This is round-off error accumulated by the solver.

p-element	P-elements are elements that do not have a pre-defined order. The solution of a p-element model requires several iterations while element order is upgraded until the difference in user-specified measures (e.g. total strain energy, RMS stress) becomes less than the requested accuracy. The name p-element comes from the p-order of polynomial functions which defines the displacement field in the element. This order is upgraded during the iterative solution.
p-adaptive solution	This refers to an option available for static analysis with solid elements only. If the p-adaptive solution is selected (in the properties window of a static study), **SOLIDWORKS Simulation** uses p-elements for an iterative solution. A p-adaptive solution provides results with narrowly specified accuracy.
Pre-load	A pre-load is a load that modifies the stiffness of a structure. A pre-load may be important in a static or frequency analysis if it significantly changes structure stiffness.
Power Spectral Density	A function describing random excitation; its argument is frequency.
Principal stress	Principal stress is the stress component that acts on the side of an imaginary stress cube in the absence of shear stresses. A general 3D state of stress can be represented either by six stress components (normal stresses and shear stresses) expressed in an arbitrary coordinate system or by three principal stresses and three angles defining the cube orientation in relation to that coordinate system.
Random analysis	The dynamic analysis of a system response to a random excitation.
Rigid body motion	A rigid body motion is the ability to move without deformation. If a structure is not fully supported, it can move as a rigid body without deformation. A structure with no supports has six rigid body motions. Rigid body motions are only allowed in Frequency study.
RMS stress	Root Mean Square stress. RMS stress may be used as a convergence criterion if the p-adaptive solution method is used.

Shell element	Shell elements are intended for meshing surfaces. The shell element that is used in **SOLIDWORKS Simulation** is a triangular shell element. Triangular shell elements have three corner nodes. If this is a second order triangular element, it also has mid-side nodes, making the total number of nodes equal to six. Each node of a shell element has 6 degrees of freedom.
Small Displacement assumption	Analysis based on small displacements assumes that displacements caused by loads are small enough to not significantly change the structure stiffness. Analysis based on this assumption of small deformations is also called a linear geometry analysis or a small displacement analysis. However, the magnitude of displacements is not the deciding factor in determining whether or not a small displacement solution will produce correct results. What matters is whether or not those displacements significantly change the stiffness of the analyzed structure.
Solid element	This is a type of element used for meshing solid geometry. The only solid element available in **SOLIDWORKS Simulation** is a tetrahedral element. It has four triangular faces and four corner nodes. If used as a second order element (high quality) it also has mid-side nodes, making the total number of nodes equal to 10. Each node of a tetrahedral element has 3 degrees of freedom.
Steady state thermal analysis	Steady state thermal analysis assumes that heat flow has stabilized and no longer changes with time.
Structural stiffness	Structural stiffness is a function of shape, material properties, and restraints. Stiffness characterizes a structural response to an applied load.
Symmetry boundary conditions	These refer to displacement conditions defined on a flat model boundary allowing only for in-plane displacement and restricting any out-of-plane displacement components. Symmetry boundary conditions are very useful for reducing the model size if the model geometry, load, and supports are all symmetric.
Thermal analysis	Thermal analysis finds temperature distribution, temperature gradient and heat flux in a structure.

Topology optimization	Finding the best material layout within a given design space, for a given set of loads, boundary conditions and constraints with goals such as minimizing mass, maximizing stiffness, etc.
Transient thermal analysis	Transient thermal analysis is an option in a thermal analysis. It calculates temperature, temperature gradient and heat flow changes over time as a result of time dependent thermal loads and thermal boundary conditions.
Ultimate strength	The maximum stress that may occur in a structure. If the ultimate strength is exceeded, failure will take place (the part will break). Ultimate strength is usually much higher than the yield strength.
Vibration analysis	An analysis of oscillations of a model about its position of equilibrium.
von Mises stress	This is a stress measure that takes into consideration all six stress components of a 3D state of stress. Von Mises stress, also called Huber stress, is a very convenient and popular way of presenting FEA results because it is a scalar, non-negative value and because the magnitude of von Mises stress can be used to determine safety factors for materials exhibiting elasto-plastic properties, such as most types of steel and aluminum alloys.
Yield strength	The maximum stress that can be allowed in a model before plastic deformation takes place.

Notes:

25: Resources available to FEA users

Readers of "Engineering Analysis with SOLIDWORKS Simulation" may wish to review the books:

- "Vibration Analysis with SOLIDWORKS Simulation 2019" (Figure 25-1)

- "Thermal Analysis with SOLIDWORKS Simulation and Flow Simulation 2019" (Figure 25-2)

These books are not introductory texts; they are written for users who are familiar with topics presented in "Engineering Analysis with SOLIDWORKS Simulation." Both are available from SDC Publications and from online booksellers.

Figure 25-1: "Vibration Analysis with SOLIDWORKS Simulation 2022".

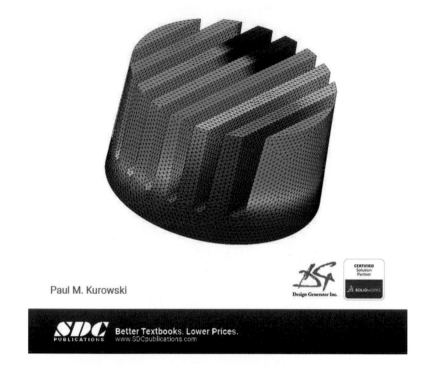

Paul M. Kurowski

Figure 25-2: "Thermal Analysis with SOLIDWORKS Simulation 2022 and Flow Simulation 2022".

Readers of "Engineering Analysis with SOLIDWORKS Simulation" may also benefit from the book "Finite Element Analysis for Design Engineers" which expands on many FEA topics.

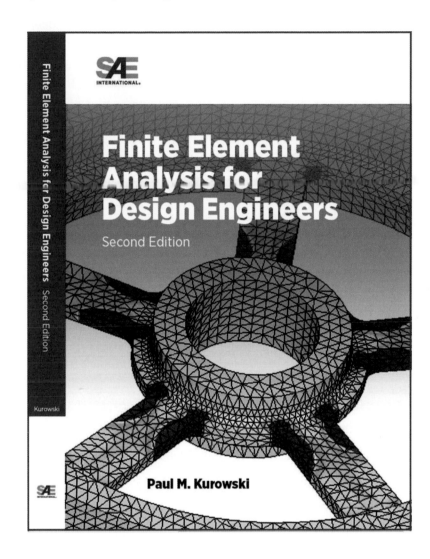

Figure 25-3: "Finite Element Analysis for Design Engineers".

"Finite Element Analysis for Design Engineers" Second Edition is available through the Society of Automotive Engineers website (www.sae.org) and from online booksellers.

Engineering literature offers a large selection of FEA related books, a few of which are listed here:

1. Adams V., Askenazi A. "Building Better Products with Finite Element Analysis," OnWord Press

2. Incropera F., Dewitt D., Bergman T., Lavine A. "Fundamentals of Heat and Mass Transfer," John Wiley & Sons, Inc.

3. Inman D., "Engineering Vibration" Prentice Hall

4. Logan D. "A First Course in the Finite Element Method," Brooks/Cole

5. Szabo B., Babuska I. "Finite Element Analysis," John Wiley & Sons, Inc.

6. Zienkiewicz O., Taylor R. "The Finite Element Method," McGraw-Hill Book Company

With so many applications for FEA, attempts have been made to create a governing body overlooking FEA standards and practices. One of the leading organizations in this field is the National Agency for Finite Element Methods and Standards, better known by its acronym NAFEMS. It was founded in the United Kingdom in 1983 with the specific objective: "To promote the safe and reliable use of finite element and related technology." NAFEMS has published many FEA handbooks such as:

- A Finite Element Primer

- A Finite Element Dynamics Primer

- Guidelines to Finite Element Practice

- Background to Benchmarks

The full list of NAFEMS publications can be found at www.nafems.org

26: List of exercises

Chapter	Part	Assembly
1	BRACKET_DEMO LUG_01* LUG_02* RING_01* CONTACT_01* VASE_01	
2	HOLLOW PLATE	
3	L BRACKET	
4	PIPE SUPPORT MISALIGNMENT*	
5	LINK	
6	TUNING FORK PLASTIC PART	
7	PIPE CONNECTOR HEATER	
8		HEAT SINK
9		HANGER
10		LOOP
11	I BEAM	
12	BRACKET	
13	RING	

* Simulation study saved with model

Chapter	Part	Assembly
14	NL002 NL002 solid* CLIP CLIP DEFORMED ROUND PLATE LINK02 BRACKET NL SPRING	LINK02
15	FLYWHEEL	
16	ROPS TUBE*	OUTRIGGER
17	VASE COVER PERFORATED PLATE ANGLE BRACKET CONTACT	
18		SDOF BUMPER
19	HD HEAD	
20	BRIDGE BRIDGE_50% BRIDGE_65% BRIDGE_80% WATER BASIN	

* Simulation study saved with model